Self-Assembled Organic-Inorganic Nanostructures

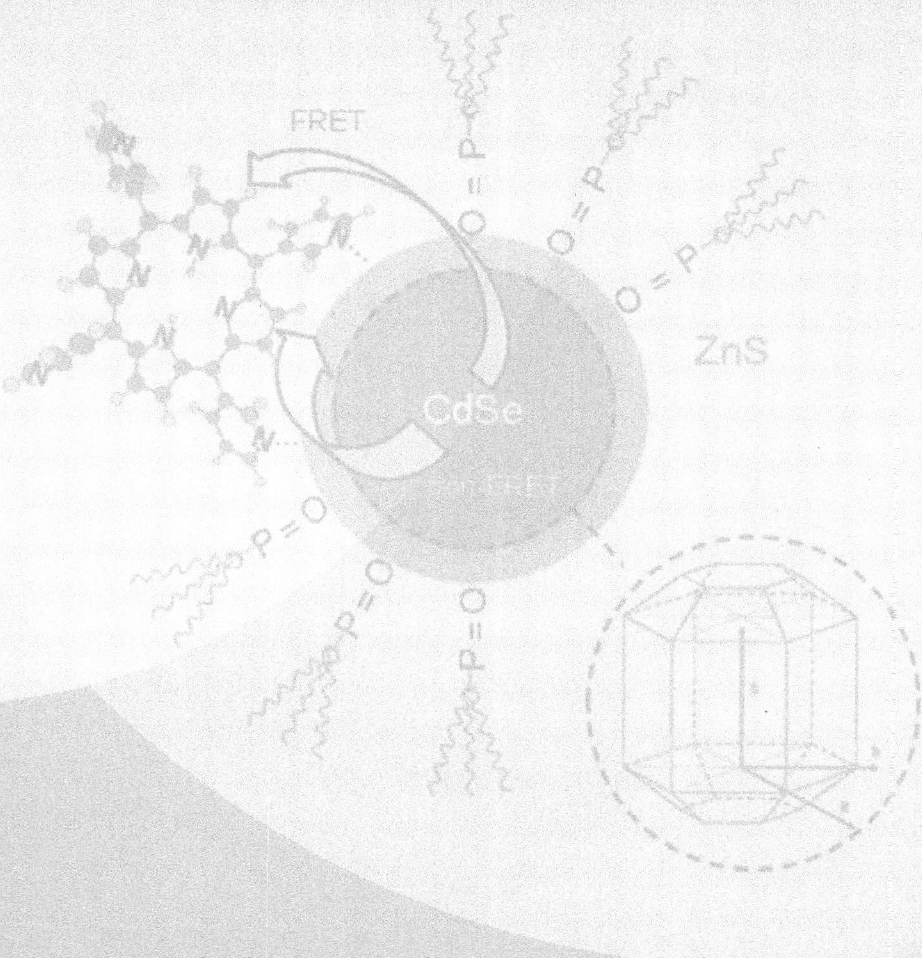

Self-Assembled Organic-Inorganic Nanostructures

Optics and Dynamics

edited by

Eduard Zenkevich

Christian von Borczyskowski

Pan Stanford Publishing

Published by

Pan Stanford Publishing Pte. Ltd.
Penthouse Level, Suntec Tower 3
8 Temasek Boulevard
Singapore 038988

Email: editorial@panstanford.com
Web: www.panstanford.com

British Library Cataloguing-in-Publication Data
A catalogue record for this book is available from the British Library.

Self-Assembled Organic-Inorganic Nanostructures: Optics and Dynamics

ISBN 978-981-4745-43-7 (Hardcover)
ISBN 978-981-4745-44-4 (eBook)

MIX
Paper from responsible sources
FSC www.fsc.org FSC® C013056

Printed and bound in Great Britain by
TJ International Ltd, Padstow, Cornwall

For Tamara and Sabine

Contents

Preface

The current state and perspectives in sciences are strongly linked to the development of novel complex materials as well as to the availability of sophisticated state-of-the-art experimental tools that enable the investigation and manipulation of the objects at various levels of organization, including single nanoobjects and biological sub-structures. The combination of organic and inorganic materials promises to make use of advantages of both types of materials, e.g., downsizing inorganic materials to the nanoscale (as is typical for state-of-the-art devices) and, e.g., nearly unlimited synthesis of functional organic compounds (even mimicking biologically relevant entities). Of special interest are those organic/inorganic entities that are formed due to self-assembly offering chemically specific and versatile formation routes according to the concepts of supramolecular chemistry. Due to the high functionality of such self-assembled nanostructures, applications are readily envisaged or already realized in nanosensorics, bio-medical applications, and photovoltaics.

Recent advancements in nanotechnology permit to produce a variety of functional colloidal semiconductor quantum dots (QDs) and QD-based nanomaterials with unique optical and physico-chemical properties which are principally different from those of bulk materials of the same composition. Science and technology of QDs and QD-based nanomaterials have to deal with (on every length scale, from the molecular to the macro) surface and interfacial phenomena that can be tuned by varying the surface and interfacial energy and by changing the specific chemical interactions with organic compounds attached to such surfaces and interfaces. It means that namely the surface chemistry related to organic/inorganic interactions plays the principal role in the formation of optical properties of QDs and QD-based nanomaterials as well as may be considered as gateway to their possible applications in optoelectronic devices and nanosensors and as optical labels and drug carriers in biomedicine.

An alternative approach to inorganic/organic self-assembled structures on the nanoscale is to scale down inorganic substrates via nanolithography resulting in functionalized nanostructures or to even make use of inherent chemical heterogeneities of an inorganic surface. Again, in both cases appropriate organic molecules self-assemble on such "artificial" or "natural" nanostructures.

In a recent Monograph, *Tuning Semiconducting and Metallic Quantum Dots: Spectroscopy and Dynamics*, we described the optical properties and photophysics of quantum dots concentrating on interfaces/surfaces and the perspectives of tuning the energies of electronic states and related dynamics. Basically, the present book provides a comprehensive description of the morphology and main physico-chemical properties of QD–dye self-assembled nanostructures and natural or lithographically generated surface inhomogeneities (with focus on results, including even some unpublished ones, obtained in our groups within the past decade, as well as some applications in the field of nanotechnology. It crosses disciplines to examine essential nanoassembly principles of inorganic nanostructures with organic molecules, excited state dynamics in nanoobjects, theoretical models, and methodologies. We show that upon nanoassembly formation, photochemical processes occurring at the interface between inorganic nanostructures and functionalized organic molecules and/or bio-objects are complex and may yield new and unexplored phenomena. Additionally, we discuss what parameters may control the photochemical and optical properties of such structures upon attachment of functionalized moieties and how such attachment onto the respective surface proceeds. Especially, we describe a solid base for the application of QD–dye based nanoassemblies in various fields of nanotechnology and biomedicine.

Organized into eight chapters, the book begins with Chapter 1, titled "Structural and Energetic Dynamics in Quantum Dot–Dye Nanoassemblies," which describes the results of basic research concerning formation principles and energetic dynamics in heterogeneous organic-inorganic QD–dye nanoassemblies, based on CdSe QDs and various dye molecules. Using a combination of ensemble (steady-state and time-resolved technique in a temperature range 77–295 K) and single-molecule spectroscopy of QDs and nanoassemblies, we show that single functionalized dye molecules act as extremely sensitive probes for studying the complex interface physics and exciton relaxation processes in QDs. Our findings discussed here and in Chapters 2 (Interrelation of Assembly Formation and Ligand Depletion in Colloidal Quantum Dots), 3 (Fluorescence Quenching of Semiconductor Quantum Dots by Multiple Dye Molecules), and 4 (Static and Dynamic Quenching of Quantum Dot Photoluminescence by Organic Semiconductors and Dye Molecules) show that surface-mediated processes dictate the probability of several of the most interesting and potentially useful photophysical phenomena observed for colloidal QDs. In fact, Förster resonance energy transfer (FRET), charge transfer, and non-FRET processes are the main reasons for QD photoluminescence quenching in QD–dye nanoassemblies. We have succeeded to quantitatively clarify that the major part of the observed QD photoluminescence quenching in QD–dye nanoassemblies, namely non-FRET processes, can be understood, on one hand, in terms of electron tunneling beyond the CdSe core under conditions of quantum confinement and, on the other hand, by the influence of ligand dynamics. Such a comparative approach is presented in this book for the first time.

Chapter 5, titled "Selected Applications of QDs and QD-Based Nanoassemblies," lists a brief history of QD applications as well as representative selected examples of how QDs and QD-based nanomaterials can be applied in photovoltaics, sensing, biomedicine and sub-diffraction imaging (the last one together with single molecule detection was awarded by the 2014 Nobel Prize in Chemistry).

Chapters 6 (Nanolithography and Decoration of Generated Nanostructures by Dye Molecules), 7 (Identification of Heterogeneous Surface Properties via Fluorescent Probes), and 8 (Selective Surface Binding of Dye Molecules on Hybrid Humidity Sensors) are devoted to the description of the interaction of dye molecules with lithographically generated or "natural" surface inhomogeneities with dye molecules.

The bibliography at the end of each chapter contains numerous leading papers, recent reviews, and books in which the readers will find specific references relevant to their subjects of interest.

Concluding, this book links interdisciplinary fundamental research (including lithography, surface chemistry, photochemistry of semiconductor QDs, and QD–dye nanoassembly formation), and selected perspectives for applications of QD–dye nanoassemblies and other organic/inorganic nanostructures. It offers an overview for graduate students, academics, researchers, and industry professionals, and anyone interested in this interdisciplinary field of nanomaterials.

Eduard Zenkevich
Minsk, Belarus

Christian von Borczyskowski
Chemnitz, Germany

September 2016

Acknowledgments

All results from our groups in Chemnitz and Minsk have been obtained by graduate students, PhD students, and post-docs. We appreciate all the work that they have contributed to the development of this field during the past 10 years.

This book contains a whole bunch of results which have not yet been published: Ines Trenkmann (Chapters 6 and 8); Danny Kowerko (parts of Chapters 1 and 2); Thomas Blaudeck (parts of Chapters 1, 3, and 4); Alexandre Stupak (part of experiments and analysis in Chapter 1); Klementina Szwaykowska (part of experiments in Chapter 1); Harald Graaf and Thomas Baumgärtel (parts of Chapter 5); Joerg Martin (parts of Chapter 7).

We would like to thank Frank Cichos (Leipzig, Germany), Dmitry and Svetlana Kilin (South Dacota, USA) as well as Alexandre Stupak and Sergei Gaponenko (Minsk, Belarus) for numerous fruitful discussions. Eugene Petrov's first time-resolved experiments on QDs and assemblies have been stimulating throughout our investigations. Thomas Blaudeck designed the graph on the front cover, which is gratefully acknowledged.

Of course, the results of the international cooperation would not have been possible without financial support throughout the years from various sources: (i) Belarusian State Program for Scientific Research "Convergence 3.2.08—Photophysics of Bioconjugates, Semiconductor and Metallic Nanostructures and Supramolecular Complexes and Their Biomedical Applications" (Chapters 2, 3, and 5); (ii) INTAS Project 03-50-4540 "Optical Active Assemblies of Colloid Quantum Dots and Tetrapyrrole Compounds: Laser Induced Relaxation Processes and Opto-Electronic Functionalities" (Chapters 2 and 3); (iii) DAAD Grant No A/08/08573 "Single Particle Spectroscopy of Individual Nanoobjects Fixed on Heterogeneous Surfaces" (Chapters 1, 2, and 3); (iv) DFG Project GR 2695/8-1 "Optical Probe for Semiconductor Nanoparticles" (Chapters 4 and 6); (v) DFG Research Unit FOR 388 "Laboratory Astrophysics" (Chapter 4); (vi) DFG Graduate College GRK 82 "Accumulation of Single Molecules to Nanostructures (Accumol)" (Chapters 6 and 8); (vii) DFG Research Unit FOR 877 "From Local Constraints to Macroscopic Transport" (Chapters 1, 2, 3, and 7).

As a whole, spectroscopy and photophysics of semiconductor QDs and QD-Dye nanoassemblies are based on more than 25 years of joint cooperation between our research groups. The scientific and personal exchange grants were strongly supported over these years by the Volkswagen-Foundation, INTAS, DAAD, and especially within the framework of the DFG Research Unit FOR 877.

All the experiments would not have been possible without the synthesis of functionalized dye molecules by the group of Frank Würthner (University of Würzburg, Germany) and Alexander Shulga (National Academy of Belarus). Alexander was the one to lay the synthetic ground for our approach to functionalized self-assembled aggregates. Alexander passed away in 2015 but would have liked to see how his efforts have been presented in this book.

Chapter 1

Structural and Energetic Dynamics in Quantum Dot–Dye Nanoassemblies

Eduard Zenkevich[a] **and Christian von Borczyskowski**[b]

[a]*Department of Information Technologies and Robototechnique, National Technical University of Belarus, Prospect Nezavisimosti 65, 220013 Minsk, Belarus*

[b]*Institute of Physics, Center for Nanostructured Materials and Analytics Technische Universität Chemnitz, Reichenhainer Str. 70, D-09107 Chemnitz, Germany*

zenkev@tut.by, borczyskowski@physik.tu-chemnitz.de

1.1 Introduction

At present, nanostructured materials, such as semiconductor quantum dots (QD) and QD-based organic-inorganic nanoassemblies with tunable morphology, have attracted exceptional interest because of their unique architectures, tailored physico-chemical properties, possessing their wide potential covering material science (Kovalenko et al., 2015; Gaponenko 2010; Nicolini, 2010; Klimov, 2010; Schmitt et al., 2011), optoelectronics (Haghi et al., 2015; De Mello Donegá, 2014; Rogers et al., 2014; Vajtai, 2013; Rogach, 2008; Gur et al., 2007; Liu et al., 2010; Wang et al., 2010; Hillhouse and Beard, 2009; Coe-Sullivan et al., 2003; Baldenebro-Lopez et al., 2013), sensorics (Frasco and Chaniotakis, 2009; Frasco et al., 2010; Jin

Self-Assembled Organic-Inorganic Nanostructures: Optics and Dynamics
Edited by Eduard Zenkevich and Christian von Borczyskowski

ISBN 978-981-4745-43-7 (Hardcover), 978-981-4745-44-4 (eBook)
www.panstanford.com

and Hildebrandt, 2012; Lemon et al., 2013; Hadar et al., 2015), biology and medicine (Klimov, 2010; He and Ma, 2014; Vo-Dinh, 2003; Shatkin, 2012; Mocatta et al., 2011; Mattoussi et al., 2012; Ji et al., 2015; Knowles et al., 2012; Liang et al., 2010; Moyano et al., 2012; Rakovich et al., 2010a; Guo et al., 2015), etc. From the basic point of view, the science and technology of QDs and QD-based materials involve on every length scale, from the molecular to the macro, surface and interfacial phenomena that can be tuned by varying the surface and interfacial energy and by varying the specific chemical interactions and chemical groups populating such surfaces and interfaces (Bullen and Mulvaney, 2006; Kalyuzhny and Murray, 2005; Knowles et al., 2012; Munro et al., 2007; Ji et al., 2008; Koole et al., 2008; Voznyy and Sargent, 2014). It is clear that surface-mediated processes dictate the pathways and efficiencies of several of the most interesting and potentially useful photophysical and photochemical phenomena observed for colloidal QDs.

In this respect, the main obstacle on the path to develop efficient QD-based nanomaterials is our limited understanding of QD surfaces, their interaction with surface-attached functionalized organic molecules, and their impact on QD excited states relaxation, including charge or energy transfer between QDs and attached functionalized molecules. Methods of optical spectroscopy (steady-state, time-resolved, and single objects detection) are fundamental to furthering our understanding of the structural, organizational, and dynamic properties of QDs and QD-based organic-inorganic nanoassemblies because a wide variety of complementary, modern non-invasive optical techniques possessing high sensitivity and specificity.

This chapter presents a comparative description of structural dynamics and exciton relaxation processes in QD–dye nanoassemblies, which would be too early given some open problems as well as some specific structural aspects. Here, we would like to characterize the non-covalent self-assembly possibilities and the dynamics in these systems, thus providing the data for a further development of well-defined multicomponent nanostructures for exploitation as artificial light-harvesting complexes, electro- and photochemical devices, nanosensors, etc.

There are several objectives of this chapter. One goal is to familiarize the reader with a specific role of functionalized dye molecules (porphyrins and perilene diimides) in the nanoassembly formation and exciton relaxation dynamics. Then we demonstrate that the QD surface is inhomogeneous with respect to the involved attachment and detachment processes, i.e., the formation of QD–dye nanocomposites is in competition with TOPO and AM capping molecules exchange dynamics. Finally, we describe that very few or even only one attached porphyrin molecule change the distribution and/or presence of dye-related surface trap states considerably, and that the "decoration" of QDs by dye molecules makes a phase transition of the QD capping ligand shell (at low temperatures) highly visible or even amplifies the phase transition. Notably, in order to obtain more insight about these processes we used a combination of bulk and single molecule/ single quantum dot experiments as a tool to precisely identify the interaction of exactly one QD with one dye molecule leading to a microscopic understanding of the QD–dye nanoassemblies formation (including ligand dynamics) and related mechanisms of PL quenching dynamics. On the basis of these well-documented understandings, one could search functional nanocomposites for possible application in various fields of nanoscience and nanotechnology.

1.1.1 QD Surface Properties and Interface Phenomena

Semiconductor QDs (e.g., CdSe or CdSe/ZnS and other II–VI compounds) represent a specific class of matter between atomic clusters and bulk materials with well-defined size-dependent tunable photophysical characteristics (Peng et al., 1997; Klimov, 2000; Rogach, 2008; Gaponenko, 2010; Talapin et al., 2010). Within the past two decades, optical properties of colloidal semiconductor QDs have been investigated intensively (Efros and Rosen, 2000; Klimov, 2000; Rogach, 2008; Michler, 2009; Gaponenko 2010; Talapin et al., 2001), including design strategies of core–shell systems (Reiss et al., 2009), ligand chemistry (Kalyuzhny and Murray, 2005; Munro et al., 2007; Koole et al., 2008; Ji et al., 2008; Morris-Cohen et al., 2010; Munro and Ginger, 2008; Knowles et al., 2012; McArthur et al., 2012; Bullen and Mulvaney, 2006; Evans et al., 2010) and surface functionalization (Querner et al., 2004; Dayal et al., 2006; Lim et al., 2008; Ren et al., 2008; Knowles et al., 2010; Knowles et al., 2012). It should be mentioned that one specific common to all QDs is a high ratio of surface area to volume relative to bulk materials. For a spherical QD with a diameter of 5 and 3 nm, approximately 40% and, correspondingly, 56% of the atoms are on the surface (Murray et al., 2000; Bischof et al., 1996; Knowles et al., 2010). Thus, colloidal QDs in solution are sensitive to various dynamic processes which are related to QD interface phenomena (where ions experience a range of degrees of coordination) that affect the optical properties and, in principle, exciton relaxation pathways (Wang et al., 2015). Some structural properties for CdSe/ZnS QDs are shown schematically in Fig. 1.1.

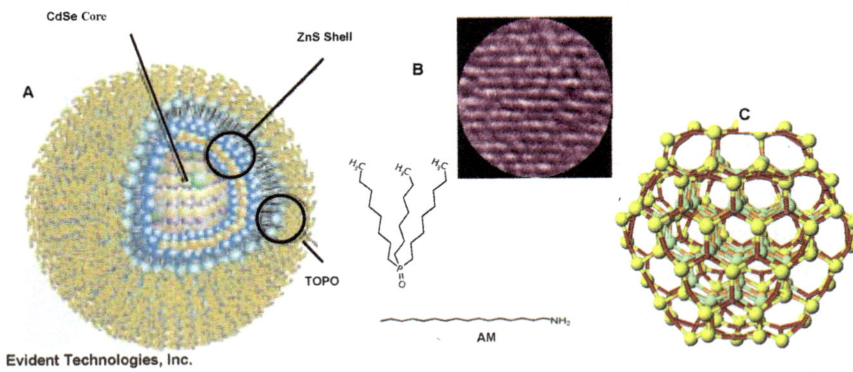

Figure 1.1 Schematic structural presentation (A), transmission electron microscope image (B) and optimized structure (C, surface saturated (Kilin et al., 2007)) of CdSe/ZnS QD. Tri-*n*-octyl phosphine oxide (TOPO) and long-chain amine (AM) ligands are also shown.

Correspondingly, in addition to size distribution, surface chemistry is the major source of heterogeneity in the optical properties of QDs both in time, which manifests itself in PL intermittency or "blinking" for single QDs, and within an ensemble of particles. It means that the heterogeneity and dynamics of QD surface complicates efforts to understand the mechanisms by which surface chemistry may influence the optical and exciton dynamic properties of QDs. In fact, PL characteristics for semiconductor QDs in solution are subject to various dynamic processes which are related to QD interface properties: (i) the adsorption of spurious molecules (Evans et al., 2010; Morris-Cohen et al., 2010; Knowles et al., 2012), (ii) attachment and detachment of protective electrostatically bound (von Holt et al., 2008; Mansur and Mansur, 2011), chemisorbed (Bullen and Mulvaney, 2006; Knowles et al., 2010) or chelating ligands (Dayal et al., 2006), and (iii) the anchoring of functionalized dye molecules (Schmelz et al., 2001; Zenkevich et al., 2005; Zenkevich et al., 2007a, 2007b; Knowles et al., 2012; Dayal et al., 2006; Lim et al., 2008; Ren et al., 2008) or protein complexes (Williard et al., 2001; Schmitt, 2010) to QD surface in hybrid nanoassemblies. In general, a surface reconstruction due to "mobile traps" (Voznyy, 2011) can be invoked by any of the ligands mentioned. Typically, changes in solvent or surface-bound ligands have been found to affect these surface traps and thereby influence steady-state (Talapin et al., 2010; Qu and Peng, 2002; Xie et al., 2005; Kalyuzhny and Murray, 2005; Smith et al., 2006; Zenkevich et al., 2005; Zenkevich et al., 2007b) and time-resolved (Underwood al., 2001; Lee et al., 2005; Petrov et al., 2005; Zenkevich et al., 2012) PL of QDs.

The large variety of functional organic molecules at hand allows for a broad scenario for modification of optical QD properties as has been successfully studied experimentally on single QD level (Empedocles and Bawendi, 1999; Müller et al., 2005; Gomez et al., 2006a; Fernee et al., 2012; Cui et al., 2014; Brown et al., 2014). Several experiments demonstrate the influence of ligand shells on surface structure (Yu et al., 2003), hot carrier relaxation (Munro and Ginger, 2008), PL quantum yield (Munro and Ginger, 2008; Ji et al., 2008) and PL energy (Brown et al., 2014; Smith and Nie, 2010). In addition, recent progress in calculations of QD structures and electronic properties has set new milestones in the understanding of QD surfaces and the impact of ligands and solvents (Puzder et al., 2004; Rempel et al., 2006; Frenzel et al., 2007; Inerbaev et al., 2009; Gomez-Campos and Califano, 2012; Kilina et al., 2009a, 2009b; Albert et al., 2011). Calculations show that ligands cause (depending on their concrete number) surface reconstruction, modify even on a single ligand base electronic states or electron-phonon coupling and hot carrier relaxation (Kilin et al., 2007; Kilina et al., 2012; Fischer et al., 2012; Califano et al., 2013). Pronounced dependencies on structure-related specific positions of surface atoms have also been identified (Fischer et al., 2012) including the mobility of individual surface atoms or ligands

(Voznyy and Sargent, 2014; Voznyy, 2011; Frantsuzov et al., 2013; Schmidt et al., 2014). In case of nanoassemblies based on QDs and organic functionalized molecules, the formation of OD-ligand delocalized (hybridized) surface states may take place which are mostly inactive in absorption spectra but have significantly influence on QD PL dynamics (Kilina et al., 2009a, 2009b; Albert et al., 2011). Thus the kind and concentration of ligands in nanoassemblies may control the optical energies of surface states and their influence on the optical properties of QDs which is important upon analysis of QD PL quenching mechanisms.

From the experimental point of view, now it is a challenge to reach on a single ligand (or dye) level a comparable sophistication of QD surface characterization in order to finally achieve tailoring electronic properties on demand. There are ample investigations on nanoassemblies formed by one or a few functionalized dye molecules attached to a QD surface, which mostly report the influence of the attached dye upon the PL intensity of II/VI colloidal QDs (Smith and Nie, 2010; Clapp et al., 2004; Zenkevich et al., 2005; Dayal et al., 2006; Kowerko et al., 2010; Gerlach et al., 2013; Dworak et al., 2014; Zenkevich et al., 2011). However, a microscopic understanding of the interaction with the QD surface is far from being at hand.

Very recently, we have succeeded to carry out systematic investigations in this direction using a combination of steady-state, time-resolved experiments on both bulk and single nanoassembly detection level (Zenkevich et al., 2012, 2015; Kowerko et al., 2009, 2010; Gerlach et al., 2013) including also low temperature experiments (Zenkevich et al., 2015). Our approach was based on the replacement of one or at most a few capping ligands by exactly one organic dye molecule since a dye allows as a kind of spectator for an efficient and direct optical access.

Based on literature data and our recent results, we discuss some specific QD–dye interactions in nanoassemblies which show that an understanding of the contribution of the surface chemistry of QDs to their electronic structure and dynamics is the key to exploit the desirable properties that should accompany quantum confinement.

1.1.2 Formation Strategies for QD–Dye Nanoassemblies

While quantum confinement is basically understood, the anchoring of functional organic molecules (dyes) to tunable wide gap semiconductor colloidal QDs using various approaches is still of considerable scientific and practical interest, because (as was analyzed above) the particular chemical composition of the surfactant shell decisively affects the photophysical properties of the nanoassembly, especially the quantum yield of QD PL. With respect to QDs, in most cases the bottom-up formation of functional nanoassemblies is used. Basically several routes have been implemented in order to realize QD–dye nanoassemblies in liquid solutions and polymeric matrices (Fig. 1.2):

(i) Blends of QD–dye moieties have been used to form photoactive devices (Ginger and Greenham, 1999; Anni et al., 2004) (see Fig. 1.2A). The use of QDs as electron acceptors in polymer blends provides some advantages to the study of photoinduced charge separation because of few reasons: (1) The energy levels of the host polymers can be tuned through chemical derivatization of the backbone chains. (2) There exists the possibility to alter the blend morphology or to introduce a controlled spatial barrier to charge transfer. (3) The energy levels of QDs can be tuned through size-dependent quantum confinement effects. Since this is a kind of macroscopic approach, not much is known about the microscopic realization of the involved nanoassemblies.

(ii) Dye molecules may be intercalated into QD capping ligand shell without real chemical interactions with QD surface (Dayal et al., 2006; Lim et al., 2008; Kowerko, 2010) (see Fig. 1.2B). This method seems to be relatively simple, but the quantitative analysis of the efficiencies of the energy relaxation channels (especially, FRET) remains to be still difficult because of unknown donor-acceptor distances and mutual orientations.

(iii) In rare cases electrostatic interactions are used for the realization of QD–organic ligand nanoassemblies (Lim et al., 2008; Halpert et al., 2009; Orlova et al., 2010). The interpretation of QD PL properties as well as the energy relaxation pathways in these nanoassemblies should take into account also the perturbations caused by electrostatic interactions.

(iv) The attachment of dye molecules has been accomplished via chemical bonds to a polymer shell covering the QD surface of QD–polymer–dye nanoassemblies (Clapp et al., 2004; Potapova et al., 2005; Morris-Cohen et al., 2010) (see Fig. 1.2C). This results in timely and spatially fixed donor-acceptor complexes, which are well-defined (Clapp et al., 2006) but with a relatively low efficiency of Foerster resonance energy transfer (FRET) (Clapp et al., 2004) since donor-acceptor distances were relatively large. Thus, more than one dye molecule per QD is needed to achieve high FRET efficiencies. QD–dye nanoassemblies of this type are considered as possible candidates for probing in vitro and in vivo biological processes (Mattoussi et al., 2012), but such systems might not be optimal sensors to detect processes on the base of a single event. On the other hand, on the base of covalent sulfur-metal linkage multiporphyrin-modified CdSe QDs have been prepared which exhibit efficient light-harvesting properties in the visible region (Kang et al., 2008) (see Fig. 1.2D).

(v) The most exploited approach to form QD–dye nanoassemblies is based on self-assembly processes (see Fig. 1.2E) being the fundamental phenomenon that generates structural organization on all scales and may be realized in solutions and solid state via various basic interactions: hydrogen bonding, coordination bonding, electrostatic and donor-acceptor interactions, and metal-ion binding (Whitesides and Grzybowski, 2002). The designed self-assembled interaction, the information necessary to initiate such a process, and the algorithm behind must be stored in the components and must be operative via selective/specific interactions (Lehn, 1990). Spontaneous self-assembly, which occurs as a result of the complementarity of superstructure components, has enabled the preparation of QD-based architectures for a wide range of possible fundamental and practical applications mentioned above.

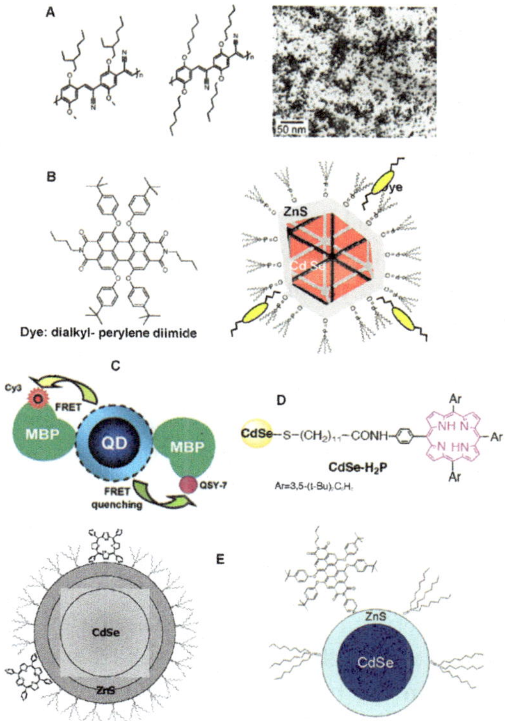

Figure 1.2 Schematic presentations of basic routes for the formation of QD–dye nanoassemblies in liquid solutions and polymeric matrices. A: Structures of some polymers blended with CdSe QDs (d_{CdSe} = 4.0 nm) and TEM images of nanoassemblies in films of 10–20 nm thickness (partly adapted from Ginger and Greenham (1999)). B: Intercalation scheme for dialkyl-perylene diimide molecules into capping ligand (TOPO) shell of CdSe/ZnS QD (partly adapted from Kowerko (2010) and von Borczyskowski and Zenkevich (2014)). C: Nanoassemblies based on DHLA coated-CdSe/ZnS QDs conjugated with polyhistidine domain His_5-MBPs labeled with two different acceptors, Cy3 (emitting acceptor) and QSY-7 (dark quencher) (partly adapted from Clapp et al. (2005)). D: Amine-modified CdSe QD (CdSe-HAD, d_{CdSe} = 4.0 nm) covalently coupled (via sulfur-metal linkage) with *meso*-aryl-substituted porphyrin molecule (partly adapted from Kang et al. (2008)). E: Formation of QD–dye nanoassemblies based on self-assembly coordination interactions of TOPO-capped CdSe/ZnS QDs and pyridyl functionalized porphyrin (left) and perylene diimide (right) molecules (partly adapted from Kowerko et al. (2009, 2010) and Zenkevich, et al. (2005, 2007a, 2007b)).

The formation of self-organized QD–dye nanoassemblies (leading to QD PL quenching in most cases) may be realized via various functional groups (Schmelz et al., 2001; Zenkevich et al., 2005; Dayal et al., 2006; von Holt et al., 2008; Mansur and Mansur, 2011; Bullen and Mulvaney, 2006; Kowerko et al., 2010; Lu et al., 2010; Ren et al., 2008; Knowles et al., 2010). The variation of QD–dye interactions (and the properties of QD–dye nanoassemblies that result from them) shows that functionalized organic dye molecules can perform all of the functions that inorganic shells can perform with respect to controlling the optical and energetic parameters of QDs in self-organized nanoassemblies (Evans et al., 2010; Morris-Cohen et al., 2010; Knowles et al., 2012; von Holt et al., 2008; Mansur and Mansur, 2011; Bullen and Mulvaney, 2006; Knowles et al., 2010; Dayal et al., 2006; Schmelz et al., 2001; Zenkevich et al., 2005; Zenkevich et al., 2007a, 2007b; Dayal et al., 2006; Lim et al., 2008; Ren et al., 2008; Talapin et al., 2010; Qu and Peng, 2002; Xie et al., 2005; Kalyuzhny and Murray, 2005; Smith et al., 2006; Zenkevich et al., 2005; Underwood et al., 2001; Lee et al., 2005; Petrov et al., 2005; Zenkevich et al., 2012). This control, of course, is achievable through the design of every functional group in the molecule, not only the adsorbing anchor substituent. The potential of the organic approach for controlling QD photophysics in self-organized QD–dye nanoassemblies lies in the power of the directed synthetic chemistry to produce a large range of electronic structures and adsorption chemistries with small synthetic adjustments. In this respect it should be mentioned that the sensitivity of the decay of the exciton to organic ligands as well as to attached dye molecules provides an opportunity to control the dynamics and fate of the exciton by controlling the properties of these ligands as well as their attachment and orientation with respect to QD surface.

However, it should be also mentioned that there are further sources of heterogeneity which influence the in solution-phase self-organized QD–dye nanoassemblies. Main items of interest are collected in Fig. 1.3. With respect to the formation of such nanoassemblies in the liquid phase (besides attachment/detachment of dye molecules and the presence/formation of various surface trap states mentioned above) the competitive dynamic interplay of dye molecule attachment and capping ligand exchange dynamics (e.g., tri-*n*-octyl phosphine oxide (TOPO) or long chain amines (AM)) play an essential role for the efficiency and stability of nanoassembly formation (Kalyuzhny and Murray, 2005; Munro et al., 2007; Ji et al., 2008; Koole et al., 2008; Blaudeck et al., 2008; Munro and Ginger, 2008; Evans et al., 2010; Morris-Cohen et al., 2010; Blaudeck et al., 2012; Zenkevich et al., 2012). Moreover, this competitive interplay of dynamics is influenced by temperature (Califano et al., 2005; Zenkevich et al., 2012, 2015) and the (dielectric) properties of the solvent (Issac et al., 2005; Blaudeck et al., 2012; von Borczyskowsk and Zenkevich, 2014). Correspondingly, in order to achieve dye-mediated control of the QD excitonic relaxation, one needs to take into account all these reasons in order to elucidate the specific mechanisms by which ligands and the surrounding may affect the QD PL efficiency as well as the main pathways of exciton relaxation in QD–dye nanoassemblies.

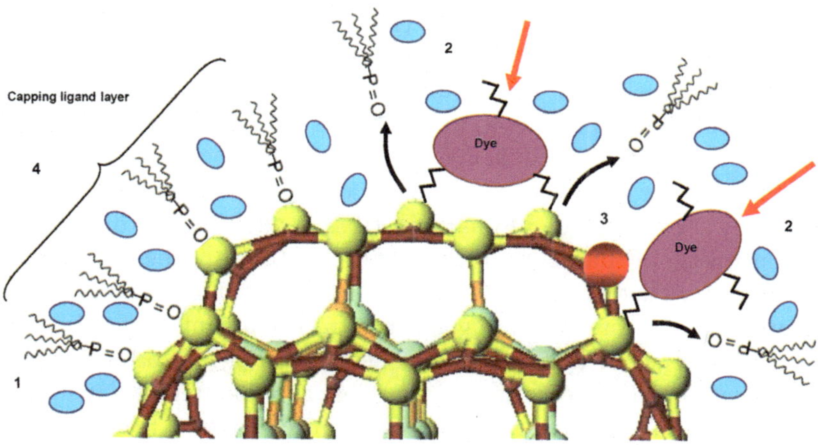

Figure 1.3 Schematic presentation of some surface phenomena realized for QD–dye nanoassemblies in solutions: (1) solvent properties (polarity and solubility of capping ligands and functionalized dye molecules); (2) dye and capping ligand exchange (concentrations C_{Dye} and C_{TOPO}, variation of complexation constants, QD surface curvature); (3) surface defects and transformations (marked by red color); (4) temperature (exciton–phonon coupling, capping ligand layer transformation, thermodynamics of complexation).

In common case, for a quantitative analysis of QD–dye interactions in nanoassemblies it is necessary to have access to both the number of surface-bound ligands per QD and the abundant ones remaining in the solution using models based on either a binomial distribution or Poisson distribution (depending on the degree of saturation of the QD surface) (Tachiya, 1975; Zenkevich et al., 2005; Song et al., 2011; Knowles et al., 2012; Morris-Cohen et al., 2012). In addition, solutions of QD–dye nanoassemblies are inherently heterogeneous in multiple parameters which may influence significantly on QD photoluminescence (PL) properties, such as: (i) the size and shape of QDs, (ii) the type and number of surface-bound dye molecules per QD, (iii) the conformation (temperature phase transition) of the ligand capping layer of the QD, (iv) the dynamic exchange of surface-bound capping ligands and attaching dye molecules, (v) the structure and composition of the surface atoms on the QD (crystallographic structure of the QD surface), (vi) energetic and structural surface defects and transformations, (vii) electronic structure of dangling bonds, etc. (Koole et al., 2008; Ji et al., 2008; Knowles et al., 2010, 2012; Blaudeck et al., 2012). A schematic diagram illustrating some surface phenomena is presented in Fig. 1.3. Thus, in light of this heterogeneity, it is evident that numerous surface-mediated processes dictate the probability of several of the most interesting and potentially useful photophysical phenomena observed for colloidal QDs. In addition, the dramatic sensitivity of the optical properties of QDs to their surface chemistry may be considered as a great potential for tuning these properties by rationally designed and attached organic molecules.

It is evident that nanoassemblies consisting out of only one QD and one dye molecule are most desirable to attempt a unique modeling of the interaction processes. In the past, a few studies have been devoted to the elucidation of thiol-type ligand exchange at relative low molar ratios between 1 and 10 (Munro and Ginger, 2008; Klimov, 2000). Recently, QD–dye nanoassembly formation has been studied under the conditions of extremely low concentrations of QDs and dye molecules and a microscopic description of QD PL quenching has been presented (Blaudeck et al., 2012; Zenkevich et al., 2012; Kowerko et al., 2010, 2012; Gerlach et. al, 2013; Knowles et al., 2010, 2012). In contrast to the number of bulk measurements on large ensembles of QDs, there have been far fewer studies on the effects of surface attached dye molecules on the photoluminescence (PL) of CdSe QDs at the single-QD level but with several attached dye molecules focusing predominantly on photoluminescence intermittency (or blinking) investigations (Hohng and Ha, 2004; Fomenko and Nesbitt, 2008; Gomez et al., 2006b; Hammer et al., 2006; Odoi et al., 2007; Cichos et al., 2007). However, a better understanding of the dye attachment effects on single QD PL is essential to interpret existing data derived from solution PL measurements. In this respect, the combination of bulk and single molecule/single nanoassembly experiments (Kowerko et al., 2010; Zenkevich et al., 2012) is a tool to precisely identify the interaction of exactly one QD with one dye molecule leading to a microscopic understanding of the formation (including ligand dynamics) and related mechanisms of PL quenching dynamics for QD–dye nanoassemblies.

As far as the quantum efficiency and PL dynamics for QDs depend drastically on the environment it is a demanding task to control the environment in a systematic and well-defined way, e.g., by changing the temperature in a controlled manner and observing the corresponding influence on PL for, at least, individual QDs, single QD–dye nanoassemblies and bulk solutions. From this point of view, experiments in a wide temperature range (2–300 K) are of urgent interest because of two reasons:

(i) Temperature- and size-dependence of the exciton decay of individual QDs provides a direct way to characterize the QD energy band gap (Munro and Ginger, 2008; Klimov, 2000; Crooker et al., 2003; Labeau et al., 2003; De Mello Donega et al., 2006), as well as the elucidation of the specificity of the interactions between QDs and organic ligands (see also Chapters 2 and 3 in Von Borczyskowski and Zenkevich (2016)).

(ii) A possible phase transition (or reconstruction) of the capping ligand shell (Wuister et al., 2004; van Driel et al., 2005; Zenkevich et al., 2012) and the rearrangement of organic ligands upon temperature changes will affect both the nanoassembly formation and QD PL.

With respect to the measured band gap energies of QDs reported results of temperature measurements and related interpretations are, however, not fully consistent (follow Chapters 2 and 3 in Von Borczyskowski and Zenkevich (2016)), while for QD–dye nanoassemblies the corresponding data are rare (Zenkevich et al., 2012, 2015). In addition, the crucial role of solvents and ligands on PL properties upon temperature variation has been demonstrated for individual QDs (van Driel et al., 2005; De Mello Donega et al., 2006), while the corresponding studies on temperature effects for QD–dye nanoassemblies have been published only recently. Very recently we have shown (Zenkevich et al., 2012, 2015), that the controlled use of dyes (at molar ratio x = [dye]/[QD] = 1:1), which quench the PL of the QD, is a useful tool to follow indirectly the thermodynamics of ligand–QD interactions at various temperatures. In addition, due to the simultaneous observation and analysis of QD PL quenching effects (intensity and decay) there is a direct access to surface-related processes as well as to the elucidation of PL quenching mechanisms.

Based on our recent results, we will demonstrate in Section 1.2 (together with the discussion of exciton relaxation dynamics) that the non-covalent self-assembly leading to the formation of QD–dye nanostructures is realized in dynamic competition with ligands, and that these ligand exchange processes are noticeably influenced by the solvent and temperature. In its turn, using a combination of spectroscopy on single nanoassemblies and nanoassemblies in bulk solutions we will characterize the influence of interface properties on the exciton dynamics in QD–dye nanoassemblies. In the remaining part of Section 1.1, we will exemplarily discuss how the formation of nanoassemblies can be identified via optical experiments.

1.1.3 Verification of QD Photoluminescence Quenching in QD–Dye Nanoassemblies

In most cases the formation of QD–dye nanoassemblies is followed by QD PL quenching, which is studied both in bulk solutions and on a single nanoassembly detection level. The observed QD PL quenching may be considered, of course, as indicator of the QD–dye nanoassembly formation but the real mechanism of this quenching should be quantitatively proven in every case. Commonly, this PL quenching is interpreted as being due to photoinduced charge (electron or hole) transfer (PCT) (Burda et al., 1999; Schmelz et al., 2001; Issac et al., 2008; Ko et al., 2011; Koposov et al., 2011; Cui et al., 2010, 2011; Ji et al., 2015a, 2015b; Tripathi et al., 2015) and/or FRET processes from the QD to the dye (Williard et al., 2001; Clapp et al., 2004, 2006; Potapova et al., 2005; Zenkevich et al., 2005; Dayal et al., 2006; Ren et al., 2008; Lee et al., 2009; Halpert et al., 2009; D' Souza et al., 2009; Kowerko et al., 2009, 2010; Boldt et al., 2011; Schmitt et al., 2011; Bean et al., 2014; Lemon et al., 2013; Hadar et al., 2015). Before analyzing some examples of these events for QD–dye nanoassemblies, it should be mentioned that from energetic background, FRET and PCT differ principally: while FRET process needs the excitation energy matching (resonance) (Foerster, 1965; Agranovich and Galanin, 1982), PCT is governed by appropriate redox conditions and reorganization energy (Marcus, 1993). We like to add also that besides photoinduced PL quenching upon assembly formation also the so-called static quenching might apply. For a deeper understanding, the reader is referred to Chapter 4 of this book.

To date, though for lot of systems ample qualitative evidence for the presence of such quenching processes is given, only a limited number of papers unravel quantitatively whether the PL quenching (full or in some cases partly, at least) can uniquely be assigned to PCT (Burda et al., 1999; Anni et al., 2004; Zhu et al., 2010; Tagliazucchi et al., 2014) or FRET for bulk solutions (Clapp et al., 2004, 2005, 2006; Zenkevich et al., 2005; Tsay et al., 2007; Jhonsi and Rengnathan, 2010; Lemon et al., 2013; Bean et al., 2014) and for single QD–dye nanoassemblies (Kowerko et al., 2009, 2010; Pons et al., 2006; Hadar et al., 2015).

Some examples of the quantitative analysis of PCT and FRET events in QD–dye nanoassemblies are shown in Figs. 1.4 and 1.5. Energetic scheme and some results of the interfacial photoinduced electron transfer CdSe/ZnS QDs → pyromellitimide (Cui et al., 2010) are presented in Fig. 1.4. Here, dependence of the photoinduced electron transfer (PET) on the solvent polarity has been investigated on the ensemble and single-assembly level. Based on steady-state, time-resolved PL and absorption measurements, it was quantitatively shown that PET rates increase when going from nonpolar solvents to highly polar ones. The results were consistently interpreted in terms of the Marcus theory. It was found also that two pathways are included in the PET process in the CdSe/ZnS-pyromellitimide nanoassemblies: the direct and ZnS layer-mediated PET from the CdSe core to pyromellitimide molecules adsorbed on the surfaces of CdSe and ZnS, respectively.

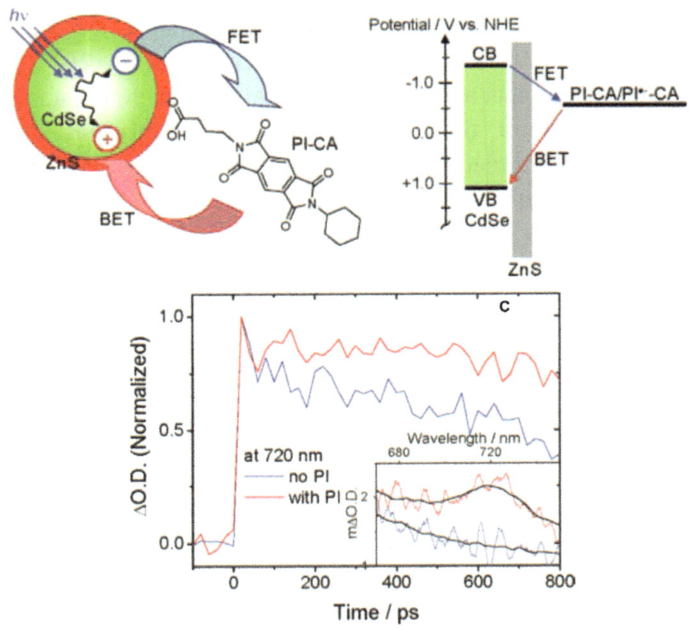

Figure 1.4 Photoinduced electron transfer (PET) in nanoassemblies based on CdSe/ZnS QD and pyromellitimide derivative with a carboxylic acid (PI-CA) anchoring group in solutions (partly adapted from Cui et al. (2010)). (Top left): Scheme of PET reactions between CdSe/ZnS QD and PI-CA. Minus and plus signs denote the formation of exciton that is photogenerated electron e^- and hole h^+, respectively. (Top right): Energy diagram for PET processes. FET and BET denote the forward and back electron transfer processes. (Bottom): Time profiles of transient absorption at 720 nm (absorption of PI-CA anion radical) at laser excitation λ = 400 nm of CdSe/ZnS toluene solutions (5 μM) in the absence (blue) and presence of PI-CA (100 μM) (red). Inset shows the transient absorption spectrum observed at delay time θ = 800 ps after the excitation pulse.

Figure 1.5 shows a scheme and principal results on FRET processes being studied for CdSe/ZnS QDs (donor) noncovalently self-assembled with maltose binding protein (MBP) appended with an oligohistidine tail and labeled with an acceptor dye

(Cy3) (Clapp et al., 2004). This type of QD–dye nanoassemblies allows accurate control of the donor-acceptor distance to a range smaller than 10 nm and provides a good model system to study FRET phenomena in detail.

Results presented in Fig. 1.5 show a good coincidence of independently measured values of FRET efficiency for various QD–Cy3 pairs being obtained from the donor PL quenching data (intensity and decay) and the acceptor fluorescence enhancement. This system has to be considered as one of the rare examples that namely FRET is the main reason of QD PL quenching in these nanoassemblies. Using the Foerster model, the authors were able to extract measurements of the average donor-acceptor spatial distance for each QD–dye pair and found that, in general, distance measurements according to the Foerster model varied proportionally with the QD core size from one set of conjugates to the other.

On the other hand, PL quenching may be induced by so-called non-FRET processes (Dayal et al., 2006; Burda et al., 1999; Zenkevich et al., 2007b; Lim et al., 2008; Orlova et al., 2010) and can be related to the involvement of QD surface states (Frenzel et al., 2007; Issac et al., 2005; Albert et al., 2011; Knowles et al., 2012) or the photoinduced self-trapping of charges in the dielectric medium of QDs environment (Guyot-Sionnest et al., 1999; Klimov et al., 2000; Issac et al., 2005; Cichos et al., 2007). It should be mentioned in this respect that all the following papers (Frenzel et al., 2007; Issac et al., 2005; Albert et al., 2011; Knowles et al., 2012) discuss surface states but not the effect of assembly formation. The latter has been discussed in our recent publications (Blaudeck et al., 2012; von Borczyskowski and Zenkevich, 2014; Zenkevich et al., 2012, 2015). A more microscopic model of non-FRET quenching and its implications and challenges will be given in Section 1.4. Moreover, the dynamically controlled attachment process may be an additional reason for PL quenching in QD–dye nanoassemblies (Bullen and Mulvaney, 2006; Ji et al., 2008; Koole et al., 2008; Munro and Ginger, 2008; von Holt et al., 2008; Schapotschnikow et al., 2009; Kilina et al., 2012). Finally, it should be added that in some rare cases the quenching of the dye-excited states has been detected upon attachment to QD conjugates (Clapp et al., 2004, 2005, 2006).

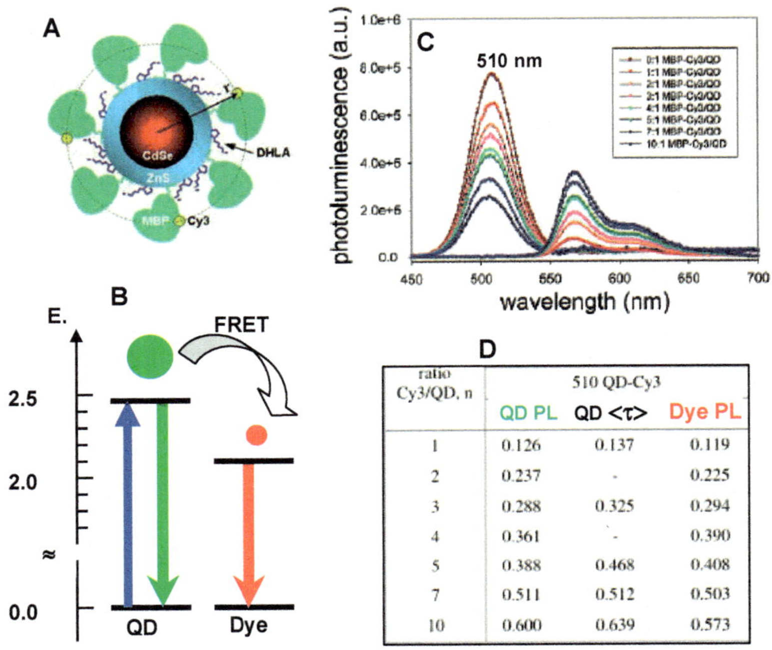

ratio Cy3/QD, n	510 QD-Cy3		
	QD PL	QD <τ>	Dye PL
1	0.126	0.137	0.119
2	0.237	-	0.225
3	0.288	0.325	0.294
4	0.361	-	0.390
5	0.388	0.468	0.408
7	0.511	0.512	0.503
10	0.600	0.639	0.573

Figure 1.5 FRET processes between CdSe/ZnS QDs (donor) and the acceptor dye (Cy3) in nanoassemblies based on QD noncovalently bond with maltose binding protein (MBP) appended with an oligohistidine tail and labeled with dye (partly adapted from Clapp et al., (2004)). A: Schematic presentation of the QD–MBP–dye nanoassembly. r is intercenter donor-acceptor distance. B: Energy diagram for FRET processes corresponding to the given D-A pair. C: Evolution of emission spectra from the QDs (λ_{max} = 510 nm) and Cy3 dyes (λ_{max} = 570 nm) in the QD–MBP–Cy3 nanoassemblies upon increasing dye-to-QD ratio n. D: Independent measured values of FRET efficiency for various QD–Cy3 pairs being obtained from the donor (QD) PL quenching data (QD PL, green column), donor QD PL decay data (QD <τ>, black column) and the acceptor (Cy3) fluorescence enhancement.

In fact, the evaluation of the microscopic reason of QD PL quenching in QD–dye nanoassemblies as well the role of FRET seems to not be a trivial question and thus needs a thorough experimental and theoretical verification for every system. Some aspects will be discussed more deeply in Section 1.4. One principal argument should be underlined in this respect. It turned out (Blaudeck et al., 2012; von Borczyskowski and Zenkevich, 2014) that in case of PL quenching in QD–dye nanoassemblies, a quantitative verification of FRET or PCT is mandatory. That these processes are the dominant reason for QD PL quenching can be only provided by a quantitative comparison of related enhancement (FRET: sensitization of the acceptor (dye) fluorescence) or quenching of the acceptor (PCT) efficiencies and the donor (QD) PL quenching. This is often overlooked in many publications resulting in incorrect assignments of processes and erroneous data evaluation. In this respect, experimental evidence for such QD PL quenching processes without a corresponding fluorescence enhancement (or charge transfer state formation in case of PCT) of dye molecules attached to QD surfaces was, for example, reported for complexes of QDs and tetramethylrhodamine-labeled streptavidin (Williard et al., 2001), but mechanisms have not been quantitatively discussed.

A quantitative comparison has been carried out by us for QD–porphyrin (Zenkevich et al., 2005, 2009; Blaudeck et al., 2008, 2012) and QD–perylene bisimide (Kowerko et al., 2009, 2010) nanoassemblies. From this, we reported for the first time that a competing process clearly distinct from PCT and FRET is responsible for dye-induced QD PL quenching in several QD–dye nanoassemblies.

Again we like to mention that we will discuss a microscopic model of these findings in more detail in Section 1.4. To outline this model is one of the central achievements of this chapter. We will approach this goal by discussing two different intensively investigated QD–dye nanoassemblies in Section 1.2. Although many of the results presented in Section 1.2 have been published in distinct publications we aim at a comprehensive description since our findings set up a mosaic to finally implement a microscopic model of non-FRET quenching which is at a same time an approach to understand surface-related properties of QDs more thoroughly (Zenkevich et al., 2015).

From methodological point of view, we also like to show what necessary steps should be realized upon analysis of QD PL quenching processes in QD–dye nanoassemblies in order to evaluate the real reasons of QD PL quenching in QD-based nanomaterials. In addition, based on our experimental data we will show that very few or even only one attached dye molecule change the distribution and/or presence of dye-related QD surface trap states noticeably.

Concluding, it is evident that properties of QD–dye nanoassemblies are not only interesting in themselves, but also may be also considered as a valuable tool to investigate surface-related phenomena in QDs on an extremely low level of surface modification thus providing the data for a further development of defined multi-component structures for various fundamental and practical applications.

1.2 Self-Assembly of Semiconductor Quantum Dots and Functionalized Dye Molecules

As we have outlined at the end of the previous section, we aim at an understanding how in general PL energetics and dynamics of colloidal QDs are related with the effects on optical properties (e.g., PL quenching (or eventually enhancement)) upon nanoassembly formation with dye (or eventually ligand) molecules. To approach this ambitious goal we have organized the following presentation of results and the related discussion in a way that we present 2 classes of nanoasssemblies, namely QD–porphyrin and QD–perylene diimide nanoassemblies. Porphyrins have been chosen because they are related with a versatile chemical functionalization both with respect to anchoring groups to facilitate attachment to the QD surface (Zenkevich and von Borczyskowski, 2002, 2012) and central atoms in the centre of the porphyrin macrocycle to allow for a considerable modification of redox potentials (Zenkevich and von Borczyskowski, 2002 and references herein). Perylene diimides have been chosen because they are ideal candidates to be detected via single molecule detection schemes (Wuerthner, 2004; Kowerko, 2010). Moreover they are very photostable, can be designed to have various functional groups and offer a variety of conformers. Respective QD–dye nanoassemblies allow for a detection of assemblies on a 1:1 QD–dye assembly instead of a statistical abundance of nanoassemblies as in the case of experiments on porphyrin-based assemblies.

1.2.1 Interacting Subunits and Self-Assembly Approach

Here, we will discuss our results being obtained for QD–dye nanoassemblies based on CdSe/ZnS QDs capped with n-trioctylphosphine oxide, TOPO, or long-chain amines, AM (the diameters d_{CdSe} of QD vary between 2.1 and 5.2 nm, two capping ZnS monolayers were applied). Schematic structures of these QDs are shown in Figs. 1.1 and 1.2E. Two classes of functionalized dye molecules have been chosen for the interaction with QD surface: meso-pyridyl substituted porphyrins, $(m\text{-Pyr})_n\text{-}H_2P$ or H_2P and highly photostable pyridyl functionalized perylene diimides, PDI. Structures of some dye molecules being used are shown in Fig. 1.2E. The main structural and optical properties of CdSe/ZnS QDs as well functionalized dye molecules are described in our earlier publications (Zenkevich, et al., 2005, 2007a, 2007b; Kowerko et al., 2009, 2010). Nevertheless, for the reader's benefit, during discussion some concrete structural and spectral information for interacting counterparts will be included in the corresponding sections of the given chapter.

The self-formation of QD–dye nanoassemblies is realized via non-covalent ligating coordination of pyridyl heteroatoms N (Zn···N-Pyr) with Zn^{2+} ions of ZnS core-shell or Cd^{2+} ions of QD core. Details of this so-called key-lock principle are described by us earlier (Zenkevich, et al., 2005, 2007; Kowerko et al., 2009, 2010) (Fig. 1.6).

In fact, using this approach on the formation of inorganic-organic heterogeneous QD–dye nanoassemblies we shared our concept with the model developed for the directed self-assembly of organic multicomponent arrays (based on porphyrin or chlorin molecules) which show tuneable photoinduced energy/electron transfer and charge separation (Chernook et al., 1996a, 1996b; Zenkevich, et al., 2003, 2009, 2014; Kilin et al., 2015).

Typically, the preparation of QD–dye nanoassemblies in non-polar or low polar liquid solutions at ambient conditions for bulk experiments is based on quantitative titration steps where into low concentrated ($C_{QD} \sim 10^{-7}$ M) QD solutions dye molecules are added sequentially in steps of 10 μL from a highly concentrated ($C_{Dye} \sim 10^{-5}$ M) stock solution up to the wanted molar ratios

$x = [C_{Dye}]/[C_{QD}]$ (where C_{Dye} and C_{QD} denote the concentration of the species), thus varying the number of dye molecules per QD. Preparation procedures (sampling, titration, etc.), details of the experimental technique, steady-state and time-resolved measurements for bulk solutions and single nanoassemblies as well as theoretical approaches are presented in corresponding publications (Zenkevich, et al., 2005, 2007a, 2007b, 2009, 2014, 2013, 2015; Blaudeck, 2007; Blaudeck et al., 2008, 2012; Kowerko et al., 2009, 2010; von Borczyskowski and Zenkevich, 2014).

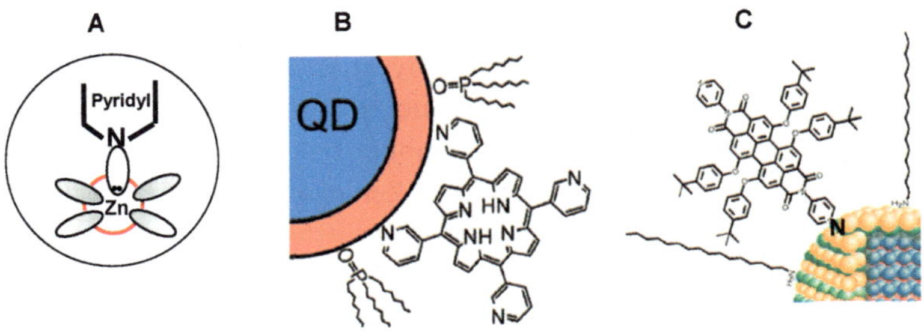

Figure 1.6 Key-lock principle (A) and schematic presentation of the interaction between CdSe/ZnS QD surface and functionalized pyridyl substituted porphyrin (B) and perylene diimide, PDI (C) molecules.

1.2.2 Nanoassemblies Based on Quantum Dots and Porphyrin Molecules

1.2.2.1 Comparative titration experiments and nanoassembly formation

In most cases, the titration of CdSe/ZnS QD toluene solution by a comparable amount of meso-pyridyl substituted porphyrin molecules leads to the QD PL quenching manifesting in the relative PL intensity decrease and decay shortening (Blaudeck et al., 2008, 2012; von Borczyskowski and Zenkevich, 2014; Zenkevich et al., 2005, 2007a, 2007b, 2009, 2011, 2014, 2015). Common spectral-kinetic features of these effects are shown in Figs. 1.7 and 1.8. Noteworthy, in all cases QD absorption remains constant, while QD PL is considerably quenched upon titration by *meso*-pyridyl substituted porphyrins.

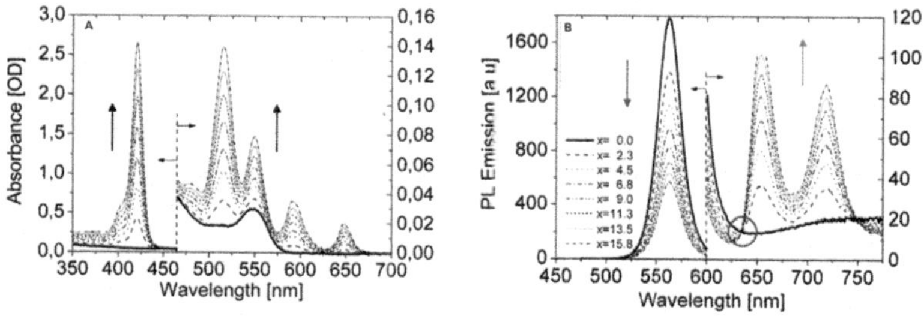

Figure 1.7 Absorption (A) and emission (B, λ_{ex} = 465 nm) spectra for CdSe/ZnS QD (diameter of CdSe core d_{CdSe} = 3.0 nm, 2 ZnS monolayers) upon titration by H_2P(m-Pyr)$_4$ molecules in toluene at 295 at sequence molar ratio $x = [C_{H2P}]/[C_{QD}]$ increase. Adapted from Blaudeck (2007).

It becomes also evident from Fig. 1.8B that at the molar ratio $x \geq 1.14$ the distribution of decay times is abruptly changed. We like to mention that this might be related to the observation that the nature of assembly formation of QD–porphyrin nanoassemblies changes qualitatively at molar ratios $x > \approx 3$ (Blaudeck et al., 2012).

It should be mentioned also that the fluorescence decay time of the H_2P molecule attached to QD surface is longer (τ = 11.7 ns in Fig. 1.8C) than the one for H_2P in the absence of QDs (τ = 8.1 ns). This effect may be caused by three reasons, namely (i) an influence of the QD on the rate constants of the porphyrin molecule in a nanoassembly since the structure is slightly changing (Zenkevich et al., 2005), (ii) a reduction of the porphyrin excited singlet state quenching by molecular oxygen due to the screening effect caused by QD having a high hydrodynamic volume like it has been observed upon interaction of porphyrin triads and pentads with molecular oxygen in non-degassed liquid solutions at 295 K (Sagun et al., 2002) and (iii) an influence of the FRET transfer time from the QD to porphyrin which will last even if H_2P emission has decayed since the intrinsic PL decay time of the QD is more than twice as long as that of porphyrin (for respective model calculations see, e.g., Rempel et al. (1998)).

Noteworthy, we observe a build-up component of ≈3 ns in the range of the H_2P fluorescence the appearance of which correlates with a corresponding increase of a PL decay component of ≈2 ns for CdSe/ZnS upon nanoassembly formation (see also for

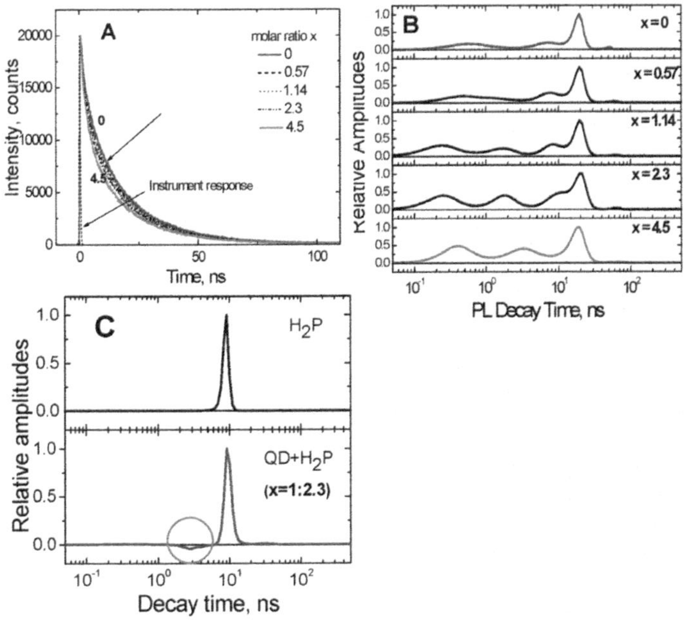

Figure 1.8 Time-resolved PL traces (A, λ_{exc} = 575 nm, λ_{em} = 585 nm), decay amplitude distributions for CdSe/ZnS QD (B, diameter of CdSe core d = 3.8 nm) and for $H_2P(m\text{-}Pyr)_3Ph$ molecules (C, λ_{exc} = 575 nm, λ_{em} = 650 nm) upon nanoassemblies formation in toluene at 295 K. The red circle in Part C shows a small build-up component (with negative amplitude) centered at about 2–3 ns, which indicates a population mechanism of a porphyrin molecule via QD in nanoassemblies. The nature of this build-up component will be discussed in Section 1.4 devoted to the analysis of the reasons for QD PL quenching in QD–dye nanoassemblies. Adapted from Zenkevich et al. (2005).

similar findings for a slightly different type of nanoassemblies in Chapter 4 of this book) which is however, somewhat shorter. As will be discussed later this close correlation allows to conclude from which component (state) the quenching originates to which spectral range, namely taking a decay of 2 ns and an increase of 3 ns results in an "intrinsic" decay time of the originating QD state of about 7 ns which nicely corresponds to the medium decay component of CdSe/ZnS prior to assembly formation. However, without given a proof (see discussion in Section 1.4.3) this time is considerably too short to be responsible for a Foerster based transfer time of 35 ns as has been calculated previously (Zenkevich et al., 2005). In principle, we suggest that the build-up component stems from population of a red shifted (charged?) QD emission induced by attached H_2P molecules.

Summarizing, all these observations prove the formation of QD–porphyrin nanoassemblies. QD PL quenching is a dynamic and static process caused by the increased non-radiative relaxation channels in QD excited states (see also detailed discussion in Chapter 4). Finally, it should be mentioned that in the concentration range C = (0.5–10) × 10^{-7} M being used in titration experiments in toluene at ambient conditions, the PL dynamic quenching due to collisions of interacting moieties within ~20 ns (QD mean PL decay time (Petrov et al., 2005, 2006; Blaudeck, 2007; Zenkevich et al., 2005)) is of low probability. Absorption spectra of the QD–porphyrin nanoassemblies are essentially a linear combination of the corresponding QD and porphyrin $H_2P(m\text{-}Pyr)_4$ ligand, with only small changes in wavelength maxima (blue shift by ~1.5 nm) and in relative band intensities Q(0,0)/Q(0,1) for the porphyrin counterpart (Zenkevich et al., 2005, 2007b; Blaudeck et al., 2012). This means that in the ground state the interaction between QD and porphyrin molecules in nanoassemblies is weak, and they basically maintain their individual identities.

Stimulated by experience with self-assembled porphyrin arrays (Chernook et al., 1996a, 1996b; Zenkevich, et al., 2003, 2009, 2014) we would like to discuss here the specificity of QD–porphyrin interactions upon exchange competition with TOPO capping molecules. The comparative analysis is based on the results of titration experiments for CdSe/ZnS QD with fixed parameters (d_{CdSe} = 2.5 nm, ZnS monolayers n = 2, C_{QD} = 4 × 10^{-7} M) but using *meso*-pyridyl substituted porphyrins $H_2P(Pyr)_n$ with changed properties such as: (i) the number of pyridyl rings from 1 to 4 including the two variants for the displacement $(m\text{-}Pyr)_2$– and $(m\text{^}Pyr)_2$–, where pyridyl rings are opposite or adjacent to each other; (ii) for the fourfold pyridyl substituted H_2P the type of N-substitution within the pyridyl ring from *meta*-(*m*) to *ortho*-(*o*) and *para*-(*p*) N position (Fig. 1.9A). It is seen from Fig. 1.9B that $H_2P(o\text{-}Pyr)_4$ and $H_2P(m\text{-}Pyr)_2$ porphyrins almost do not quench the QD PL (curves 1, 2), whereas the quenching is strongest for $H_2P(p\text{-}Pyr)_4$ and $H_2P(m\text{-}Pyr)_4$ ones. Within the

$H_2P(m\text{-Pyr})_4$ manifold, there is a systematic increase of the quenching efficiency on the number (n) of *meso*-pyridyl substituents. It stems from the absence of QD PL quenching in the case of $H_2P(o\text{-Pyr})_4$ porphyrin, that despite the flexibility of pyridyl rings with respect to the H_2P macrocycle, a parallel orientation of the macrocycle with respect to the QD surface is obviously not favored. On the other hand, an orientation of the porphyrin macrocycle nearly perpendicular to the QD surface would be easily possible for *meta* and *para* N positions in pyridyl substituted porphyrins (see Figs. 1.2E and 1.6B). Theoretical calculations of the optimized geometry for nanoassemblies based on CdSe/ZnS QD and $H_2P(m\text{-Pyr})_2$ molecule (having adjacent *meta*-pyridyl rings) predict also perpendicular placement of the porphyrin macrocycle with respect to QD surface (Kilin et al., 2007).

In addition, Fig. 1.9B shows that the QD PL quenching efficiency and thus the probability to form QD–porphyrin nanoassemblies is decreased with a decreasing number of pyridyl rings. Assuming that the probability to form an assembly is linearly proportional to the number of pyridyl rings, we obtain a rescaling of the QD PL quenching efficiency for every porphyrin molecule ($x_{pyr} = x \cdot N/4$ (Zenkevich et al., 2005)), as depicted in Fig. 1.9C. The overall result is a kind of "master" curve for the quenching efficiency. Correspondingly, all of the quenching curves besides those for $H_2P(m\text{-Pyr})_1$, $H_2P(m\text{-Pyr})_2$, and $H_2P(o\text{-Pyr})_4$ are shifted toward one single curve. In case that only one pyridyl ring can be anchored effectively, the agreement with the master curve becomes less satisfactory. The stability of a two-point interaction will be at least a factor of 2 stronger than a one-point interaction, as can be deduced from the pronounced mismatch of the (scaled) one-point interaction curves for $H_2P(m\text{-Pyr})_1$, $H_2P(m\text{-Pyr})_2$ as compared to the master curve. On the other hand, the difference between curves 6 [$H_2P\text{-iso}(m\text{-Pyr})_2Ph$] and 5 [$H_2P(m\text{-Pyr})_2Ph$] (for porphyrins with three pyridyl anchoring groups) compared to curves 7 [$H_2P(m\text{-Pyr})_4$] and 8 [$H_2P(p\text{-Pyr})_4$] (for porphyrins with four pyridyl anchoring groups) is much less. The latter may be explained by more complex dynamics for QD–porphyrin interactions upon nanoassembly formation including rebuilding probability (depending on the number of pyridyl rings), binding energy, the influence of competing TOPO molecules and some other entropic factors.

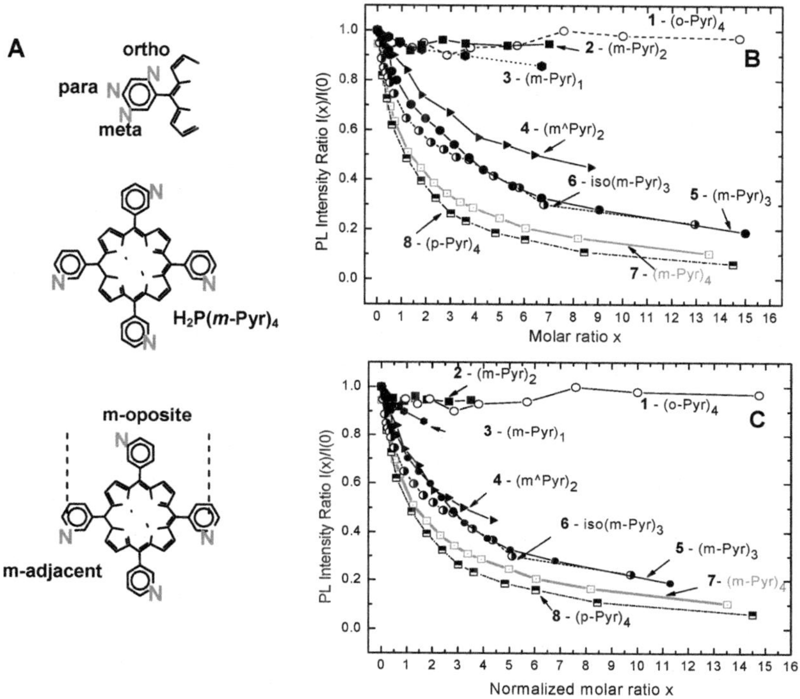

Figure 1.9 A: Chemical structures, abbreviation and positions of nitrogen for *meso*-pyridyl substituted porphyrins. B: Relative photoluminescence intensity dependence $I(x)/I(0)$ for CdSe/ZnS QD (d_{CdSe} = 2.5 nm, two ZnS monolayers, $C_{QD} = 4 \times 10^{-7}$ M) on the molar ratio ($x = [C_{H2P}]/[C_{QD}]$) for various porphyrin molecules. C: PL intensity ratio $I(x)/I(0)$ for the same QDs as function of the normalized (to the number of pyridyl groups) molar ratio x and various porphyrin molecules. Porphyrins being used are the following: 1-H_2P(*o*-Pyr)$_4$; 2-H_2P (*m*-Pyr)$_2$(Ph)$_2$; 3-H_2P(*m*-Pyr)$_1$; 4-H_2P(*m*^Pyr)$_2$; 5-H_2P(*m*-Pyr)$_3$; 6—H_2P(iso)-(*m*-Pyr)$_3$; 7-H_2P(*m*-Pyr)$_4$; 8-H_2P(*p*-Pyr)$_4$. Adapted from Zenkevich et al. (2005).

Concluding, the QD PL quenching efficiency and thus the probability to form QD–porphyrin nanoassemblies scales with the number of pyridyl rings having access to the QD surfaces, being the strongest one in the case of the fourfold pyridyl substituted H_2P molecules, namely H_2P(*m*-Pyr)$_4$ or H_2P(*p*-Pyr)$_4$ (see curves 7 and 8 in Fig. 1.9B) (Zenkevich et al., 2005).

1.2.2.2 Conditions for nanoassembly formation

We will discuss the details of the attachment of porphyrin molecules to QD surface (such as surface-related processes, solvent and temperature influence, complexation and ligand exchange, temporal effects, etc.) for nanoassemblies based on TOPO-capped CdSe/ZnS QDs and $H_2P(m\text{-}Pyr)_4$ molecules. Here, with respect to items outlined in Fig. 1.3, we would like to pick out main results and conclusions based on our experimental data and theoretical considerations and marked below by bold numbers.

(1) Poisson distribution in QD-dye nanoassemblies. For nanoassemblies, upon numerical analysis of QD PL quenching in order to evaluate the corresponding complexation constants K_C and quenching rates one should take into account dye exchange dynamics (depending on capping ligands and dye concentrations as well as on solvent properties) and the number of attached porphyrin (H_2P) molecules per QD. Without consideration of the dynamic equilibrium, it is reasonable to assume that the PL quenching rate for a given QD with n attached chromophores is increasing. As has been proposed earlier (Tachiya, 1975) as well as in our (Zenkevich et al., 2005) and in recent other publications (Song et al., 2011; Knowles et al., 2012) QD–dye formation may be described by a Poisson distribution

$$P(n) = x^{n}\exp(-x)/n!, \tag{1.1}$$

where x is the average number of chromophores per one QD (estimated from a molar ratio ($x = [C_{H_2P}]/[C_{QD}]$), n is the number of attached chromophores on a given QD. Correspondingly, the QD PL intensity ratio $I(x)/I(0)$ may be written as

$$\frac{I(x)}{I(0)} = \frac{\exp(-x)\cdot\Sigma x^{n}}{n!(1+nk_q/k_D)} \quad \text{(Tachiya, 1975) or} \tag{1.2}$$

$$\frac{I(x)}{I_0} = \sum_{n=0}^{\infty}\frac{k_D}{k_D+nk_Q}x^{n}\frac{\exp(-x)}{n!} \quad \text{(Zenkevich et al., 2005),} \tag{1.3}$$

where k_D is the total QD PL intrinsic decay rate and k_Q the QD PL quenching rate. Using this approach, the complexation constant for QD–porphyrin nanoassemblies was found to be no less than $K_C \sim 3 \times 10^7\ M^{-1}$ for $H_2P(m\text{-Pyr})_4$ molecules and TOPO-capped CdSe/ZnS QDs (d_{CdSe} = 3.0 nm and two ZnS monolayers of l = 1 nm) (Zenkevich and von Borczyskowski, 2009, 2014). The details of this approach including also the analysis of time resolved kinetics of QD PL decays are presented also in our recent publication (Zenkevich et al., 2015).

(2) **Influence of QD concentration**. QD PL quenching upon titration by porphyrins scales inversely with QD concentration (Blaudeck et al., 2012). The decrease for later times is an obvious result of the ongoing PL quenching for QD which reflects the particular dynamic processes at the QD surface. The decrease of QD PL decay upon formation of QD–porphyrin nanoassemblies seems to be broadly distributed reflecting the presence of different quenching processes, namely those already inherent in the pure QD sample and those imposed by addition of the H_2P resulting in assembly formation (Zenkevich et al., 2005, 2015; Blaudeck et al., 2008, 2012).

(3) **Complexation and temporal ligand exchange on QD surface**. It is also clear that the attachment of functionalized porphyrin molecules to QD surface is realized in competition (like dynamic equilibrium) with capping TOPO molecules. Considering space-filling molecular entities for such nanoassemblies, in the case of competitive exchange of TOPO capping molecules by attaching porphyrin ligands of one H_2P molecule may replace about 2–3 TOPO molecules or, alternatively, fills a free volume corresponding to 2–3 TOPO molecules. Like in experiments recently described (Koole et al., 2008), in our case the ligand exchange of the native TOPO ligands of CdSe/ZnS QDs by porphyrins is not a simple first-order process. It is seen from Fig. 1.10A that PL of CdSe/ZnS QDs in nanoassemblies shows long-time development (within time scale from 60 s to minutes and hours (Blaudeck et al., 2012)) in the presence of $H_2P(m\text{-Pyr})_4$ molecules. This is not so surprising since the CdSe/ZnS QDs are initially ligated by TOPO molecules, which have different binding constant and exchange kinetics as compared to porphyrins. The experimentally observed "instantaneous"

formation of assemblies within less than 60 s can be explained by the assumption of a few easily accessible ("vacant") surface sites to allow for dye molecule attachment (Blaudeck et al., 2008, 2012).

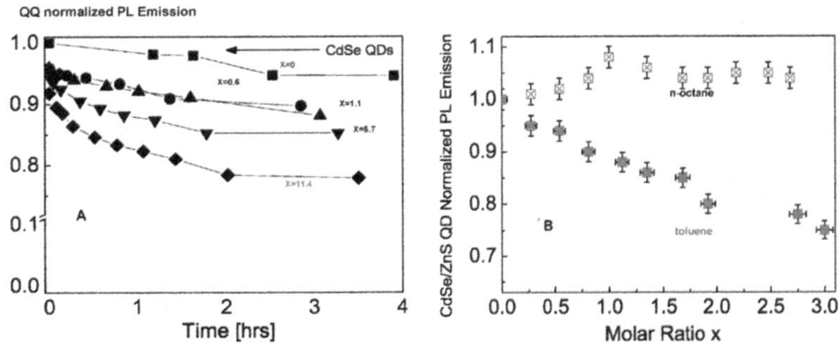

Figure 1.10 (A) PL temporal changes for CdSe/ZnS QDs in nanoassemblies with $H_2P(m\text{-}Pyr)_4$ molecules at different molar ratios $x = [C_{H2P}]/[C_{QD}]$. The values on the left axis $I(x)/I_0$ were normalized to the PL of the individual sample prior titration. (B) Normalized PL intensities of CdSe/ZnS QDs upon titration of $H_2P(m\text{-}Pyr)_4$ molecules as a function of the molar ratio $x = [H_2P]/[QD]$ in *n*-octane and toluene. Adapted from Blaudeck et al. (2012).

Obviously, the observed dependence of QD PL on the molar ratio x = [Dye]/[QD] requires a model that includes two functions: a binding isotherm, and a function that describes the response of the PL to ligands once they are bound at their equilibrium surface coverage (Koole et al., 2008). Both functions are sensitive to the kind of ligand. The exchange of a dye molecule (porphyrin) with capping ligands (TOPO in our case) is defined by a "reaction"

$$(\mathrm{TOPO})_n : \mathrm{QD} + \mathrm{Dye} \rightarrow (\mathrm{TOPO})_n + \mathrm{Dye} : \mathrm{QD},$$

with the equilibrium constant

$$K_x = [(\mathrm{TOPO})_n] \times [\mathrm{Dye{:}QD}]/[(\mathrm{TOPO})_n{:}\mathrm{QD}] \times [\mathrm{Dye}].$$

In case of ligand exchange and assuming that, prior to ligand exchange, the surface of the QD is completely covered with the "native" TOPO ligand, the corresponding evaluation of the equilibrium constants K_x (according to PL quenching experiments and theoretical DFT calculations) has been recently described (Knowles et al., 2010). These considerations have been also taken into account upon estimation of the complexation constants of $K_C \sim 3 \times 10^7$ M^{-1} for the formation of QD–porphyrin nanoassemblies in toluene at ambient conditions (see item 1 of this Section). Correspondingly, the rapid exchange time for CdSe/ZnS QD–porphyrin nanoassemblies may be caused by two reasons (Zenkevich et al., 2015): (i) weaker one-point bond strength of TOPO molecules ($K_C \sim 10^5$ M^{-1} (Moreels et al., 2007; Koole et al., 2008)) with respect to the two-point interaction of *tetra-meso*-phenyl substituted porphyrin molecules with QD surface, (ii) faster kinetics due to the relatively small volume occupied by a TOPO molecule in comparison with the porphyrin.

According to the whole set of the experimental findings and the comprehensive analysis of the obtained results presented by us in (Blaudeck et al., 2012), the influence of the ligand exchange dynamics on the QD PL quenching efficiency may be outlined in the following way. Typically, the initial sample preparation is accompanied by diluting the original QD–toluene solution followed by a new equilibrium of TOPO coverage caused by a relatively low concentration of TOPO in the sample. This implies an overall reduction of the QD surface coverage by TOPO ligands resulting in a decrease of the QD PL quantum yield due to the formation of more unsaturated surface states with different emission efficiencies in comparison with near band edge or excitonic states (Frenzel et al., 2007; Ji et al., 2008). In addition, the reduction of the QD surface coverage by capping ligands may be accompanied by further attachment of quenchers, either those which are already inherently present or are added during a titration procedure. The time scale of such processes is in our experiments typically between 200 and 2000 s and depends critically on the experimental conditions such as the type of QD, TOPO, (unknown) impurities and/or dye concentration. However, besides processes on a timescale below 2000 s, a further reorganization of the surfactant shell (accompanied by increased quenching of

QD PL) occurs on much longer time scales which can be still observed. This reorganization may be explained by the formation of new TOPO structures such as surfactant islands on the QD surface or TOPO micelles in the solution. It is not excluded also that long time QD PL changes might be related to an oxidative degradation of the QD surface, found, for instance, for amine ligands and PDI molecules (Koole et al., 2008; Ji et al., 2008; Kowerko et al., 2009, 2010; Kowerko, 2010).

(4) **Solvent influence on QD–porphyrin nanoassembly formation**. The dynamic equilibrium between capping TOPO ligands and attaching porphyrin molecules as well as QD PL quenching processes depend critically on the kind of the solvent. Addition of polar acetone into toluene solutions containing QD–porphyrin nanoassemblies increases significantly the PL quenching (Zenkevich et al., 2005). According to recent calculations (Schapotschnikow et al., 2009) this effect may be connected with ligand binding specificity which takes place via electrostatic forces. On the other hand, our experiments show (Blaudeck et al., 2012), that QD PL quenching upon titration by $H_2P(m\text{-}Pyr)_4$ molecules in non-polar n-octane is essentially smaller at the same molar ratios with respect to that found for toluene solutions (Fig. 1.10B). It is clearly seen that QD PL quenching is only observed in toluene, whereas the QD PL intensity stays more or less constant in *n*-octane indicating that in the later case the ligand shell is stabilized against porphyrin molecules. This is easily understood since the solubility of TOPO is largely reduced in *n*-octane while that for porphyrin molecules is increased as compared to toluene. This will shift the complexation constants K_C for TOPO and $H_2P(m\text{-}Pyr)_4$ molecules towards each other. In that case the ligand shell remains intact effectively preventing the porphyrin attachment to QD surface.

(5) **Role of dye electronic properties in QD PL quenching in nanoassemblies**. As was discussed in Section 1.2.1, the formation of QD–dye nanoassemblies is realized via non-covalent ligating coordination. It means that functionalized organic molecules (porphyrins or perylene diimides in our case) should have the specific anchoring groups (like

pyridyl rings). Indeed, no QD PL quenching effects and, correspondingly, no nanoassembly formation were observed when using various porphyrins without anchoring pyridyl rings (such as tetraphenylporphyrin (Zenkevich et al., 2012, 2015)). On the other hand, even in the case of the attachment of organic molecule to QD surface QD PL quenching should depend on electronic properties of the ligating counterpart. Figure 1.11A shows a comparison of QD PL quenching induced by titration with pyridine and with mono-pyridyl-substituted porphyrin molecules $(m\text{-Pyr})_1\text{-}H_2P$ containing only one meso-pyridyl ring which is at least necessary for the QD attachment. In the case of titration by pyridine (a hole acceptor like amines or thiols (Schapotschnikow et al., 2009; Bullen and Mulvaney, 2006; Holt et al., 2008)) a negligible QD PL quenching is observed even at 100 times increased relative molar ratio.

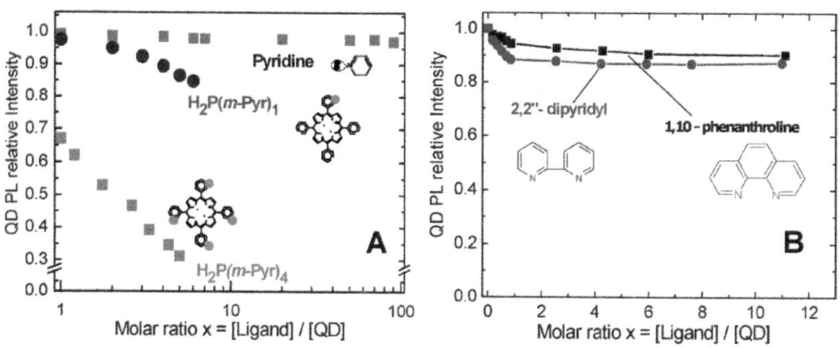

Figure 1.11 Comparison of PL quenching of colloidal CdSe/ZnS QDs by various molecules upon titration in toluene at 295 K. (A) d_{CdSe} = 2.15 nm, n_{ZnS} = 2, λ_{exc} = 465 nm, λ_{reg} = 530 nm. (B) d_{CdSe} = 2.4 nm, n_{ZnS} = 2, λ_{exc} = 465 nm, λ_{reg} = 542 nm. Adapted from Blaudeck et al. (2012) and Zenkevich et al. (2007a).

It is seen also that for molar ratios x = 1–8, the PL quenching of QDs by mono-pyridyl substituted porphyrin $H_2P(m\text{-Pyr})_1$ is essentially stronger than for pyridine. Some reasons of that concerning the competition between ligands are discussed in item 3 of this Section. Additionally, one should conclude that the strong QD PL quenching in QD–porphyrin nanoassemblies shows that electronic properties of the π-

conjugated porphyrin macrocycle having meso-pyridyl rings may play a specific role in the localization of the electron–hole pair at the QD surface (Zenkevich et al., 2007a). Namely, the porphyrin chromophore itself is of central importance although the specific electronic structure is of much less influence (Zenkevich et al., 2007b). As it will be discussed later, the results of QD PL quenching in QD–perylene diimide nanoassemblies (Kowerko et al., 2009, 2010) support this statement.

For comparison, it was shown for CdSe/ZnS–diazaperylene nanoassemblies that the strong (~60%) QD PL quenching is due the photoinduced charge (electron or hole) transfer (Schmelz et al., 2001). On the other hand, Fig. 1.11B shows that titration of CdSe/ZnS QD solutions by 1,10-phenanthroline or 2,2′-dipyridyl molecules does not show a significant QD PL quenching, though the spatial arrangement and the distance between the two anchor nitrogen atoms in the later molecules are the same as for diaza-perylene. It should be mentioned that with respect to diazaperylene molecules, absorption and fluorescence spectra of 1,10-phenanthroline and 2,2′-dipyridyl molecules are considerably shifted toward the UV range. Correspondingly, photoinduced charge (electron or hole) transfer upon excitation of QD should be ruled out in the later case (Fig. 1.12). It is not excluded also that for nanoassemblies CdSe/ZnS–diazaperylene the formation of new types of low-lying surface states (traps) may take place which are characterized by a rapid non-radiative relaxation. Concluding, all these results show that for every nanoassembly electronic properties of the ligand may play an essential role in the formation of the non-radiative channels for QD exciton relaxation.

(7) **Low-temperature capping ligand phase transition in QD–porphyrin nanoassemblies**. As discussed in Chapter 3 of Von Borczyskowski and Zenkevich (2016), the elucidation of exciton relaxation pathways for alone QDs and QD–dye nanoassemblies upon temperature lowering is the necessary step for understanding the photophysics and photochemistry of quantum-confined nanoassemblies. Upon temperature lowering, in contrast to rigid matrixes (polymers, resins, glasses, etc.), for QDs and QD–dye nanoassemblies dissolved in typical liquid solvents some specific spectral and

energetic effects may take place caused by a phase transition of the capping ligand layer (De Mello Donega et al., 2006; Wuister et al., 2004; Zenkevich et al., 2012; Zenkevich et al., 2015).

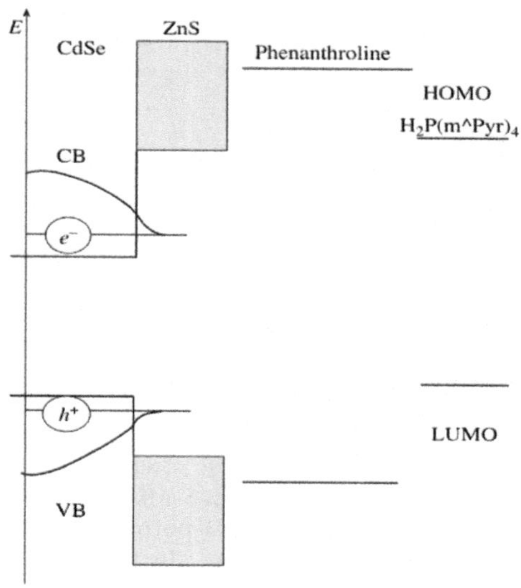

Figure 1.12 Schematic diagram of the relative positions of electron (e^-) and hole (h^+) states in CdSe/ZnS and HOMO and LUMO energy positions for organic dyes. For the CdSe core, the quantized states of an electron–hole pair in the valence band (VB) and conduction band (CB) as well as the wave functions of e^- and h^+ are shown. Adapted from Zenkevich et al. (2007a).

A set of experimental results for TOPO capped CdSe/ZnS QDs (d_{CdSe} = 3.0 nm, n_{ZnS} = 2) and various QD–porphyrin nanoassemblies performed in a methylcyclohexane/toluene (6:1) mixture in a temperature range from 77 K to 300 K is shown in Fig. 1.13. The detailed analysis of these results together with time-resolved QD PL data has been presented in our recent publications (Zenkevich et al., 2012; Zenkevich et al., 2015) as well as partly in Chapters 2 and 3 of von Borczyskowski and Zenkevich (2016). Here, we underline main ideas following from low temperature experiments what should be taken into account upon microscopic understanding of the nanoassembly formation (including ligand dynamics) and related mechanisms of PL quenching dynamics for QD–porphyrin nanoassemblies as well as for other structures of the same type.

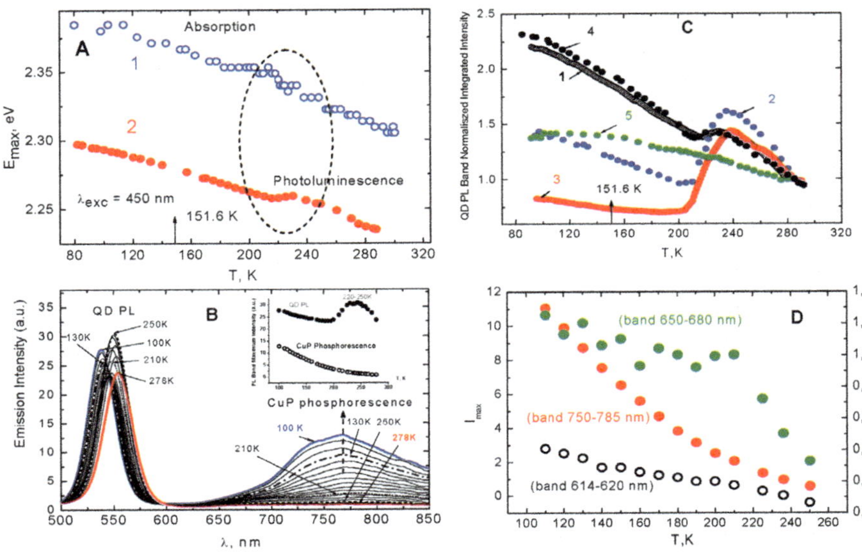

Figure 1.13 Temperature dependence of absorption and PL properties for TOPO capped CdSe/ZnS QDs (d_{CdSe} = 3.0 nm, two ZnS monolayers) and QD–porphyrin nanoassemblies in methylcyclo-hexane/toluene (6:1) mixture. The glass transition temperature (151.6 K) of the solvent mixture is indicated by solid arrow. (A): Temperature dependence of the first absorption peak (1, mean weighted data for two Gaussians lines) and PL maximum (2, λ_{exc} = 450 nm) for QDs. Dashed ellipse indicates temperature range at which optical properties change (showing "kink"). (B): Temperature dependence of PL spectra (λ_{exc} = 450 nm) for nanoassemblies based on CdSe/ZnS QDs and CuP(m-Pyr)$_4$ (1:1). The insert shows the temperature dependence of normalized PL intensity for the short wavelength (~530–570 nm) band and long wavelength band (>600 nm) between 77 and 285 K. The short wavelength band belongs to QD PL, the increasing long wavelength band belongs to CuP(m-Pyr)$_4$ phosphorescence. (C): Temperature dependence of QD PL band normalized integrated intensities (λ_{exc} = 450 nm) for nanoassemblies based on TOPO-capped CdSe/ZnS QDs (d_{CdSe} = 3.0 nm, two ZnS monolayers) and various porphyrin molecules at molar ratio $x = [C_{H2P}]/[C_{QD}]$ = 1. All curves are normalized to the integrated band intensity of QD PL at 285 K. Experimental temperature dependences are shown for the same type of CdSe/ZnS QD under various conditions and attached porphyrins: (1) QDs; (2) QD+-H_2P(m-Pyr)$_4$; (3) QD+CuP(m-Pyr)$_4$; 4) QD+tetraphenylporphyrin (having no pyridyl rings); (5) QD in dried PMMA film on a quartz plate. (D) Temperature

dependence of three Gaussian components as a result of fitting long wavelength bands which are especially pronounced in the presence of CuP. See discussion in the text. Adapted partly from Zenkevich et al. (2012, 2015).

It is seen from Fig. 1.13A upon temperature lowering, the maximum of the band-edge absorption band as well as the maximum of PL spectra for alone QDs are blue shifting. In addition, the PL line width (FWHM) becomes essentially narrower, while the PL intensity becomes higher (Zenkevich et al., 2012; Zenkevich et al., 2015). The most interesting result is that upon temperature lowering a non-monotonous temperature dependence of the respective peak energies becomes evident for absorption and PL data. Interestingly, this non-monotonous behavior (which we named "kink," for simplicity) is observed for TOPO-capped CdSe/ZnS QDs between 200 and 240 K, far from the glass transition temperatures for methylcyclohexane (146.7 K) and toluene (180 K), respectively, as well as for a methylcyclohexane/toluene (6:1) mixture (151.6 K) (Marcus, 2002). With T lowering the "kink" occurs at slightly different temperatures for the first excitonic absorption band (~218 K) and the PL band (~223 K), respectively (T range of the "kink" is shown by dashed ellipse in Fig. 1.13A).

In addition, a similar "kink" is observed in the PL temperature dependence for CdSe/ZnS QDs capped by long-chain amines but shifted to ~237 K as compared to ~223 K for TOPO-capped CdSe/ZnS QDs (see Fig. 1.19) (Zenkevich et al., 2012). It should be mentioned that unusual "luminescence anti-quenching" at a well defined temperature of $T \sim 250$ K (like "kink" in our experiments) has been observed for CdTe or CdSe/ZnS QDs initially dissolved in liquid toluene, and has been related to a phase transition in the surfactant capping layer (De Mello Donega et al., 2006). This phase transition of the surfactant was explained by surface relaxation and/or surface reconstruction which are strongly dependent on the type of capping ligand molecules (TOPO or aminoethanethiol) and, in turn, may cause a spatially energetic reordering of trap states. Notably, we have shown for the first time (Zenkevich et al., 2012) that, in addition to PL, the QD band-edge absorption may be sensitive also to such a phase transition of the surfactant capping layer at temperatures close to ~218 K. The appearance of the "kink" may be connected with some strain-induced deformations of the ZnS shell caused by the capping ligand layer at low *T*. In this respect, we believe that slightly different "kink" temperatures for the first excitonic absorption band (~218 K) and the PL band (~223 K) for TOPO-capped CdSe/ZnS QDs, may be explained by the different nature of the corresponding transitions in absorption (allowed excitonic ones) and in emission (trap dominated at low *T*) (Zenkevich et al., 2012).

In its turn, as far as a trap state emission dominates at low temperatures, these trap states or their distribution might be influenced by dye molecules attached to the QD surface due to replacing some of the surfactant (TOPO in our case) molecules (Kaluzhny and Murray, 2005; Zenkevich et al., 2012). The results of the corresponding temperature experiments for nanoassemblies based on TOPO capped CdSe/ZnS QD of one type and various porphyrin molecules at a relative molar ratio of $x = [C_{H2P}]/[C_{QD}] = 1$ are collected in Figs. 1.13B and 1.13C. In the case of CuP(m-Pyr)$_4$ being attached to QD, upon temperature lowering long wavelength emission bands between λ_{max} = 614–850 nm appear (Fig. 1.13B). The longer spectral range (>850 nm) is limited by the spectral range of the spectrometer. These long wavelength emission bands belong partly (λ > ≈750–785 nm) to the phosphorescence of CuP(m-Pyr)$_4$ (Kim et al., 1984). It is evidently seen that CuP(m-Pyr)$_4$ phosphorescence intensity is continuously increasing upon temperature lowering (insert in Fig. 1.13B,D, inset) what is typical for Cu-containing porphyrins. However, the bands (see Fig. 1.13D) between 650–680 nm (and 614–620 nm) show also a strong (weak) sudden change of PL intensity at the phase transition temperature. We tentatively assign the 650–680 nm band to CuP-induced deep traps of the QD, whereas the 614–20 nm band is probably due to remaining (uncomplexed) QD emission. See also discussion later on.

The PL intensity for QD in QD–CuP(m-Pyr)$_4$ nanoassemblies shows a "kink," which, however, is now much more pronounced (–23%) than in the absence of CuP molecules. Comparative results presented in Fig. 1.13C show that porphyrin molecule attachment to QD surface enhances QD PL decrease at the phase transition temperature, and the effect is strongest for CuP molecules. We suggest that CuP exhibits the largest effect, since either internal molecular charge transfer states influence the energy distribution of CdSe/ZnS QD trap states considerably or

TOPO molecules may be strongly coordinated at low temperatures to the central Cu ion, which might increase the disorder of the surfactant capping shell (Zenkevich et al., 2015).

Interestingly, it is seen from Fig. 1.13C (curve 4) that tetraphenylporphyrin without *meso*-pyridyl anchoring groups does not enlarge the QD PL decrease at the "kink" (because of self-assembly absence in this case as was discussed earlier in Section 1.2.2.1), and the "kink" is absent for QDs in a dried rigid PMMA film on a quartz plate (curve 5). In the later case, the film rigidity suppresses the reorganization of the TOPO capping layer (De Mello Donega et al., 2006), thus weakening a spatially energetic reordering of QD trap states. It is of interest also, that no "phase transition" of the TOPO capping layer is observed for the same nanoassemblies obtained by spin coating from toluene onto a quartz substrate followed by evaporation of the solvent under ambient conditions (Zenkevich et al., 2015). This is probably due to the fact, that the TOPO capping layer is (almost) completely removed or that the remaining ligands can freely re-orientate due the absent of an embedding solvent.

Very recently on the base of time-resolved PL spectroscopy in combination with spectral deconvolution we have argued (von Borczyskowski and Zenkevich, 2014; Zenkevich et al., 2015) that CdSe/ZnS QDs undergo a phase transition in the sense, that the "freezing" ligand shell exerts strain on the ZnS shell thus creating trap states with reduced PL quantum yields. Thus, surface properties of QDs are a complex interplay of the nature of the dye molecule and the topography of the ligand layer across a temperature range of 77–290 K. Moreover, this phase transition changes also the CdSe/ZnS absorption, which might be explained by a modification of the CdSe core or ZnS layer structure. In solution, the solvent molecules adsorbed on the QD surface can reduce the total surface free energy, and the surface can undergo reconstruction when the solvent molecules leave the QD surface. In turn, dye attachment creates new and/or more trap states, which obviously quench the PL very effectively. Phase transition temperature and the influence of dye-induced trap states depend critically on the type of the capping ligand. In the result, such surface transformations may change the nature and the energy position of surface states observed below and above the phase transition temperature.

In this respect, it is interesting to note that the Gaussian deconvolution of the long wavelength emission band especially expressed for QD–CuP nanoassemblies shows that the temperature dependences for at least two components differ significantly (see Fig. 1.13D): the longest wavelength component (λ_{max} ~ 750–785 nm) does not show any kink upon temperature lowering, while the short wavelength component with λ_{max} ~ 650–680 nm does show such a kink just in the vicinity of the phase transition temperature for TOPO-capped CdSe/ZnS QDs. It is reasonable to relate the longest wavelength component with the phosphorescence of CuP which is not sensitive to the phase transition of TOPO capping layer. On the other hand, the temperature specificity of the component with λ_{max} ~ 650–680 nm indicates that this emission belongs to QD intra band states. It means that Cu-porphyrin molecule being attached to a QD surface induces (especially at temperatures below 230 K) deep intraband states (traps) accompanied by quenching of the near band edge exciton PL. This idea coincides with results of (Baker and Kamat, 2010) showing that the interaction of 3-mercaptopropionic acid with CdSe QD creates emissive deep trap sites (Se vacancies) by exchange of surface bound amines. In that case the enhanced intensity of deep traps as well as the corresponding quenched band edge emission of QD are readily tuned by controlling the concentration of mercaptopropionic acid.

A comprehensive analysis of temperature dependent spectral-kinetic properties and interface phenomena for given CdSe/ZnS QDs is presented in Chapter 2 of Von Borczyskowski and Zenkevich (2016). In addition, the presence of phase transition-type changes have been found also for QDs capped by long chain amines (Zenkevich et al., 2015). Most importantly, our experimental findings and qualitative description can be closely related to recent calculations of ligand-induced modifications of surface states of QDs (Kilina et al., 2009a, 2009b, 2012; Albert et al., 2011; Fischer et al., 2012; Califano et al., 2013).

Thus, the above results and discussion demonstrate clearly that capping ligand and dye molecules noticeably influence the optical properties at ligand controlled phase transitions. Our findings highlight that single functionalized heterocyclic molecules can be considered as extremely sensitive probes for the complex interface physics and dynamics of colloidal semiconductor QDs. In this respect, the observed QD PL changes (for alone QDs and QD–dye nanoassemblies) upon temperature lowering (especially for liquid solutions) should be analyzed not only from the common approach (including thermally activated carrier trapping (Dayal and Burda, 2007)), and/or the electron-phonon coupling decrease (Klimov, 2000; Morello et al., 2007; De Mello Donega et al., 2006; Liptay and Ram, 2006; Liptay et al., 2007; Fernee et al., 2008) but take into account also the influence of surface states formed by ligand attachment and depending on QD surface transformations upon temperature changes. Briefly, it means that the spatially energetic reconstruction of trap states caused by the phase transition of capping ligand layers has impact even on the QD core structure and exciton–phonon coupling. The reader may find the detailed and additional information in this respect in Chapters 2 and 3 of Von Borczyskowski and Zenkevich (2016).

1.2.2.3 Quantitative studies of QD PL quenching and porphyrin fluorescence sensitization in QD–porphyrin nanoassemblies

It follows from experimental results presented in Section 1.2.2.1 (Figs. 1.7 and 1.8) that the formation of QD–porphyrin nanocomposites leads to the increase of the non-radiative relaxation channels in the excited states of QD namely while the fluorescence of the attached porphyrin is not quenched. These features are typical for the energy transfer events (via FRET or other mechanisms (Clapp et al., 2004, 2005, 2006; Zenkevich et al., 2005; Tsay et al., 2007; Jhonsi and Rengnathan, 2010; Lemon et al., 2013; Bean et al., 2014)) and for single QD–dye nanoassemblies (Kowerko et al., 2009, 2010; Pons et al., 2006; Hadar et al., 2015). In this respect, the main question is: in what extent namely FRET QD $\rightarrow$ Dye is fully responsible for QD PL quenching in nanoassemblies. Usually in the FRET case, the direct verification of the energy transfer process as a real reason of PL quenching is the comparison of the experimental values of FRET efficiencies via the donor (QD) PL quenching and the sensitization of the acceptor (porphyrin) fluorescence. Such a quantitative comparison has been carried out by us for QD–porphyrin (Zenkevich et al., 2005, 2009; Blaudeck et al., 2008, 2012) and QD–AM–perylene bisimide self-assembled nanostructures including both bulk solutions and single particle detection level (Kowerko et al., 2009, 2010). Below, we quantitatively analyze the results (shown in Figs. 1.7 and 1.8) being obtained for nanoassemblies based on TOPO-capped CdSe/ZnS QDs (d_{CdSe} = 3.0 nm, 2 ZnS monolayers, $C_{QD} = 4 \times 10^{-7}$ M) and $H_2P(m\text{-}Pyr)_4$ molecules in toluene at 295 K.

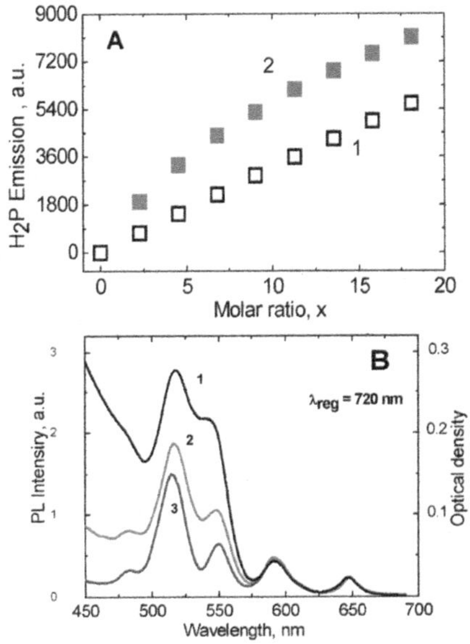

Figure 1.14 Porphyrin $H_2P(m\text{-}Pyr)_4$ fluorescence intensities in toluene at two experimental conditions. A: Integrated fluorescence intensity of individual $H_2P(m\text{-}Pyr)_4$ molecules (1) and enhanced integrated fluorescence intensity of these molecules in the presence of CdSe/ZnS QDs (d_{CdSe} = 3.00 nm/2 ZnS monolayers) (2) at the same molar concentrations of porphyrin and excitation conditions (λ_{exc} = 465 nm). B: Comparison of the absorption spectrum (1) and porphyrin fluorescence excitation spectrum (2, λ_{reg} = 720 nm, corrected for excitation spectral response) for a solution of CdSe/ZnS QD with $H_2P(m\text{-}Pyr)_4$ at molar ratio x = 4.0; (3) fluorescence excitation spectrum for alone porphyrin at the same $H_2P(m\text{-}Pyr)_4$ concentration and experimental conditions as for case (2). A remaining QD emission in the porphyrin fluorescence spectral range has been subtracted. Spectra (1), (2) and (3) are normalized with respect to the intensity of Q(0,0) porphyrin band (λ = 647 nm). Adapted from Zenkevich et al. (2011).

For QD–porphyrin nanoassemblies, the direct comparison of QD PL quenching results and sensitization data for porphyrin fluorescence was carried out using a complete set of titration points. Figure 1.14A shows that at the same experimental conditions (λ_{exc} = 465 nm, the donor QD strong absorption) upon increasing the molar ratio x in solutions of QD–porphyrin nanoassemblies the integrated fluorescence intensities of H_2P(m-Pyr)$_4$ molecules are noticeably larger than those being measured for individual porphyrin molecules. This finding reflects the existing of the sensitization (enhancement) effect that is typical for FRET phenomena. In addition, the fluorescence excitation spectrum of H_2P(m-Pyr)$_4$ ligand in nanoassembly (λ_{reg} = 719 nm) also shows the existence of QD→porphyrin FRET process (Fig. 1.14B).

Nevertheless at a given molar ratio x, there is no full coincidence between absorption and fluorescence excitation spectra for QD–porphyrin nanoassemblies. Thus, the realization of FRET with a quantum efficiency of 100% is not the case for the nanoassembly under study. It means that competing non-FRET processes may realize in such nanoassemblies.

In order to evaluate a real contribution of FRET into the whole QD PL quenching in nanoassemblies at every titration step, we measured QD PL quenching efficiency using equation

$$E_{\text{QUENCH}} = 1 - I(x)/I_0, \tag{1.4}$$

where I_0 is QD integrated PL intensity without porphyrin molecules in solution, and $I(x)$ is that for QD–porphyrin solutions at given molar ratio x.

In its turn, FRET efficiencies Φ_{FRET} were obtained via porphyrin fluorescence sensitization effect using an approach described in (Zenkevich et al., 1986; Clapp et al., 2004; Knowles et al., 2010; Zenkevich et al., 2005; Lakowicz, 2006)

$$E_{\text{FRET}} = [\varepsilon_A(\lambda_{ex})F_{AD} - F_A]/[\varepsilon_D(\lambda_{ex})F_A \cdot n], \tag{1.5}$$

where F_{AD} corresponds to the acceptor (A, porphyrin) fluorescence intensity measured at λ_{em} (651 nm, $Q(0,0)$ band maximum of porphyrin while exciting at the strong donor absorption λ_{ex} = 465 nm). F_D corresponds to A emission at the same λ_{em} and λ_{ex} in the absence of QD (donor). n is the number of A molecules per D, ε_A

and ε_D are molar extinction coefficients of A and D, respectively, at the excitation wavelength λ_{ex}.

The comparison of QD PL quenching data and FRET efficiency values Φ_{FRET} found via the porphyrin sensitization are shown at every titration step in Fig. 1.15. It follows from this figure that for every molar ratio x, Φ_{FRET} values of ~6–14% (found from the acceptor sensitization data, curve 2, right scale) are significantly smaller than those estimated from QD PL quenching efficiency [$\Phi_{QUENCH} = 1 - I/I_0$ ~50–80%, curve 1, left scale].

The same differences between Φ_{FRET} (sensitization) and Φ_{QUENCH} (quenching) values have been observed for all QD–porphyrin nanoassemblies based on $H_2P(m\text{-}Pyr)_4$ porphyrin molecules and CdSe/ZnS or CdSe QDs of various sizes (Zenkevich et al., 2005; Blaudeck et al., 2008, 2012). Thus, the obtained results demonstrate clearly that FRET is not a dominant reason of QD PL quenching in QD–porphyrin nanoassemblies. Below we will discuss some other mechanisms which may be responsible for the observed QD PL quenching upon attachment of functionalized dye molecules.

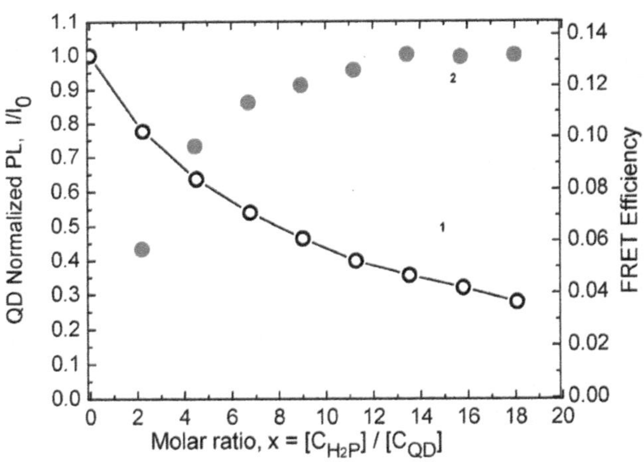

Figure 1.15 Comparison of QD PL quenching I/I_0 (1, left scale) and Φ_{FRET} values (2, right scale) estimated via the porphyrin fluorescence sensitization for QD–porphyrin nanoassemblies based on TOPO-capped QD CdSe/ZnS (d_{CdSe} = 3.0 nm, two ZnS monolayers) and $H_2P(m\text{-}Pyr)_4$ molecules as function of the molar ratio $x = [C_{H2P}]/[C_{QD}]$ in toluene at 295 K. Adapted from Zenkevich et al. (2011).

1.2.3 Nanoassemblies Based on QDs and Perylene Diimide Molecules

Another class of dye molecules for self-assembled nanoaggregates are pyridyl functionalized perylene diimide (PDI) molecules. In this case we have chosen CdS/ZnS with long chain amine (AM) ligands to allow for intercalation of appropriate anchor groups of PDI molecules (see Figs. 1.2E and 1.6C). Several reasons for choosing of PDI molecules may be pointed out: (i) alternative investigation (verification) of non-FRET processes; (ii) investigation of the influence of the ligand-related phase transition; (iii) variation of the anchoring conditions for PDI molecules as compared to porphyrins which allows for different geometric orientations of dyes with respect to the QD surface (Blaudeck et al., 2008, 2012; von Borczyskowski and Zenkevich, 2014; Zenkevich et al., 2005, 2007, 2009, 2011, 2014, 2015; Kowerko et al., 2009, 2010); (iv) most importantly, an additional unique information via single QD and single dye molecule detection in an individual nanoassembly, which becomes feasible (contrary to porphyrin dyes) due to the high quantum yield of fluorescence and large photostability of PDI molecules (Wuerthner, 2004; Kowerko et al., 2009, 2010; Kowerko, 2010; Zenkevich et al., 2012); and (v) investigation of the influence of the solvent polarity on the competition between FRET and non-FRET quenching processes down to the single assembly level.

1.2.3.1 Nanoassembly formation and manifestation of surface-related and temperature effects

Typical experimental spectral-kinetic findings showing the formation of QD–AM–PDI nanoassemblies in toluene at ambient temperature (Figs. 1.16 and 1.17) are very similar to those found for QD–porphyrin nanoassemblies in liquid solvents (presented in Section 1.2.2.1). Figure 1.16 shows that upon nanoassembly formation, QD–AM PL intensities change while spectral band shapes and position in the energy scale remain constant. In this case, during titration, the formation of nanoassemblies does not change QD–AM absorption properties while QD PL intensity is essentially quenched (Fig. 1.16) followed by the PL decay shortening.

It should be noted that absorption and fluorescence spectra of five PDI dyes (with various number and position of anchoring groups) used for the attachment to QD–AM surface are almost identical, but the comparative QD–AM PL quenching results show noticeable difference (Kowerko et al., 2010). The strongest quenching is observed upon QD–AM–DPP nanoassembly formation (pink curve in Fig. 1.17B). Quenching is quantitatively similar for TPP and DTPP and somewhat stronger than for PP and DAP. At x = 1 the ratio I_x/I_0 becomes 0.88, 0.82 and 0.66 for PP, DTPP and DPP, respectively (shown by dashed lines in Fig. 1.17B).

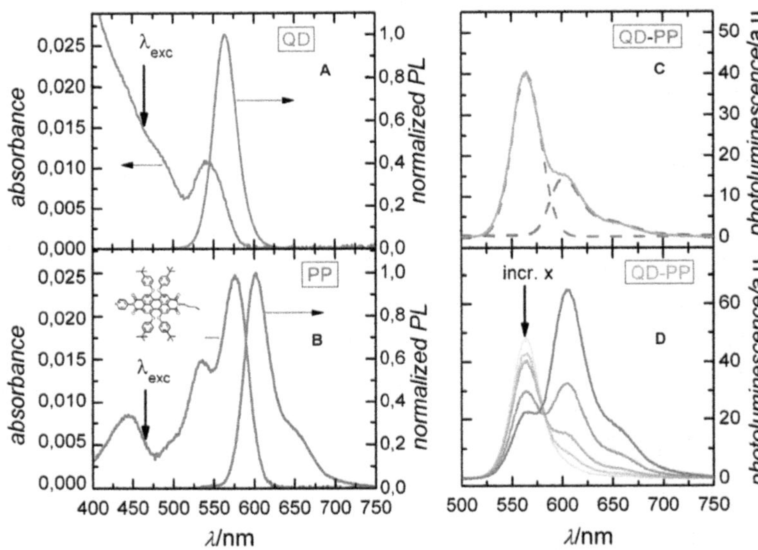

Figure 1.16 Spectral properties of alone amine capped QDs (QD–AM, d_{CdSe} = 3.2 nm, two ZnS monolayers) and (pyridyl)$_1$-PDI molecules (PP, structure is shown in insert) as well as QD–AM–PDI nanoassemblies in toluene at 295 K. (A) Absorption and emission spectra (λ_{exc} = 465 nm) spectra of 0.1 μM QD–AM. (B) Absorption and emission spectra (λ_{exc} = 465 nm) of 0.55 μM PP. (C) Photoluminescence of QD–AM–PP nanoassemblies at x = 1. Broken lines present a deconvolution of the whole PL spectrum into two components belonging to QD–AM and PP, respectively. (D) Photoluminescence quenching of QD–AM upon titration by PP at molar ratios x = 0.2, 0.5, 0.9, 2.5, and 5 (line color from orange to brown). Adapted from Kowerko et al. (2010).

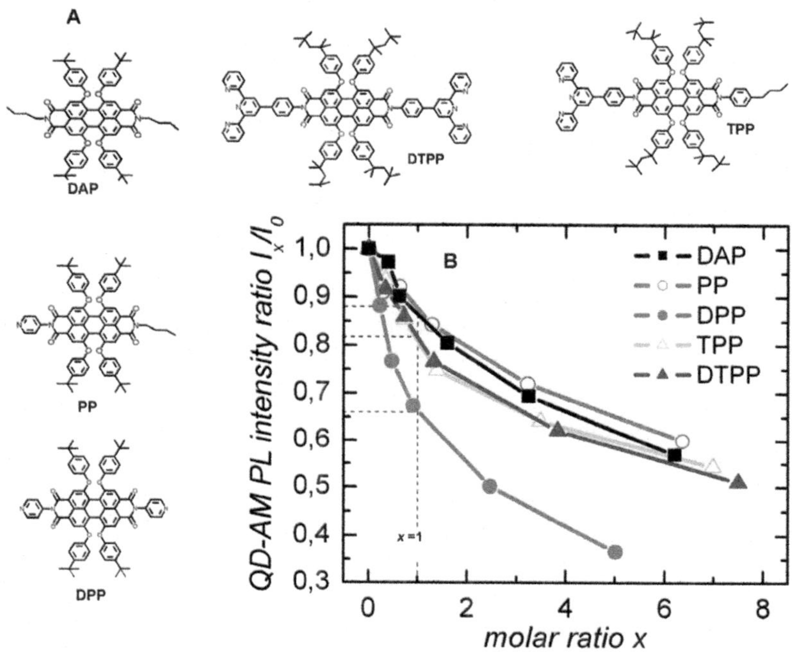

Figure 1.17 Chemical structures of perylene diimides PDI (A) and relative photoluminescence intensity dependence I_x/I_0 (B) for amine capped CdSe/ZnS quantum dots, QD–AM (d_{CdSe} = 3.2 nm, two ZnS monolayers) in toluene at 295 K. (A): Abbreviations for perylene diimides correspond DAP: dialkyl-DPI; PP: $(pyridyl)_1$-PDI; DPP: $(pyridyl)_2$-PDI; TPP: $(ter\text{-}pyridyl)_1$-PDI; DTPP: $(pyridyl)_2$-PDI. (B): Vertical dashed line corresponds to molar ratio $x = [C_{PDI}]/[C_{QD\text{-}AM}]$ for all samples. Adapted partly from Kowerko et al. (2010).

1.2.3.1.1 Intercalation

The unexpected result is, that DAP molecules (without any functional pyridyl group) quench the QD–AM PL almost to the same extent (black line) as compared to that found for PP molecules (red line). In case of DAP, we apt to believe that the attachment might be realized for dyes without pyridyl groups by intercalation of the aliphatic tale of DAP into the corresponding ones of the AM layer. A similar way for the formation of self-assembled QD–dye nanostructures has been proposed in (Dayal et al., 2006; Lim et al., 2008).

The details of the interaction of PDI dyes with the QD surface are not known, but it is reasonable to assume that they can be sorted into two closely related classes, namely DPP-type one-pyridyl and DTPP-type two-pyridyl coordination (the third lone pair orbital can for steric reasons not attach to the QD surface) via nitrogen lone pairs or electrostatic interaction between partial negative charge densities at the N-atom of pyridyl with surface Zn^{2+} ions (see, e.g., Bullen and Mulvaney (2006), Koole et al. (2008), Holt et al., 2008, Ji et al. (2008), and Schapotschnikow et al. (2009)). A third class of attachment principles is namely the DAP-type intercalation into the AM ligand shell. PP and TPP are mixed cases since they are subject to a competition between pyridyl-type coordination and intercalation. DAP did not cause quenching in case of TOPO ligated QD (Kowerko, 2010). This is reasonable since TOPO has a conical-type structure (see Fig. 1.6) which does not easily allow for an intercalation of DAP, PP or TPP, respectively.

At first glance it seems evident that the functional pyridyl group itself plays a major role in the quenching mechanism for QD–AM PL. Nevertheless, like for QD–porphyrin nanoassemblies (see Fig. 1.11A) quenching by pyridine is negligible for QD–AM PL even at molar ratios as high as $x = 105$ (Kowerko, 2010). On the contrary, strong quenching has been observed for uncapped CdSe QDs (Zenkevich et al., 2005). Also thiol-type dyes are strong PL quenchers in case of CdSe but only weak quenchers in case of CdSe/ZnS (Bullen and Mulvaney, 2006; Munro, et al., 2008; Koole, et al., 2008; Holt, et al., 2008; Ji, et al., 2008; Schapotschnikow et al., 2009). In this case it has been suggested that hole transfer has a strong impact on the amount of CdSe quenching. In case of pyridyl-type dye attachment it may be suggested that it is necessary to attach "bulky" molecules to the QD surface in order to efficiently quench QD PL. However, like for QD–porphyrin nanoassemblies (Zenkevich et al., 2007a, 2007b), the detailed electronic structure seems to be less important in QD–AM–PDI case. This is also evident from single molecule experiments carried out in our group (Kowerko, 2010): upon permanent bleaching of the dye neither the QD PL intensity, nor the PL decay time are increased, which would be the case if CT or FRET would be the major quenching mechanisms (Ren et al., 2008).

1.2.3.1.2 Determination of assembly geometry

As has been studied for QD–porphyrin interactions (discussed in Section 1.2.2.1), using PDI molecules with two different pyridyl-type functional groups (namely DPP, and DTPP, the corresponding structures are shown in Fig. 1.17A), it is possible to realize geometrically well defined but different QD–PDI nanoassemblies, which permit to determine the dependence of dynamic QD–dye interactions upon variation of the spatial arrangement of the dye molecule with respect to QD surface (e.g., distance and orientation) (Kowerko et al., 2009, 2010; Kowerko, 2010). In fact, it follows from experimental data obtained for AM-capped CdSe/ZnS QDs (three ZnS monolayers, total diameter of 5.1 nm) that QD–AM PL quenching is stronger for DTPP molecules with respect to that found for the DPP at the same molar ratio *x*, though the spectral properties of DTPP and DPP molecules are nearly the same. The observed QD PL quenching difference can be related to various geometrical displacements (e.g., different attachment angles for pyridyl and terpyridyl anchoring groups) for these two molecules on the QD surface (Fig. 1.18A).

In the case DTPP molecules a parallel placement of the chromophoric part (optical transition dipole moment) on the QD surface is reasonable, since the attachment of the terpyridyl fragment to the QD surface includes the coordination of the *ortho*-pyridyl groups, which can only be realized in a flat and slightly drilled configuration of the terpyridyl fragment. A direct experimental consequence of this coordination to the surface is a spectral red shift observed for DTPP in parallel to an enlarged removal of capping AM (Kowerko et al., 2009). Thus, for DPP and DTPP molecules with respect to the orientation to the QD surface, the centre-to-centre donor-acceptor distance is calculated to be R_{DA}(DPP) = 3.6 nm and R_{DA}(DTPP) = 4.75 nm, respectively. This implies that FRET (if it is one reason of QD PL quenching) should be stronger for DPP than for DTPP, which is quite opposite to what is observed experimentally. However, taking into account the steric restrictions discussed above the FRET efficiency was calculated as a function of the orientation φ of the long molecular axis with respect to the QD surface as is shown in the inset of Fig. 1.18A. In the result, Φ_{FRET} was estimated to be 0.85 for DTPP at $\varphi = 0°$ and 0.68 for DPP at $\varphi = 90°$ which is in appropriate agreement with the experimental results.

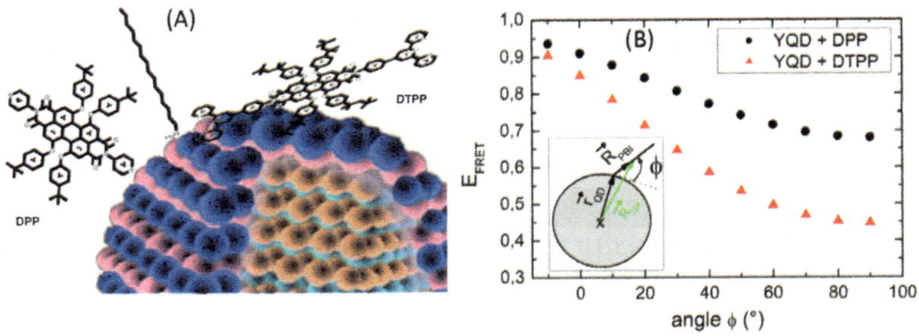

Figure 1.18 Scheme of surface attachment of DPP and DTPP molecules to a surface of AM-capped CdSe/ZnS QD (A) and calculated FRET QD → Dye efficiency for DPP and DTPP as a function of angle ϕ (B). The insert represents a scheme of geometrical model being used in calculations. From geometrical constraints the orientation for DTPP molecule with respect to QD surface corresponds to $\phi \approx 0°$ and for DPP molecule to $\phi \approx 90°$, respectively. Simple calculations of FRET efficiency (Φ_{FRET}) have been carried out using the standard Foerster model $\Phi_{FRET} = [1 + (R_{DA}/R_0)^6]^{-1}$, where R_0 corresponds to the Foerster transfer radius and R_{DA} is the centre-to-centre donor-acceptor distance. While almost all relevant parameters are nearly the same with respect to DPP and DTPP, they differ considerably in values of $R_{DA} = R_{PDI}/2 + r_{QD}$, where $R_{PDI}/2$ corresponds to the distance of the coordinating point to the centre of PDI chromophore, while r_{QD} is the CdSe/ZnS QD radius. Adapted from Kowerko (2010).

It should be noted that the FRET efficiency is quite low (even lower as compared to some of the QD–porphyrin assemblies) with respect to the overall PL quenching effect for all QD–PDI nanoassemblies being studied in toluene at ambient conditions (Kowerko et al., 2009, 2010). Thus, in spite of that relative FRET efficiencies helped to identify two different geometries of PDI molecules with respect to the QD surface, these differences may be completely masked by low FRET efficiencies experimentally observed for these nanoassemblies.

1.2.3.1.3 Influence of temperature on assembly formation and PL quenching

Finally, we varied the temperature to investigate the impact of dye molecules and ligands on the phase transition already discussed for QD–porphyrin nanoassemblies. We performed experiments both for AM-capped quantum dots QD–AM and QD–PDI nanoassemblies. Also in case of amine capping ligands, we observe a similar phase transition type behavior as is shown in Fig. 1.19.

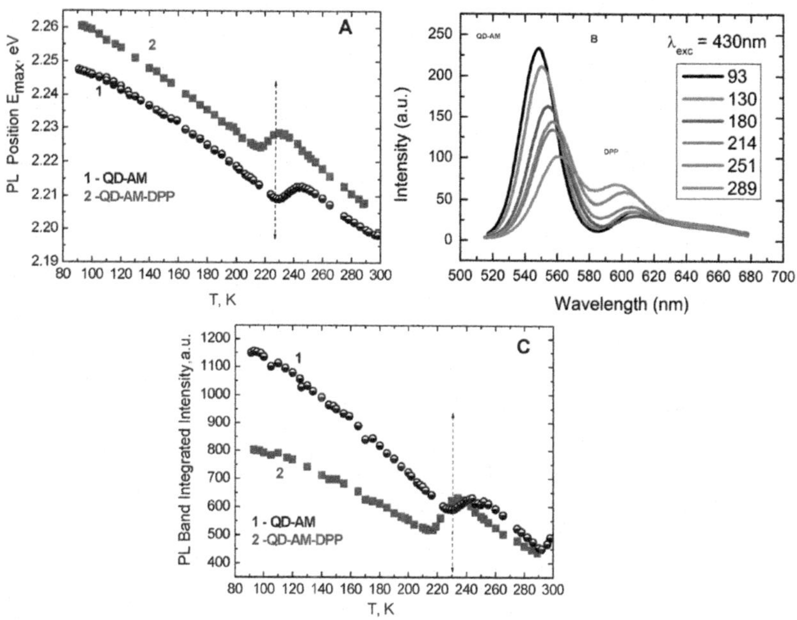

Figure 1.19 Temperature dependence of (A) PL band energies E_{max}, (B) PL spectra, and (C) PL band integrated intensities for alone CdSe/ZnS-AM QDs (1, black points, d_{CdSe} = 3.2 nm, two ZnS monolayers) as well as for QDs in (2, blue points) QD–AM–DPP nanoassemblies at molar ratio $x = [C_{DPP}]/[C_{QD}] = 1$ in methylcyclohexane-toluene (6:1) mixture at λ_{exc} = 430 nm. Fits in parts (B) and (C) have been obtained by assuming only one Gaussian component. The mean temperature of a phase transition assigned to the capping AM layer ($T_{crit} \approx 230$ K) is indicated by a dashed arrow. Adapted partly from Zenkevich et al. (2012, 2015).

When defining the temperature at the mid-intensity of the "kink" as the apparent phase transition temperature, this temperature is about 5 K higher for QDs as compared to QD–dye nanoassemblies. In this case the "kink" is more pronounced than in case of TOPO ligands and is shifted to 235–240 K as compared to 230 K. Notably, DPP (the structure is shown in Fig. 1.17A) does not show any "kink" in the absence of QDs (Zenkevich et al., 2012, 2015). It is seen that the attachment of DPP molecules enhances the decrease of QD PL intensity and shifts the phase transition to lower temperatures. The comparison of data presented in Figs 1.19 (for QD–DPP nanoassemblies) and 1.13 (for QD–porphyrin ones) reveals that in the case of TOPO ligands QD PL changes at the phase transition temperature are more pronounced with respect to those observed for AM ligands upon attachment of functionalized dye molecules. At the same time, the QD–DPP nanoassembly formation does not influence the phase transition temperature T_{crit}. We tentatively explain this difference by the fact that DPP removes upon surface attachment more ligands than H_2P does (Kowerko et al., 2010; Blaudeck et al., 2012) thus having a larger impact on the phase transition temperature.

The other reason of the observed differences in the range of phase transition for TOPO- and AM-capped QDs may be connected with the different CdSe core size (d_{CdSe} = 3.2 nm for QD–AM and d_{CdSe} = 3.0 nm for QD–TOPO). In fact, our results are in agreement with earlier observations showing the noticeable dependence of the phase transition effects on the type of ligands (De Mello Donega et al., 2006), and, additionally, supports the suggestion that a phase transition of ligands couples to the optical properties of QDs.

1.2.4 Role of the Solvent Polarity in Competition between FRET and Non-FRET Quenching Processes for QD PL in Nanoassemblies

As was shown in Section 1.2.2.1, the dynamic equilibrium between capping TOPO molecules and functionalized porphyrins upon attachment to QD surfaces as well as QD PL quenching processes (including FRET and non-FRET) for QD-porphyrin nanoassemblies depend critically on the kind of the solvent being smaller for low polar solvents (*n*-octane versus toluene). Addition of polar acetone into toluene solutions containing QD–porphyrin nanoassemblies increases significantly the QD PL quenching (Zenkevich et al., 2005), which indicates the involvement of charged QD states as far as ligand binding realizes via electrostatic forces (Schapotschnikow et al., 2009). Such effects were also observed for various TOPO-capped CdSe/ZnS or CdSe QDs (Blaudeck et al., 2008, 2012; Zenkevich et al., 2007a, 2007b, 2009; von Borczyskowski and Zenkevich, 2014) as well as in QD-AM-PDI nanoassemblies based on AM-capped CdSe/ZnS QDs (Kowerko et al., 2009, 2010; Kowerko, 2010). To investigate this finding in more detail we present detailed studies of solvent variation on TOPO capped QDs and making use of the possibility to study the related influence on the single molecule, single QD level.

In most cases QD–porphyrin and QD–PDI nanoassemblies have been studied in the relatively low-polar toluene (ε = 2.38). Here, for bulk solutions containing TOPO-capped CdSe/ZnS QDs and DTPP molecules, we will discuss in more detail what happens with QD PL quenching and competing FRET and non-FRET processes when using a solvent with considerably reduced polarity (ε < 2.3) like TEHOS (tetrakis-2-ethylhexoxy-silane having long aliphatic chains). The related structure is shown in Fig. 1.20). The focus on polarity is motivated by the finding that blinking processes in single QDs depend strongly on solvent polarity (Cichos et al., 2007; Issac et al., 2005; Schmidt et al., 2014) and is also discussed intensively in Chapters 4 and 6 of Von Borczyskowski and Zenkevich (2016).

Before proceeding, we like to emphasize that FRET efficiencies obtained for single QD-DTPP nanoassemblies in TEHOS are essentially higher ($\Phi_{FRET} \sim 0.7$) (Gerlach et al., 2013) with respect to those ($\Phi_{FRET} \leq 0.1$) for the same QD–DTPP nanoassemblies both from ensemble (in toluene) and single nanoassembly detection (on a SiO_2 surface) (Kowerko et al., 2009, 2010) as well as for QD-porphyrin nanoassemblies (in toluene, discussed in Section 1.2.2.3) (Zenkevich et al., 2005, 2007a, 2007b; Blaudeck et al., 2008, 2012). Experimental results on single QD–DTPP nanoassemblies in TEHOS will be discussed in more detail later on in Section 1.3. As was mentioned above, low FRET efficiencies may be related to the presence of competing non-FRET processes, which open a quenching pathway other than the non-radiative energy transfer by merely attaching a dye molecule to the QD surface (Zenkevich et al., 2007, 2009, 2015; Blaudeck et al., 2008, 2012; Kowerko et al., 2009, 2010; Kowerko, 2010; von Borczyskowski and Zenkevich, 2014). This will be discussed separately in Section 1.4.

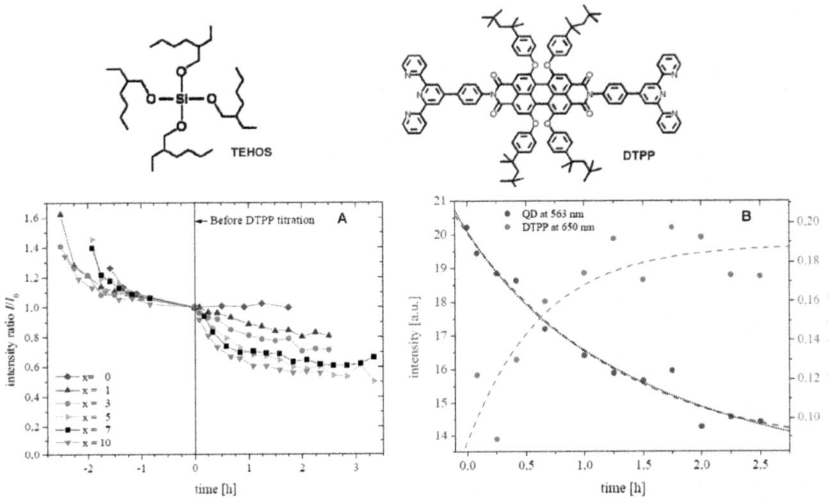

Figure 1.20 Temporal changes of relative intensities I/I_0 for TOPO-capped CdSe/ZnS QDs upon titration by DTPP molecules in TEHOS (λ_{exc} = 412 nm). (A): Titration of with various relative DTPP molar ratios x at t = 0. The first spectra are taken 1 min after titration. Before t = 0 all samples come nearly to an equilibrium with x = 0 in each case. (B): Time dependent PL intensities of CdSe/ZnS QDs and DTPP at x = 7. The first DTPP spectra have been taken about 1 min after titration. Lines present bi-exponential fits for CdSe/ZnS QD and mono-exponential one for DTTP. Chemical structures of TEHOS and DTTP molecules are shown in the top. Adapted from Von Borczyskowski and Zenkevich (2014).

Changes of the PL intensity ratio I/I_0 at waiting time after sample preparation for CdSe/ZnS PL upon titration by DTPP in TEHOS at various molar ratios x are presented in Fig. 1.20. In these experiments λ_{ex} = 412 nm was used, at which QD absorption is 100 times stronger compared to that for DTPP. It means that the direct excitation of DTPP molecules is negligible, and the formation of excited DTPP molecules is realized for these conditions via FRET QD → DTPP. It is seen in Fig. 1.20A, that the QD PL quenching immediately after the respective titration step is very small but increases within in time (hours) as a function of x.

Interestingly, quenching times become shorter with increasing *x*. Figure 1.20B also shows that in addition to a temporal quenching of QD PL intensity (fitted by two exponential decay functions) a comparable increase of DTPP fluorescence (fitted by 1 exponential function) takes place on the same time scale at a molar ratio $x = 7$.

The analysis of fitting results obtained for the whole data set at various molar ratios *x* during the observation time shows (von Borczyskowski and Zenkevich, 2014) that one of the QD PL quenching times is very close to that for DTPP fluorescence increase. This time correlation is observed at every molar ratio *x*. Therefore it is reasonable to assign this component of the QD PL quenching to FRET QD → DTPP, whereas the second component is related to non-FRET quenching. A quantitative consideration of DTPP fluorescence enhancement (sensitization) shows that on average ~40% of the QD PL quenching for these QD–DTPP nanoassemblies in TEHOS (bulk solutions) can be attributed to FRET (von Borczyskowski and Zenkevich, 2014). This high FRET efficiency reflects the same tendency showing that for single QD–DTPP nanoassemblies in TEHOS FRET is even more effective with $\Phi_{FRET} \sim 0.7$ (Gerlach et al., 2013).

Like in the case QD–porphyrin nanoassemblies in toluene, it possible to study the dependence of QD PL quenching dynamics on the type of dye molecule for QD–PDI nanoassemblies in THEOS. Figure 1.21 shows the comparative results of titration experiments obtained for DAP and DTPP molecules at two molar ratios $x \approx 3$ and 10, correspondingly.

On the basis of the whole data set comparatively obtained during the observation time for QD–DTPP and QD–DAP solutions in TEHOS at the same molar ratios (x = 1, 3, 5, 7, 10) the following conclusions have been evaluated: (i) time-dependent QD PL quenching is shortened for both DTPP and DAP dyes upon *x* values increase; (ii) at each value of molar ratio *x*, QD PL quenching dynamics for DAP are by about a factor of 2–3 faster than those for DTPP; (iii) the final degree of quenching is somewhat larger for DTPP as compared to DAP (von Borczyskowski and Zenkevich, 2014). It is seen from Fig. 1.17 that DAP molecules (without any pyridyl group) quench the QD–AM PL almost to the same extent like PP molecules (having anchoring pyridyl group) (Kowerko, 2010). This result was explained by DAP intercalation into the ligand shell probably in an upright position of the long molecular axis with respect to the QD surface. In contrast, as it is seen in Fig. 1.18A, for DTPP molecule a parallel displacement on QD surface is preferable. To realize such an orientation more capping ligands (TOPO or AM) will have to be replaced in case of DTPP as compared to DAP, which will take a longer time as it follows from the experimental data. However, the energy transfer will be stronger for DTPP since the Foerster-related FRET efficiency is larger for a parallel as compared to a perpendicular orientation (Kowerko et al., 2009, 2010; Kowerko, 2010) as is also found in the experiment.

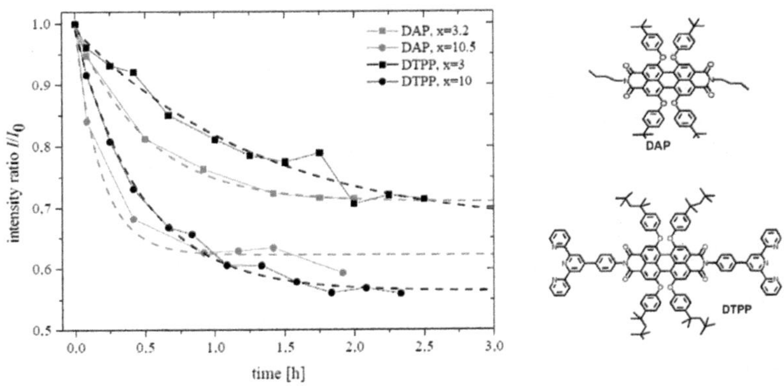

Figure 1.21 Temporal changes of relative luminescence intensities I/I_0 for DAP (red) and DTTP (black) upon attachment to CdSe/ZnS QDs (each at two molar ratios $x \approx 3$ and 10, correspondingly). Broken lines correspond to mono-exponential fits. Dyes structures are shown on the right TEHOS at RT.

We suggest that the differences in formation dynamics when comparing solvents like toluene and TEHOS are due to the different polarities which control competing ligand and dye solubility in the respective solvents. In TEHOS, (polar) ligands prefer to remain on the QD surface while (non-polar) PDI molecules prefer to remain in TEHOS. The opposite holds in toluene. Nevertheless, finally assemblies are formed which counts for an increased lowering of total solvation energies. An additional argument is related to differences in viscosity since TEHOS is highly viscous at room temperature, which will also influence (diffusion controlled) assembly formation.

Besides formation dynamics obviously dependent—at least partly—on the solvent polarity, we will in a next step discuss the influence of the solvent (polarity) on PL quenching and FRET efficiencies. Already earlier in this chapter, we have shown that an increase/decrease of solvent polarity enhances/diminishes PL quenching. Since PL decay times of QDs do not strongly depend on solvent polarity (Schmidt et al., 2014) FRET efficiencies should be nearly independent on the solvent polarity. Thus all solvent dependent PL quenching has to be cast on non-FRET processes (Kowerko et al., 2010). Moreover, recent experiments on blinking dynamics of QDs have shown (Cichos et al., 2007), that an increase of solvent polarity (increase of the dielectric constant ε) results in an increase of the lengths of times during which PL is "switched" off while at the same time PL on -times are shortened (Issac et al., 2005). Qualitatively this can be described by the influence of charges or polarization of QD emission. The question now is whether non-FRET PL quenching acts in the same way, namely an increase (or even creation) of charge-influenced QD states via nanoassembly formation.

In order to study the dependence of FRET efficiency on the solvent polarity a series of comparative experiments has been carried out for QD–DTPP nanoassemblies (at $x = 5$ and 7 for TOPO-capped CdSe/ZnS QDs) diluted in a toluene/TEHOS mixture of various relative volume ratios R (von Borczyskowski and Zenkevich, 2014). The comparative analysis of time dependent luminescence spectra (Fig. 1.22, left) and intensities (right) obtained after titration with DTPP at $x = 5$–7 leads to two evident conclusions: (i) the time scales of QD PL quenching and DTPP fluorescence increase are much faster for the mixture of the solvents ($R = 50{:}50$) as for TEHOS samples discussed before, and (ii) the DTPP fluorescence first increases strongly during about 9 min and thereafter decreases again on a much longer time scale. The letter process is accompanied by a red shift of the DTPP fluorescence band maximum by about 5 nm (see Fig. 1.22A). This red shift equals the shift of the fluorescence maximum when measuring DTPP in TEHOS or in toluene. This result may be explained by a time dependent separation of the two solvent components in such a way, that just after preparation, QD–dye nanoassemblies are predominantly solvated by TEHOS component and later on by a more toluene rich environment. This finding is supported by the observation that FRET efficiency decreases as is expected for a nanoassembly in toluene. For a solvent mixture at R = 20:80 DTPP fluorescence properties are more closely to those in TEHOS, while they are more close to a dye fluorescence in toluene for R = 80:20. Correspondingly, we find a strong "overshoot" of luminescence intensity for R = 20:80, which is completely lost for R = 80:20.

Additionally, a comparison of Figs. 1.22A,C clearly shows that the DTPP fluorescence is very low at both solvent mixture ratios R = 100:0 and R = 0:100 even after long waiting times. In contrast, strong DTPP emission at R = 50:50 is immediately present after titration and further increases in time. Comparing the related time scales of increasing FRET (as manifested by increasing DTPP fluorescence intensities at various ratios R, Fig. 1.22B) reveals, that they are similar for QD PL quenching time scales. In the case of pure toluene (R = 100:0) the PL quenching (and DTTP enhancement) main time component is shorter than 1 min, the time needed to register the spectra. On the other hand, the maximum of PL quenching for pure TEHOS (R = 0:100) is about 0.5 which is very similar as for R = 100:0, but with the difference that the related FRET is very weak in the later case. This fact may be explained by the following way. At R = 0:100 a long time is needed to form QD–dye nanoassemblies, but FRET is very effective. In the opposite case of the relative volume ratio (R = 100:0), nanoassemblies are much more readily formed, but non-FRET processes are more effective than FRET as has also been shown in (Kowerko et al., 2010; Kowerko, 2010).

At present, a qualitative model for QD–dye interactions leading to QD PL quenching may be set up following ideas discussed in Section 1.2.2.2 (Blaudeck et al., 2012). In fact, non-FRET quenching can be assigned to a competition of dye-induced ligand detachment with subsequent assembly formation, which gives rise both to FRET and formation of specific surface states (related to remaining dangling bonds). These surface states give rise to non-FRET quenching due to specific sites, probably localizing charges and thus quenching the PL. Decreasing of polarity will destabilize charge localization (Schapotschnikow et al., 2009; Issac et al., 2005; Cichos et al., 2007).

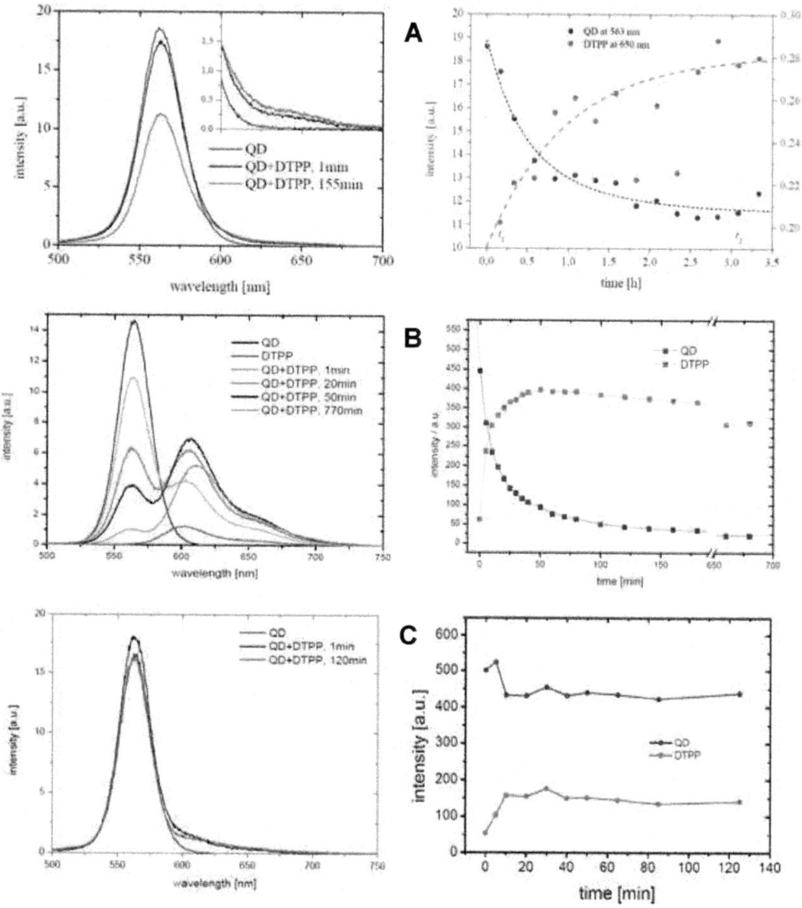

Figure 1.22 Dependence of time dependent luminescence spectra (left) and time dependent luminescence intensities (right) for QD–DTPP nanoassemblies (based on CdSe/ZnS QDs) on the solvent mixture ratio R = toluene/TEHOS: (A): R = 0:100 (x = 7); (B): R = 50:50 (x = 5); (C): R = 100:0 (x = 5). The first spectra have been taken about 1 min after titration. At excitation conditions (λ_{ex} = 412 nm) QD absorption was 100 times stronger compared to that for DTPP. Correspondingly, the direct excitation of DTPP molecules is very low, thus the formation of excited DTPP molecules may be realized only via FRET QD $\rightarrow$ DTPP.

On the other hand ligand (TOPO) detachment is less favorable in non-polar media, since the polar ligands have a low solubility. For this reason there is not enough space for large dye molecules to attach the QD surface. Hence, the nanoassembly formation needs a very long time. Therefore FRET events are observed only after a long waiting time but non-FRET quenching is relatively weak. For R = 100:0 the nanoassembly effective formation takes place within less than 1 min, since ligands are readily detached from the QD surface. At the same time surface states are created leading to the effective non-FRET QD PL quenching. It follows from the above results that the optimum of the solvent mixture ratio for effective FRET on a short time scale is close to R = 50:50. Even at R = 70:30 most of the PL quenching results in FRET. This observation leads to the conclusion that in the solvent mixture not merely the number of detached ligands has to be taken into account, but also the specific structure of the near solvent shell (solvate) surrounding the QD surface. Since TEHOS has long aliphatic chains structure (shown in Fig. 1.20), these chains probably intercalate with the aliphatic TOPO chains forming a stable TEHOS shell intercalated with the ligands. This shell is only broken up at sufficiently high toluene concentrations >70%. Only then the TEHOS shell is nearly absent and toluene may easily penetrate to the QD surface and stabilizes the non-FRET quenching states.

Concluding this part, titration experiments carried out in a toluene/TEHOS mixture of various relative volume ratios R demonstrate, that the kinetics of nanoassembly formation is complex but nevertheless provide detailed information not only on dynamic processes but also on the incorporation of QDs or nanoassemblies into complex and heterogeneous self-organizing environments. Additional experimental information showing the formation of new band-edge states stems from temperature dependent QD PL properties for similar nanoassemblies in organic glasses (Zenkevich et al., 2012, 2015) and will be discussed later on. Although most of the properties are inherent to the QDs themselves, dye molecules act as reporters or even initiate reorganization processes especially of QD surface states or within the ligand shell. Such well defined experiments provide insights, why colloidal QD chemistry is often critically dependent on small amounts of "impurities" during wet-chemical synthesis.

1.3 Single Nanoassembly

Typically, single molecule spectral experiments are usually exploited to investigate the distribution or manifold of conformations and interactions with the environment. It should be noted that number of experiments on the identification of single nanoassemblies is limited (Potapova et al., 2005; Ren et al., 2008; Hohlbein and Huebner, 2005; Roy et al., 2008), but even in these cases a complete quantitative determination of the involved quenching and transfer processes is not reported. In this Section, we will predominantly demonstrate our results being obtained for single nanoobjects (QDs, dyes and nanoassemblies) showing that this approach may be considered as a tool to precisely identify the interaction of exactly one QD with one dye molecule. In this case it is possible to receive an unique (in some cases) information on a microscopic understanding of the formation (including ligand dynamics) and related QD PL quenching dynamics in QD–dye nanoassemblies. Additionally, a comparison of results being obtained for the same bulk and single nanoassembly allows unraveling the contributions of different quenching mechanisms which are of importance for the identification of dynamic processes for QD–dye dye and/or QD-capping ligand systems in general.

1.3.1 Quantum Dot Blinking Statistics in QD–Porphyrin Nanoassemblies

In Section 1.2, it was shown for ensemble that the attachment of functionalized dye molecules to a QD surface manifests itself in noticeable QD PL quenching as well as in related complex interface dynamics accompanied by non-radiative relaxation dynamics for the exciton. Fortunately, QD PL quenching is also detected in single nanoassembly experiments. Especially blinking statistics are valuable tool to investigate various aspects of excited state dynamics. Figure 1.23 shows the comparison of blinking statistics for two samples, namely spin coated toluene solution at 295 K containing either TOPO capped CdSe/ZnS QDs or the respective QDs and porphyrins, some of them forming nanoassemblies. Both samples contain the same initial QD concentration ($C_{QD} = 1.8 \times 10^{-9}$ M) and are excited within the QDs first excitonic absorption band. Abundant nanoassemblies self-assembly at a molar ratio $x = [C_{H2P}]/[C_{QD}] = 10$, at which the QD PL quenching is about 40% in ensembles in toluene solution (Zenkevich et al., 2005).

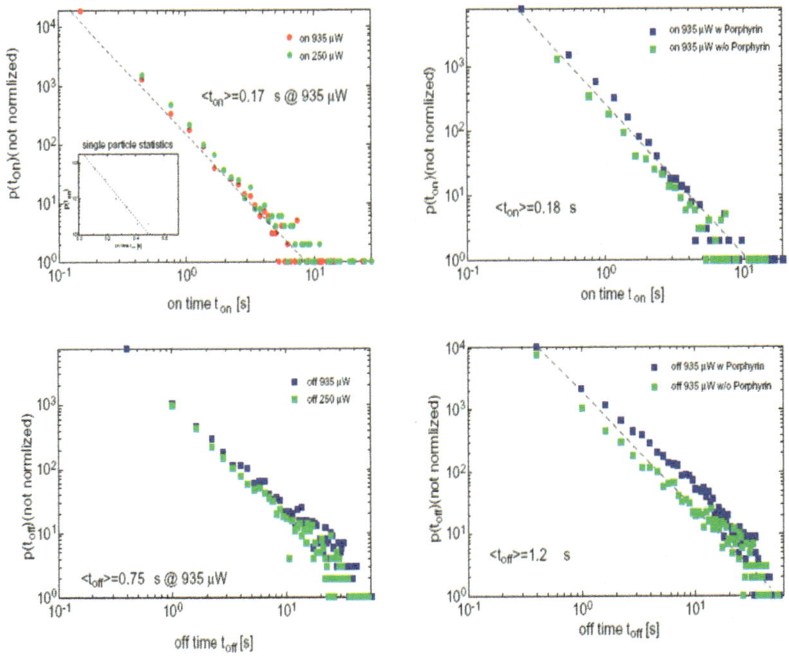

Figure 1.23 Comparative blinking statistics for a bunch of single TOPO-capped CdSe/ZnS QDs (a) and single QD–porphyrin nanoassemblies (b) upon two laser excitation power (250 and 935 μW) at $\lambda_{exc} = 514$ nm and 295 K. Sample properties: $d_{CdSe} = 3.2$ nm, three ZnS monolayers, $C_{QD} = 1.8 \times 10^{-9}$ M, $H_2P(m\text{-}Pyr)_4$ molecules (see structure in Fig. 1.9), molar ratio $x = [C_{H2P}]/[C_{QD}] = 10$. Samples have been prepared by spin coating from toluene solution onto quartz surface at 295 K. The presented data were averaged for at least 20 individual objects from alone and porphyrin-assembled QDs. Average "on"- and "off"-times are indicated in each graph. The inset in the top left figure shows the statistic of one single QD. Adapted from Zenkevich et al. (2012).

The comparison of results presented in Figs. 1.23A,B shows that for both cases blinking statistics is described by a power law distribution for "on-" and "off-"times. Dark QD states are usually explained by charged QDs (Issac et al., 2005; Schmidt et al., 2014), and the heterogeneity (power law behavior (Cichos et al., 2007)) is inherent to broadly distributed (de-)population processes of the dark state. In case of the presence of nanoassemblies, values for $<t_{on}>$ (0.18 s) do not change with respect to those measured for QD in the absence of dye molecules ($<t_{on}> = 0.18$ s). In contrast, a substantial increase of mean $<t_{off}>$ times is observed for QD with partly attached porphyrin molecules (1.2 s in comparison to 0.75 s). This elongation of dark periods is equivalent to PL quenching.

These findings are considered as a proof of QD–porphyrin interactions (that is the formation of QD–porphyrin nanoassemblies) leading to QD PL quenching also on a single assembly level. Unfortunately, because of low quantum efficiency of fluorescence as well as low photostability of porphyrin molecules it is impossible to detect the single porphyrin emission in comparative experiments with single nanoobjects.

1.3.2 PL Quenching, Geometry of Nanoassemblies and Conformational Stability of PDI Molecules on QD Surfaces

As we have discussed in Section 1.2.3, perylene diimide (PDI) dyes are ideal molecules for single molecule detection. Moreover, we have shown that they can be appropriately functionalized to form nanoassemblies with QDs. Here, for nanoassemblies based on amine-capped QDs and perylene diimide dyes we demonstrate spectral-kinetic observations on the single assembly level which may give additional information concerning distinct nanoassembly geometries and conformational mobility of PDI molecules attached to the QD surface. In the context of complementary ensemble experiments, such conformation dynamics obtained by single object spectroscopic approach may be used as probe for the surface coverage by ligands close to the dye (Kowerko et al., 2009, 2010; Kowerko, 2010). In this respect, it should be noted that additional details of nanoassembly formation and QD–dye interactions as well as chemical surface phenomena studied by time-resolved single molecule spectroscopy will be discussed also in Chapters 2 and 4 of this book.

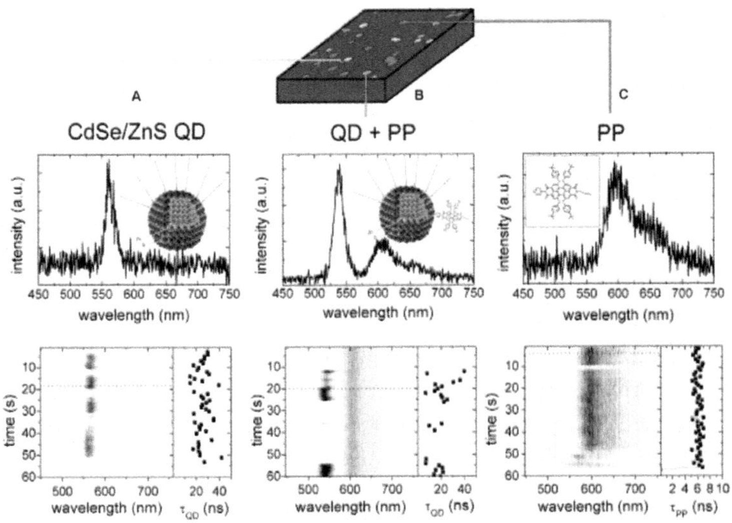

Figure 1.24 Emission spectra (λ_{exc} = 465 nm; P = 0.6 kW·cm^{-2}) of single amine-capped CdSe/ZnS QD (three ZnS monolayers, a total diameter 5.1 nm, first exciton absorption λ_{max} = 542 nm) and (pyridyl)$_1$-PDI molecule (PP). Spectra are related to various confocal spots of the spin-coated sample. It was possible to detect PL spectra (1 s binning time) of single QD (A), single PP (C), and single nanoassemblies with PP and QD spectra simultaneously (B). Typical luminescence intensities, spectral positions and decay times are shown as a function of observation time of 60 s and longer. Material quality was checked with a homebuilt fluorescence wide field microscope. Single particle sample preparation was done by spin coating a 5 × 10^{-11} M solution of the QD–dye mixture onto a Si/SiO_2 (100 nm thick SiO_2) surface. A confocal scan of a spin coated sample (x = 1 in solution) is shown on top. Adapted from Kowerko (2010).

The experimental approach to identify single nanoassemblies has been outlined in Chapter 4 of Von Borczyskowski and Zenkevich (2016). Emission spectra related to various statistically formed single nanoobjects are presented in Fig. 1.24. It is evident that the spectra can be divided into three different types belonging to various nanoobjects: QD PL (CdSe/ZnS, three ZnS monolayers, total diameter 5.1 nm, (A)); PP dye fluorescence (C), and the sum of the two spectra assigned to QD–dye nanoassembly (B). Essentially, all types of spectra are characterized by typical single quantum object fingerprints, namely emission intermittency (blinking) (Cichos et al., 2004, 2007) and spectral fluctuations (shown at the bottom of Fig. 1.24). The experimental emission decay times averaged over 500 ms are shown as a function of observation time on the right side of each series of spectra. In case of PP a nearly mono-exponential fluorescence decay varying between 4 and 6 ns is detected for different single molecules, while the decay time for QDs is fluctuating between 50 and 5 ns (experimental lower limit in time resolution).

It follows from Fig. 1.24B that at $t \approx 57$ s the fluorescence of the given PP molecule is bleached while the QD remains emitting because of higher photostability in comparison with an organic dye. It is seen also that PL band of the QD in a single nanoassembly is blue shifted during observation time which might be connected with photooxidation of the QD core and might be enhanced by the attached dye molecule (Van Sark et al., 2002; Kowerko et al., 2009; Zenkevich et al., 2012). Experimental findings show that the fluorescence of single PP molecules decays almost mono-exponentially.

On the other hand, the PL decays of single QDs (both for free or assembled ones) are clearly multi-exponential as was found for a lot of many similar QDs (Schlegel, et al., 2002; Zhang et al., 2006; Fisher et al., 2004; Kowerko et al., 2009; Schmidt et al., 2014). It should be noted also that PL average decay times $<\tau_D>$ of single QDs (fitted by a stretched exponential function) show a decrease with increasing observation time. This behavior is related to the spectral blue shift of the PL. Some aspects of this finding are discussed in Chapter 2. The average decay time for single QDs was measured to be $<\tau_D> \approx 20.5$ ns (based on measurements for 20 QDs), while for QDs in a single nanoassembly QD–PP this PL decay was found to be shortened to $<\tau_D> \approx 16$ ns (based on measurements for 18 nanoassemblies). The shortening of the decay time is clearly related to the observed decrease of the QD PL intensity upon nanoassembly formation. However, the shortening of time cannot be directly related to FRET since QD PL decay is intrinsically multi-exponential. Notably, while in a given sample the fluorescence intensity of PP molecule is nearly indistinguishable between PP and assembled PP, the QD PL intensity of single QD–PP assemblies is quenched on average by 50% as compared to a QD in the same sample (detected at a different position on the surface). On the other hand, for ensembles in solution the corresponding QD PL quenching efficiency for abundant QD–PP nanoassemblies (at $x = 1$) was measured to be 12%. This implies, that 1 PDI molecule quenches the PL by as much as 50%. Moreover, from the reduced quenching efficiency in ensemble experiments, we conclude that either not all PDI molecules are assembled (only one-fourth based on the different quenching efficiencies) or quenching efficiencies are different in distinct environments. A combination of both is probably a more realistic approximation. Taken together, these facts prove that QD PL quenching is also observed on a single nanoassembly level and reflects directly the existence of an additional non-radiative channel for QD exciton relaxation in these nanoobjects.

It should be mentioned that the spectral distribution of the fluorescence band (shape and position) for many single PP molecules is different with respect to the one for many single QD–PP nanoassemblies. This might be related to a reduced conformational mobility of PP molecules on a QD surface (Kowerko et al., 2009). It is known (Fron et al., 2008; Kowerko et al., 2009a), that individual PP molecules have different fluorescence spectra due to various conformations of the bay groups (see Fig. 1.17) resulting in a broad distribution of the fluorescence maxima. The corresponding distribution of conformations and thus the fluorescence spectra depends on the local environment such as the embedding matrix. In Fig. 1.25 typical distributions of fluorescence maxima for PP molecules are presented for PP on a SiO_2 surface (Fig. 1.25A), in a QD–PP nanoassembly (Fig. 1.25B) and in a PMMA film (Fig. 1.25C), respectively. In comparison, the formation of QD–PP nanoassembly results in PP conformations characterized by the short wavelength emission range being more consistent with the distribution of fluorescence energies observed for the rigid PMMA matrix.

We conclude, that the limited spectral range of nanoassemblies reflects a reduced conformational distribution of bay group orientations. Nevertheless, as is evident from Fig. 1.25A, fluorescence spectra of individual PP molecules show a broad distribution which reflects that spectral fluctuations still occur. Interestingly, the ligand shell is obviously more of the type of a rigid shell inhibiting free orientational fluctuations of PP molecules which is interesting with respect to the above discussed phase transition of ligand shell.

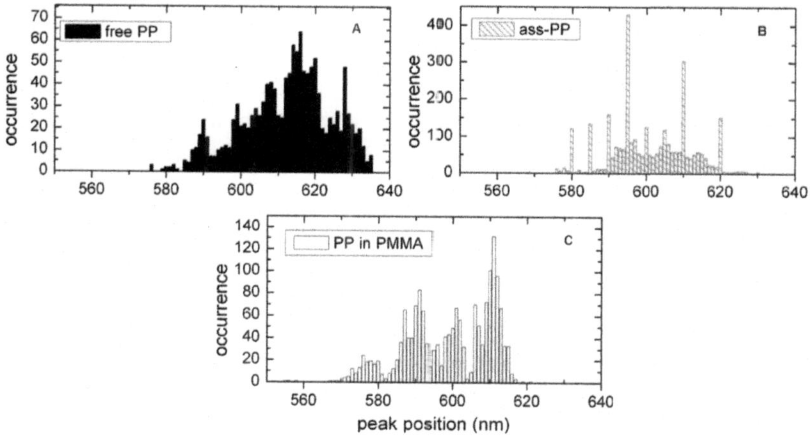

Figure 1.25 Distributions of fluorescence maxima for single PP molecules on SiO_2 surface (A), PP in single QD–PP nanoassemblies (B), and PP in a spin-coated PMMA film (C). Fluorescence maxima have been obtained by fitting the PP fluorescence spectra by two Gaussian lines. Adapted from Kowerko et al. (2009) and Kowerko (2010).

Since the PL intensity of the QD is reduced on average by 50% for each identified single nanoassembly (Kowerko et al., 2009, 2010), the corresponding dye-induced quenching rate should be (assuming $n = 1$) nearly equal to the (average) decay rate of the QD. A close inspection of the time resolved QD PL decay of single nanoassemblies provides evidence for the shortening of the QD PL decay time. An average decay time of the multi-exponential QD PL can be estimated using a stretched exponential fit (with $0 < \beta \leq 1$) according to

$$I(t) = I_0 \cdot \exp\left(-\left(\frac{t}{\tau}\right)^{\beta}\right). \tag{1.6}$$

Thus, an average decay time τ_D for single QDs and single QD–PP nanoassemblies can be calculated according to

$$\tau_D = \frac{\tau}{\beta}\Gamma\left(\frac{1}{\beta}\right). \tag{1.7}$$

Calculated τ_D values for single QDs and single "QD–PP" nanoassemblies as a function of the observation time are shown as black circles in Fig. 1.24A,B. The spectral fingerprints of 90 QDs from three different samples have been used to compare and statistically analyze the decay times of free and assembled QDs (Kowerko et al., 2010). It follows from this analysis that the related quenching efficiency of one dye molecule per QD is obviously broadly distributed between weak and near complete PL quenching.

Additionally, the calculation of the mean value of τ_D by averaging over all nanoassemblies leads to the value $\tau_D^A \approx 13.1$ ns for the decay time for QD PL in single QD-PP nanoassemblies while it is $\tau_D^D \approx 21.3$ ns for isolated QDs. Correspondingly, an average effective quenching rate k_{DA} can be estimated as

$$k_{DA}\frac{1}{\tau_{DA}} = \frac{1}{\tau_D^A} - \frac{1}{\tau_{DD}} = 2.9 \times 10^{-7} s^{-1}. \tag{1.8}$$

This value is very close to $k_q = 3.5 \cdot 10^{-7}$ s^{-1} obtained in bulk experiments for the corresponding QD–PP nanoassemblies in titration experiments at $x = 1$ in toluene (Kowerko et al., 2009, 2010). Thus, it may be concluded that the reasons of QD PL quenching processes (including low efficiencies of FRET) are very similar in ensemble and single nanoassembly experiments. More concretely, both experimental approaches being applied for QD–porphyrin and QD–PDI nanoassemblies evidently show the existence of strong non-FRET quenching pathways competing effectively with FRET.

Statistical data presented in Fig. 1.25 show that the spectral distribution of PP fluorescence is blue-shifted upon nanoassemblies formation. It means that those conformations of PP are favored, for which the bay groups are most extended as far as it is known (Fron et al., 2008; Kowerko et al., 2009, 2009a) that such situation corresponds to short fluorescence wavelengths. In this case, PP bay groups are as much within the molecular plane as possible. Such a situation is reasonable since PP has to be intercalated into the ligand shell which will be more easily accomplished for a nearly flat molecule, since less ligands have to be excluded from the QD surface. Correspondingly, the distribution is shifted from a "free" PP molecule on SiO_2 surface to a "matrix-isolated" type such as in a PMMA film (see Fig. 1.25C). In addition, spectral fluctuations are enhanced for assembled PP as compared to alone PP molecules in a rigid PMMA film (Kowerko et al., 2009). From this it follows, that the ligand shell imposes more flexibility than in a PMMA film allowing for remaining conformational changes of PP molecules. An alternative explanation is that the curved QD surface allows for more conformational flexibility than on a SiO_2 substrate. The high flexibility of the PP phenoxy side groups on the QD surface is only enabled in absence of steric hindrance.

The presented single molecule data reveal that for various QD–dye nanoassemblies, different relative orientations of the PP dyes with respect to the QD surface may be realized, for which the acceptor molecule may have different spectral properties. It means that relative efficiencies of FRET QD→Dye may differ significantly for each nanoassembly depending on the conformation of the dye attached to the QD surface. As an example, Fig. 1.26 shows the related spectral overlaps and calculated FRET efficiencies in the frame of the Foerster model (Foerster, 1965) for QD–PP nanoassemblies taking into account the range of spectral distributions for PP conformations (Kowerko, 2010). The corresponding quantitative description will be presented in Section 1.4 devoted to the analysis of this model to QD–dye interactions. It is clearly seen that FRET transfer parameters related to different QD–dye nanoassemblies with different PDI conformations vary essentially.

On the basis of above results we present the following conclusions: (i) a nearly perpendicular geometry for PP molecules is established, (ii) the ligand density close to PP has to be low, (iii) the identification of PP conformations may be used to confirm the geometry of the assembly which is useful for the analysis of FRET calculations and experimental data, and (iv) the comparative analysis of the dye conformation serves as probe for the ligand density and dynamics on single quantum dots. Concluding, single molecule data provide more specific information on the structure of the QD–PP nanoassemblies.

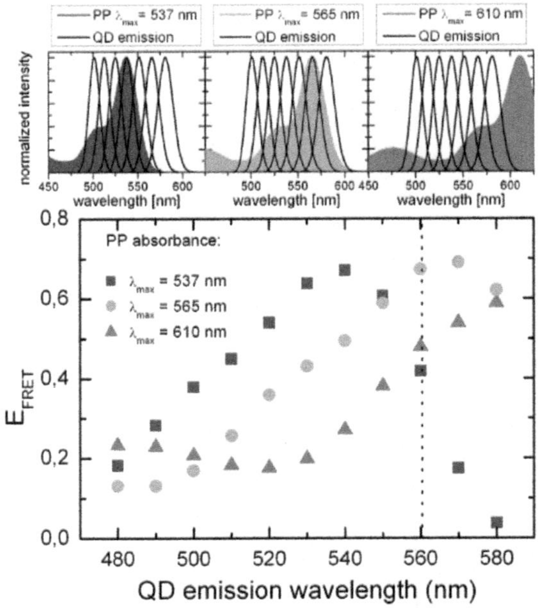

Figure 1.26 (Top) Scheme of spectral overlap for three PP conformations with the absorption maximum at 537, 565 and 610 nm, respectively, and QD PL between 500 and 580 nm. (Bottom) FRET efficiencies calculated for the respective conformations as a function of the QD emission wavelength. Parameters of TOPO-capped CdSe/ZnS QD (three ZnS monolayers, total diameter 5.1 nm, first exciton absorption λ_{max} = 542 nm) have been used for calculations. Adapted from Kowerko (2010).

1.3.3 Time Dependent Fluctuations of FRET in Single QD–Dye Nanoassemblies

It was shown in Section 1.2.4 that for bulk solutions containing QD–dye nanoassemblies based on TOPO-capped CdSe/ZnS QDs and DTPP molecules and dissolved in non-polar TEHOS, QD PL quenching caused by non-FRET processes is effectively suppressed. In the result, the efficiency of FRET QD→DTPP increases significantly. In this respect, single molecule detection is an excellent tool to evaluate details of such process (distance and orientation dependence, role of spectral overlap, etc.). More interestingly, for single nanoassemblies it is possible to detect FRET events during the observation time in order to evaluate time dependent fluctuations of the geometry or relative energies of interacting counterparts. Some results in this direction (Gerlach et al., 2013) will be the subject of this Section.

Samples were prepared by dipcoating the silicon substrates (with 100 nm thick thermal oxide, Crystec, Berlin) into solutions of 0.3% TEHOS in hexane with highly diluted TOPO-capped CdSe/ZnS QDs and DTPP molecules (at molar ratio $x = 1$) resulting in 10 nm thick films of TEHOS with nanomolar concentrated CdSeZnS QDs and DTPP. Spectra (at $\lambda_{exc} = 488$ nm) were investigated with a home-built laser scanning confocal microscope described in detail elsewhere (Krause et al., 2011a).

Intensity fluctuations of the luminescence of single QDs and DTPP molecules as a function of an observation time of several minutes are presented in Fig. 1.27. The fluctuating luminescence intensity is shown at the side of each time trace of the series of spectra. CdSe/ZnS shows strong intensity fluctuations known as blinking or luminescence intermittency (Cichos et al., 2007), while blinking is almost negligible for DTPP (Lang et al., 2005). Typically, strong intensity intermittency in combination with spectral fluctuations (spectral diffusion) is quite obvious (Krause et al., 2011b). The luminescence of CdS/ZnS and DTPP is bleached after 180 and 77 s, respectively. Bleaching of the QD is in most cases preceded by a strong blue shift of the photoluminescence which is tentatively assigned to photooxidation of the CdSe core (see also Chapter 2) (Van Sark et al., 2002; Kowerko et al., 2009; Zenkevich et al., 2012).

It was found that about 10% of the luminescent spots show spectra of both constituents (like it has been observed for solutions of ensembles of the same nanoassemblies, presented in Section 1.2.4). Two typical examples (A and B) are shown in Fig. 1.28. In general, the time dependent spectral features are quite similar to those shown above separately for the QD and the dye in Fig. 1.27, besides the enhanced blinking of DTPP. A closer inspection, however, reveals that the respective luminescence intensities are strongly (positively) correlated, as can be directly seen, when plotting the corresponding intensities on top of each other as shown at the bottom of Fig. 1.28. It is evidently seen that in both cases A and B the strong blinking of QDs is imposed on DTPP, which now blinks very often synchronously with the QD. In addition, when DTPP is photobleached (or detached) from the

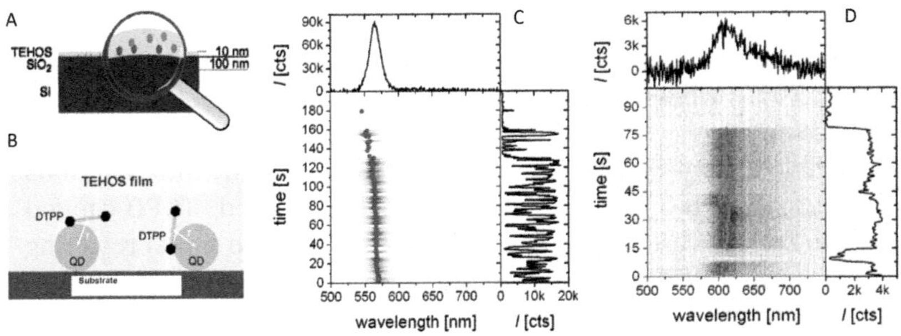

Figure 1.27 Scheme of sample preparation (A) and two different orientations of the DTPP dye on TOPO-capped CdSe/ZnS QDs (B) as well as typical time dependent spectra of CdSeZnS (C) and DTPP (D) with 1 s integration time (excitation via the 488 nm line of an Ar-ion laser). The structure of DTPP molecule is shown in Fig. 1.20. B: For steric reasons DTPP can be attached via the terpyridyl groups at only one end of the dye. DTPP is plotted along the optically active transition dipole moment. C, D: The respective over the total observation time integrated spectra are shown on top of each spectral trace. The integrated luminescence intensity is depicted on the right side of each time trace. Samples were laser scanned to detect luminescent spots, the size of which is determined by the focus of the laser. At those spots then time series of fluorescence spectra were taken with an integration time of 1 s. Luminescence intensities were obtained via integration of the spectra across the relevant spectral range of the respective constituents. Adapted from Gerlach et al. (2013).

QD (at 77 s) the QD PL increases on average by about a factor of 3. There are only three times at which intensity correlation fails, namely at 56, 66, and 72 s, respectively, for which DTPP fluorescence decreases while the QD photoluminescence increases. This can be assigned to one of the rare blinking events of DTPP. The overall interpretation is that the PL of the QD is quenched due to FRET according to QD→DTPP and that DTPP fluorescence is predominantly caused via FRET. It should be noted in this respect, that the PET from the QD to DTPP can be excluded in this case as far as the formation of charged QD DTPP counterpart in nanoassembly should result to complete quenching of DTPP which is not observed in the given experiments. Contrary to this, DTPP fluorescence is strongly decreased namely when the QD does not emit.

The comparison of blinking correlation shown at the bottom of Fig. 1.28 for two different single nanoassemblies (A and B) reveals a strong correlation of luminescence intensities in both cases. However, it is not the same for these two different nanoobjects. According to considerations presented recently (Gerlach et al., 2013) this can be explained by different orientations of DTPP molecule on the QD surface in the two nanoassemblies (see Scheme B in Fig. 1.27), and, correspondingly, various conditions with respect to laser excitation at 488 nm (favorable one for the left orientation shown in Scheme B).

The detailed quantitative consideration of FRET in single QD–DTPP nanoassemblies has been described in our paper (Gerlach et al., 2013). The main ideas may be presented as follows. Figure 1.29 shows the correlation of normalized DTPP and QD luminescence intensities and the luminescence intensity ratio $R = I_{DTPP}/I_{QD}$ for selected time ranges related to the time traces in Fig. 1.28A. It is seen that the correlation clearly depends on the observation time and changes by more than a factor of 3 at different times. A detailed analysis (with normalized intensities for each

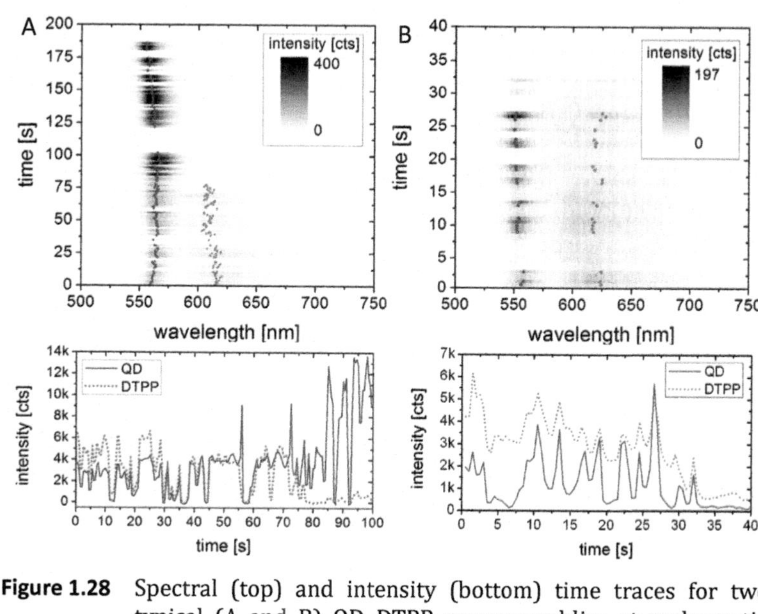

Figure 1.28 Spectral (top) and intensity (bottom) time traces for two typical (A and B) QD–DTPP nanoassemblies at molar ratio $x = 1$ in 10 nm thick films of TEHOS. Adapted from Gerlach et al. (2013).

constituent) reveals that the linear correlation of intensities is quite constant during the first 30 s, but changes in the course of time. This can be interpreted as being due to time dependent changes of the geometry of the nanoassembly or the conformation of the DTTP molecule attached to the QD surface.

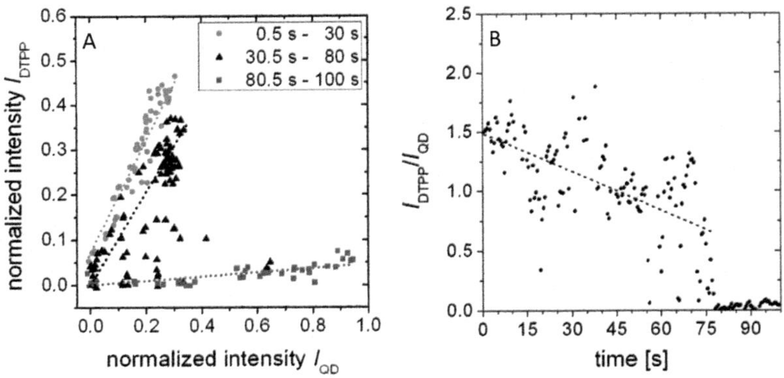

Figure 1.29 (A) Correlation of normalized DTPP and QD luminescence intensities for selected time ranges related to the time traces for QD–DTPP single nanoassembly A (Fig. 1.28A). Intensities have been normalized to the respective maximum of each time trace for each constituent. (B) Luminescence intensity ratios as a function of observation time. Adapted from Gerlach et al. (2013).

For a single QD–DTPP nanoassembly, the FRET efficiency can be calculated using quenching data for the donor (QD) PL intensity I_{QD} according to following equation (Lakowicz, 2006):

$$\Phi_{FRET} = 1 - \frac{I_{QD\text{-}DTTP}}{I_{QD}}. \tag{1.9}$$

It is found from Fig. 1.28A that DTPP is photobleached for times longer than 77 s. Thus, taking the maximal PL intensity I_{QD} in that range and the maximum of the intensity $I_{QD\text{-}DTTP}$ for a QD–DTPP nanoassembly within the time when DTPP is not photobleached (Fig. 1.28A, bottom), one can estimate the FRET efficiency to be $\Phi_{FRET} \approx 0.69$ (according to Eq. (1.9). This result can be compared to the situation at 57 and 72 s when DTPP is "off" due to blinking providing I_{QD} and compare it to $I_{QD\text{-}DTPP}$ when DTPP is "on" again. According to Eq. (1.5) this results in $\Phi_{FRET} \approx 0.55$. Both values are in reasonable agreement with $\Phi_{FRET} \approx 0.85$ calculated within the Foerster model (Kowerko et al., 2010) not taken into account the existence of some competing non-FRET quenching processes (see Section 1.2.4). Nevertheless, it should be noted that obviously, strong variations of Φ_{FRET} are observed for on one and the same assembly, since I_{QD} depends on quenching processes other than FRET.

Results presented in this section clearly demonstrate that single assembly experiments are an unsurpassed tool to investigate subtle conformational changes and the related impact on interchromophoric dynamics. In fact, due to their intrinsic flexibility, DTPP molecules can realize a large variety of conformations on the experimental time scale of seconds to minutes. For this reason, FRET becomes dependent on observation time, which can be followed quite in detail.

1.4 Quantitative Analysis of Non-Radiative Relaxation Pathways for Quantum Dots in Nanoassemblies

It follows from literature data (cited in Introduction, Section 1.1.3) including our investigations described in previous Sections that in many cases (may be all for self-assembled nanoaggregates) low FRET efficiencies in QD–dye nanoassemblies may be related to the presence of competing non-FRET processes of various nature: (i) photoinduced self-trapping of charges in an dielectric medium of the QDs environment (Guyot-Sionnest et al., 1999; Klimov et al., 2000; Issac et al., 2005; Cichos et al., 2007); (ii) dynamically controlled attachment/detachment processes and ligand exchange (Bullen and Mulvaney, 2006; Ji et al., 2008; Koole et al., 2008; Munro and Ginger, 2008; von Holt et al., 2008; Schapotschnikow et al., 2009; Kilina et al., 2012); (iii) involvement of QD surface states (Frenzel et al., 2007; Issac et al., 2005; Albert et al., 2011; Knowles et al., 2012; McBride et al., 2013; Zenkevich et al., 2015); (iv) temperature and solvent polarity effects (De Mello Donega et al., 2006; Wuister et al., 2004; Zenkevich et al., 2005, 2012, 2014, 2015; Blaudeck et al., 2012; Gerlach et al., 2013); (v) charge tunneling through the capping (ZnS) barrier under the conditions of quantum confinement (Blaudeck et al., 2008, 2012;

von Borczyskowski and Zenkevich, 2014; Kowerko et al., 2009, 2010; Kowerko, 2010). The manifestation of some of these processes has been described in previous Sections. Here, with respect to QD–dye nanoassemblies, we like to discuss quantitatively the application of the Foerster model, the manifestation of charge tunneling effects as well as surface and interfacial phenomena that may be tuned by tailoring the surface state energy and by varying the specific electronic interactions with dye molecules attached to the QD surface.

At the end we will present a model, which schematically describes how FRET and non-FRET contributions are interrelated without the necessity to identify the microscopic nature of non-FRET processes like those summarized above. Overall, we will repeatedly make use of a combination of ensemble and single quantum object experiments. Doing so, we are aware of the fact that weak ergodicity braking will not allow for a quantitative comparison of both types of experiments (Zenkevich et al., 2015). Nevertheless, we claim that an essential break-through in the understanding of many features of QD PL energies and dynamic processes in complex and heterogeneous QD–environment systems can only be realized via a careful combination of findings from these two at first glance so different approaches.

1.4.1 QD PL Quenching via Foerster Resonant Energy Transfer in QD–Dye Nanoassemblies

At the moment, there have been numerous studies of energy transfer from semiconductor QDs to various acceptor organic ligands. Experimental results (Javier et al., 2003; Clapp et al., 2004, 2005; Zenkevich et al., 2005; Blaudeck et al., 2008, 2012; Sitt et al., 2013; Halivni et al., 2012; Kowerko et al., 2009, 2010; Boldt et al., 2011; Lemon et al., 2013; Stewart et al., 2012; Sadhu et al., 2013; Mattoussi et al., 2012; Gerlach et al., 2013; Beane et al., 2014; Hadar et al., 2015) and theoretical considerations (Basco et al., 1999; Kilin et al., 2007) suggest nanocrystalline materials and organic-inorganic hybrids are best described in the weak coupling regime. The application of the weak coupling model arises due to the dissimilarity in the wave functions of the molecular orbitals in organic systems and band-like orbitals in QDs that prevents traditional strong interactions. This allows the application of

Foerster model for the fluorescence resonant energy transfer (FRET) (Foerster, 1965; Agranovich and Galanin, 1982) which typically is used at the limit of "localized oscillators" as a result of the electric dipole approximation.

Notably, the most important remark is that these considerations make sense if the other QD PL quenching channels are vanishing. In this case, FRET from a photoexcited donor (D) to a lower-energy acceptor (A) takes place due to the coupling of the dipole moment of an exciton in a D chromophore with absorbing transition in nearby A chromophores. The application of Foerster theory relies on the approximation that the distance R_{DA} between interacting D and A dipoles is larger than the length $|l|$ of the transition dipoles themselves. The necessary estimations may be derived using well-known expressions (Atkins and de Paula, 2002) for oscillator strength f and transition dipole moment $|\mu|$ of the corresponding low-energy electronic transitions of interacting subunits:

$$f = 4.33 \times 10^{-9} \cdot \varepsilon_{\max}(\nu)\Delta\nu_{1/2}, \tag{1.10}$$

$$f = (8\pi^2 m_e \nu |\mu|^2)/3he^2, \tag{1.11}$$

where $\varepsilon_{\max}(\nu)$ is the molar decimal extinction coefficient at the maximum of the absorption band and $\Delta\nu_{1/2}$ the corresponding spectral width. Making use of the experimental data known for CdSeZnS QDs and dye molecules (porphyrins and perylene diimides) we obtain that the effective length of the transition dipoles $|l| \leq 1\text{–}2.0$ Å for dyes and $|l| \approx 2$ Å for QDs under study, while estimations of intercenter D-A distances for the optimized structures of QD–dye nanoassemblies give values $R_{DA} \approx 26\text{–}36.6$ Å (Zenkevich et al., 2005; Kilin et al., 2007; Blaudeck et al., 2007; Kowerko, 2010). It means that in this case the point dipole-dipole approximation is still valid.

FRET rate constant k_{FRET} for an isolated D-A pair at distance R_{DA} is expressed by

$$k_{FRET} = \frac{1}{\tau_D^0} \cdot \frac{R_0^6}{R_{DA}^6}. \tag{1.12}$$

where τ_D^0 is the emission decay of the individual donor, R_{DA} is the intercenter of the interacting dipoles. R_0 is the critical Foerster radius which may be calculated as

$$(R_0^{\text{theor}})^6 = \frac{9000 \ln 10 k^2 \phi_D^0}{128\pi^5 n^4 N_A} \times \int_0^\infty f_D(\nu)\varepsilon_A(\nu)\frac{d\nu}{\nu^4} \qquad (1.13)$$

where N_A = 6.022 × 10^{23} M^{-1} is Avogadro number, τ_D^0 is the D emission quantum efficiency in the absence of FRET, n is the refraction index of the solvent. Orientation factor k^2 = [Cos(μ_D, μ_A) – 3cos(μ_D, r_{DA})·cos(μ_A, r_{DA})]^2, where (μ_D, μ_A) is the angle between transition dipole moment vectors of the D and A subunits, (μ_D, r_{DA}) and (μ_A, r_{DA}) denote the angles between the dipole vectors of D and A and the direction D→A, respectively. For a random static rigid distribution of interacting dipoles k^2 = 0.476 (Agranovich and Galanin, 1982) (this is the case for QD–dye nanoassemblies). Spectral overlap integral

$$J(\nu) = \int_0^\infty f_D(\nu)\varepsilon_A(\nu)\frac{d\nu}{\nu^4} \qquad (1.14)$$

is calculated on the basis of experimental acceptor (H_2P or PDI) molecules absorption and donor (QD) PL spectra, where $\varepsilon(\nu)$ corresponds to the A molar decimal extinction coefficient, f_D (ν) corresponds to the D quantum emission spectrum normalized to unity by area on a wave number scale. Figure 1.30 shows calculated dependences of spectral overlap, Foerster critical radius R_0, donor-acceptor distance R_{DA}, FRET efficiency, and reciprocal FRET rate as a function of the QD size represented by the respective PL emission wavelength and one type of porphyrin $H_2P(m\text{-}Pyr)_4$ molecule (Blaudeck, 2007).

If FRET is considered as the dominant reason of QD PL quenching in nanoassemblies, the theoretical FRET efficiency $\Phi_{\text{FRET}}^{\text{theor}}$ for pair (1:1) interaction QD–dye can be calculated as

$$\Phi_{\text{FRET}}^{\text{theor}} = \left[1 + \left(\frac{R_{DA}}{R_0^{\text{theor}}}\right)^6\right]^{-1} \qquad (1.15)$$

When a finite number of acceptors is attached to QD surface, the excitonic energy is spread among *n* acceptors on one QD with equal probability. In other words, the likelihood of an exciton transfer to particular acceptor molecules scales inversely proportional to the total number of *n* acceptors A per donor D. Thus, Eq. (1.15) becomes as

$$\Phi_{\text{FRET}}^{\text{theor}}(n) = \left[1 + \left(\frac{R_{\text{DA}}}{n \cdot R_0^{\text{theor}}}\right)^6\right]^{-1}. \tag{1.16}$$

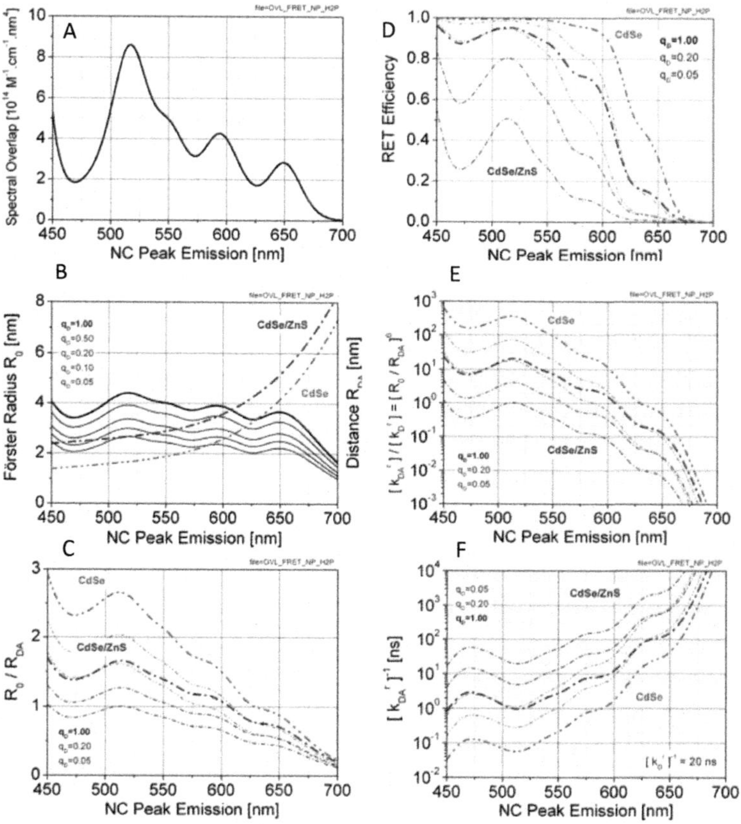

Figure 1.30 (A) Spectral overlap integral between CdSe//ZnS QDs (or nanocrystals, NC) of various sizes (PL emission) and $H_2P(m\text{-}Pyr)_4$ molecule. (B) Calculation of the Foerster radius R_0 for FRET QD→Porphyrin. (C) Ratio between the Foerster radius R_0 and the donor-acceptor distance R_{DA}. (D) FRET efficiency Φ_{FRET}^{theor}. (E) FRET rate k_{FRET} in units of the intrinsic radiative rate k_D of the QDs. (F) Reciprocal FRET rate k^{-1}, calculated for an intrinsic radiative rate of the QDs of k_D = 20 ns. Legend: (dash-dot-dot) CdSe QDs, (dash-dot) CdSe/ZnS QDs. The $H_2P(m\text{-}Pyr)_4$ molecules are coordinated by two pyridyl groups to the QD surface (see Fig. 1.6B). The PL quantum efficiency of the QDs is treated as a free parameter. Adapted from Blaudeck (2007).

With these physical backgrounds and for educative reasons it is possible to analyze comparatively FRET parameters for various QD–dye nanoassemblies (Blaudeck, 2007; Kowerko, 2010; Gerlach et al., 2013). Figures 1.30 and 1.31 present such a comparison for CdSe and CdSe/ZnS QDs of various sizes and one type of porphyrin $H_2P(m\text{-}Pyr)_4$ molecules (Blaudeck, 2007).

A further example of the theoretical analysis based on the evaluation of Eq. (1.16), illustrates the effect of acceptor labeling (e.g., number of attached porphyrin molecules to the QD surface) on the FRET efficiency (Fig. 1.31). These calculations yield that if FRET is the only reason of QD PL quenching in nanoassemblies, already upon attachment of one single acceptor molecule, the maximal achievable FRET efficiency may be reached up to 50–70% for various CdSe/ZnS QDs (curves 3–5 in Fig. 1.31). Additionally, theoretical estimations show that for the nanoassembly containing CdSe/ZnS QD ($d_{CdSe} \geq 3$ nm, 2 ZnS monolayers), FRET should be realized within a time scale > 10 ns.

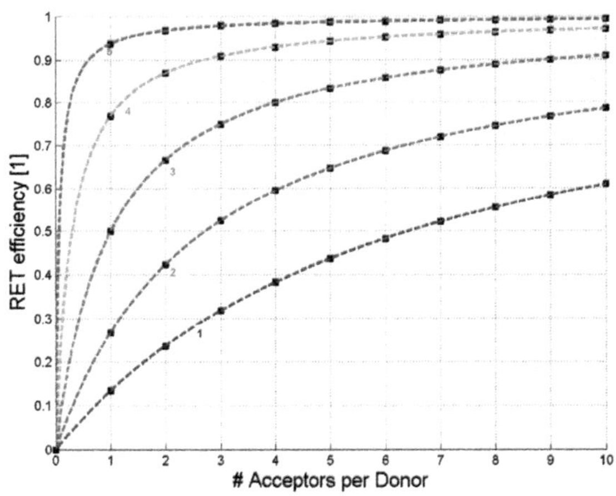

Figure 1.31 Calculated dependence of FRET efficiency Φ_{FRET}^{theor} on the number of acceptors molecules, ($H_2P(m\text{-}Pyr)_4$), per donor QD according to Eq. (1.16) for Foerster radius R_0 = 2.75 nm and various donor-acceptor distances depending on QD structure (CdSe or CdSe/ZnS with two ZnS monolayers) and size: R_{DA} = 1.75 nm (1), 2.25 nm (2), 2.75 nm (3), 3.25 nm (4), and 3.75 nm (5). Adapted from Blaudeck (2007).

With respect to QD–porphyrin nanoassemblies, the FRET dynamics have also been studied by ab initio electronic structure calculations combined with a molecular dynamic approach (Kilin et al., 2007). The geometric structure, electronic energies, and transition dipole moments were derived by density functional theory (DFT) and are utilized for calculating the Foerster coupling between the excitons of the QD and the porphyrin molecule (Fig. 1.32). Theoretical calculations of the optimized geometry for nanoassemblies based on CdSe/ZnS QD and $H_2P(m\text{^}Pyr)_2$ (having adjacent *meta*-pyridyl rings) predict as experimental findings a perpendicular position of the porphyrin macrocycle with respect to the QD surface (see Fig. 1.9).

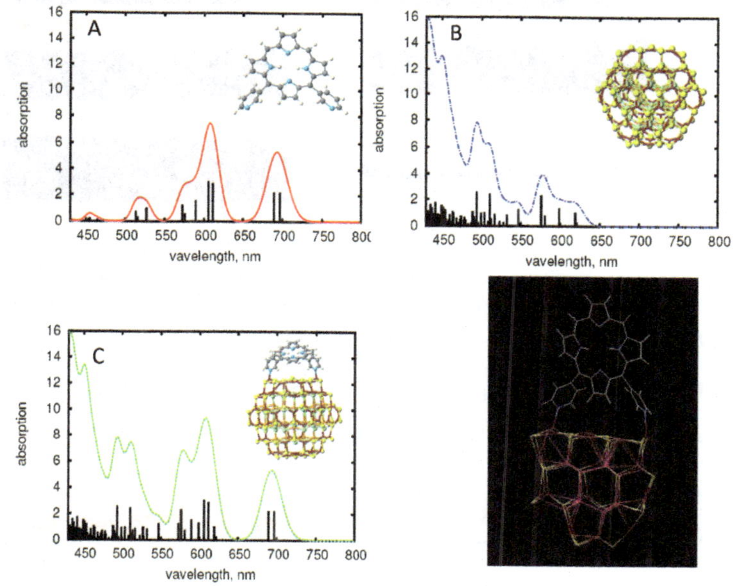

Figure 1.32 Calculated absorption spectra and structures of the dipyridyl substituted porphyrin (B), the core-shell CdSe/ZnS QD (B) and the QD-porphyrin nanoassembly (C). The QD contains 222 atoms, including 66 atoms in the $Cd_{33}Se_{33}$ core and 156 atoms in the $Zn_{78}S_{78}$ shell. Both the core and the shell are of wurtzite structure of bulk CdSe and ZnS upon geometry optimization (HyperChem 7.0, DFT). The porphyrin molecule contains two adjacent *meso*-pyridyl rings, $H_2P(m\text{-}Pyr)_2$. The optimized geometry of the nanoassembly based on CdSe/ZnS QD ($Cd_{33}Se_{33}$ core and $Zn_{78}S_{78}$ shell, wurtzite structure) and $H_2P(m\text{^}Pyr)_2$ molecule has been obtained using molecular dynamic calculations within the HyperChem software package and DFT simulations performed with the VASP code. Adapted from Kilin et al. (2007).

The dynamics of the energy transfer between CdSe/ZnS QD and porphyrin molecule (in experimentally relevant nanosecond timescale) was described by the reduced density matrix formalism (Schreiber et al., 1999; Prezhdo, 1999; Cina et al., 2003). It has been shown that the calculated direction and rate of the irreversible exciton transfer are determined by the initial photoexcitation of QD, the QD–porphyrin coupling and the interaction to the electronic subsystem with the vibrational environment (Fig. 1.33).

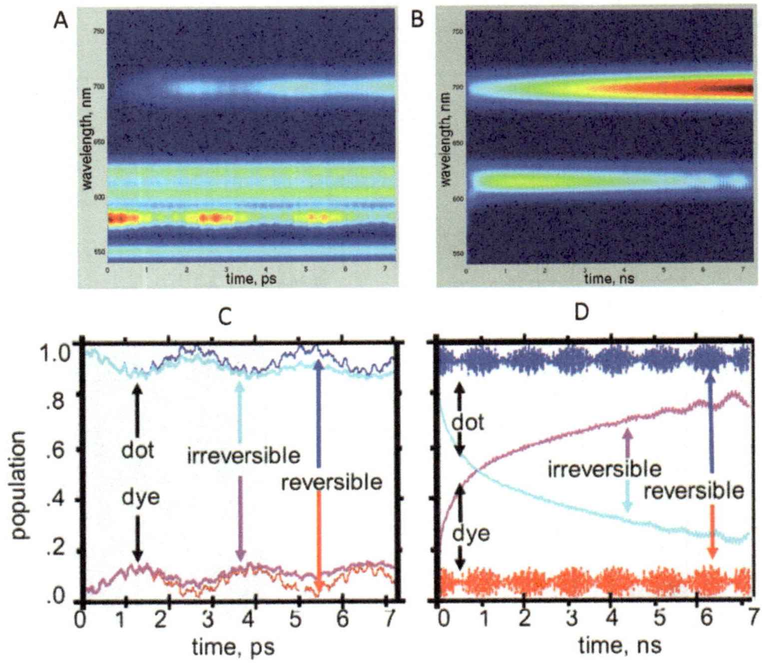

Figure 1.33 Calculated (in the frame of the reduced density matrix formalism) time resolved fluorescence signal as a function of time and wavelength (A, B) as well the localization of the exciton on QD and the porphyrin molecule as a function of time (C, D) for nanoassemblies based on $Cd_{33}Se_{33}/Zn_{78}S_{78}$ QD and represent both the irreversible dynamics including electron-phonon coupling (cyan and magenta) and the reversible dynamics without the phonon bath (blue and red). Adapted from Kilin et al. (2007).

It follows from these calculations that the initial excitation transfer from the QD to the porphyrin occurs in a resonance fashion and remains on the porphyrin due to irreversible relaxation induced by a thermalized phonon bath. As a result, the emission of the QD is quenched giving rise to the emission of the porphyrin. According to this approach in the case of a small $Cd_{33}Se_{33}/Zn_{78}S_{78}$ QD with attached porphyrin molecule, FRET QD $\rightarrow$ porphyrin should occur on a nanosecond time-scale (up to ~7 ns). As we mentioned above, for larger QDs ($d_{CdSe} \approx 4.0$ nm, three ZnS monolayers) estimations of experimental FRET times give values $t_{FRET} = 1/k_{FRET} \approx 30$ ns. In the result, if one does not take into account the possible competition of other reasons for QD PL quenching, the real-time atomistic details of this quenching in QD–porphyrin nanoassemblies may be fully described in the frames of the Foerster model.

Obviously (as shown above), in the frame of the Foerster model FRET efficiency should depend on the spectral overlap of the donor emission and acceptor absorption. Correspondingly, if QD PL quenching in nanoassemblies is caused presumably by FRET, this quenching should depend also on the spectral overlap integral $J(\nu)$ (the so-called resonance conditions, (Foerster, 1965)). Such a comparison has been carried out for QD–porphyrin nanoassemblies (Zenkevich et al., 2007a; Zenkevich and von Borczyskowski, 2009). For a given TOPO-capped CdSe/ZnS QD, PL quenching results obtained with H_2P(m-Pyr)$_4$ molecules have been compared with those measured with CuP(m-Pyr)$_4$ and THP(m-Pyr)$_4$ molecules, thus changing the overlap integral values by 2–2.5 times with respect to H_2P(m-Pyr)$_4$. Nevertheless, Fig. 1.34 (A and B) shows that upon increasing the molar ratio x, QD PL quenching is nearly the same for H_2P(m-Pyr)$_4$ and CuP(m-Pyr)$_4$ as well as for H_2P(m-Pyr)$_4$ and THP(m-Pyr)$_4$, respectively. It means that the contribution of FRET (QD $\rightarrow$ porphyrin) to the total QD PL quenching in QD–porphyrin nanoassemblies containing various porphyrin molecules is not the only and even not a major reason.

On the other hand, in the hypothetical case of a photoinduced charge (hole or electron) transfer process in QD–dye nanoassemblies the fluorescence of the dye should be also quenched. However, fluorescence parameters (efficiency φ_F and decay time τ) for H_2P(m-Pyr)$_4$ and PDI molecules upon complexation with QDs do not show quenching effects and remain practically the same with respect to those measured for individual constituents under the same conditions (Zenkevich et al., 2005; Blaudeck et al., 2008,

2012; Zenkevich and von Borczyskowski, 2009; Kowerko et al., 2009, 2010). In addition, for a given TOPO-capped CdSe/ZnS QD, the comparative titrations experiments have been carried out for $H_2P(m\text{-}Pyr)_4$ molecules as well as for $THP(m\text{-}Pyr)_4$ (hole acceptor, $E^{red}_{1/2}$ = – 1.08 V), $H_2P(m\text{^}Pyr)_2(5FPh)_2$ (electron acceptor, $E^{red}_{1/2}$ = −0.85 V), and $H_2P(m\text{^}Pyr)_2(Anthraquinone)_2$ (having two electron accepting meso-anthraquinones) (Zenkevich et al., 2007a). Nevertheless, it is seen from Fig. 1.34C that practically near the same QD PL quenching curves are observed for all types of porphyrins. Thus, non-dependence of QD PL quenching efficiency on redox properties of porphyrin molecules and the absence of the porphyrin fluorescence quenching in QD–porphyrin nanoassemblies rule out the dominant role of typical photoinduced charge transfer processes.

On the basis of QD PL quenching results, experimental and theoretical FRET analysis as well as the absence of a typical manifestation for photoinduced charge transfer processes we conclude that in QD–dye nanoassemblies low FRET efficiencies are due to the presence of competing non-FRET processes, which open a quenching pathway other than energy transfer by merely attaching a dye molecule to the QD surface. These processes will be discussed in more detail in the next Section 1.4.2.

Nevertheless, we finally show that even low FRET efficiencies (experimentally found for QD–porphyrin nanoassemblies in toluene at ambient temperature) may increase significantly the photochemical activity of these assemblies. Taking into account physical backgrounds for FRET and non-FRET processes in QD–dye nanoassemblies we investigated the interaction of QDs and QD–porphyrin nanoassemblies with molecular oxygen in toluene at ambient temperature. For this analysis we have chosen CdSe/ZnS QD (d_{CdSe} = 3.0 nm, two ZnS monolayers) and nanoassemblies based on this QD and $H_2P(m\text{-}Pyr)_4$ molecules at a molar ratio x = 4 (spectral and QD PL quenching results are presented in Figs. 1.7, 1.14, and 1.15). As we have shown at x = 4 for these nanoassemblies (see Section 1.2.2.3), experimental FRET efficiencies found via direct sensitization of porphyrin fluorescence are in the range of Φ_{FRET} = 0.12–0.14 (Blaudeck et al., 2008, 2012).

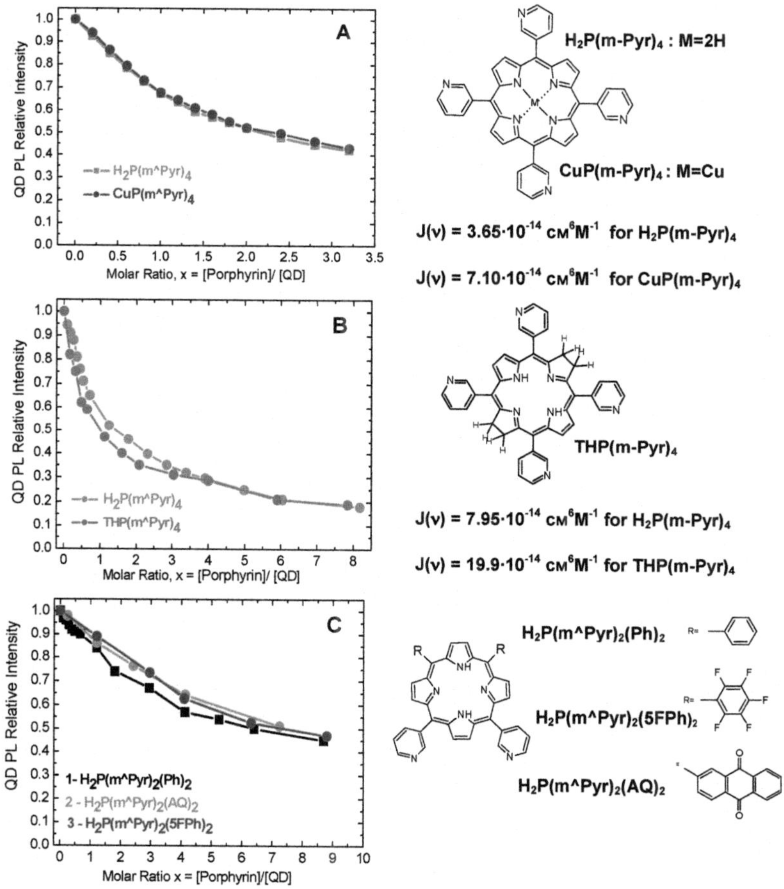

Figure 1.34 PL quenching for TOPO-capped CdSe/ZnS QDs upon titration (in toluene at 295 K) for various attached porphyrin molecules. (A) d_{CdSe} = 2.6 nm, two monolayers ZnS, λ_{exc} = 440 nm, λ_{reg} = 554 nm, $H_2P(m\text{-}Pyr)_4$ and $CuP(m\text{-}Pyr)_4$. (B) d_{CdSe} = 2.5 nm, 2 monolayers ZnS, λ_{exc} = 465 nm, λ_{reg} = 522 nm, $H_2P(m\text{-}Pyr)_4$ and $THP(m\text{-}Pyr)_4$. (C) d_{CdSe} = 2.5 nm, two monolayers ZnS, λ_{exc} = 465 nm, λ_{reg} = 522 nm, porphyrins with adjacent pyridyl rings–$H_2P(m\text{\^{}}Pyr)_2(Ph)_2$, $H_2P(m\text{\^{}}Pyr)_2(5FPh)_2$, and $H_2P(m\text{\^{}}Pyr)_2(AQ)_2$. Chemical structures of the corresponding porphyrin molecules are shown on the right. Spectral overlap integrals $J(\nu)$ for a given QD and various porphyrins have been calculated using Eq. (1.13). Adapted partly from [Zenkevich et al., 2007a, 2007b).

Typically in liquid solutions at ambient temperature, the photosensitized generation of singlet oxygen ($^1O_2$) in a long-living $^1\Delta_g$ electronic state is realized via the interaction of a potential light-activated photosensitizer (QD or porphyrin molecule) with molecular oxygen, O_2. In the result, the excited states of the photosensitizer (PS), particularly the longer-lived triplet T_1-state of porphyrins (Dzhagarov et al., 1987; Zenkevich et al., 1996; Zenkevich and von Borczyskowski, 2012, 2015), are efficiently quenched by O_2. According to Wigner spin conservation rule (transitions between terms of the same multiplicity are spin-allowed (McNaught and Wilkinson, 1997)), the quenching of the T_1-state of the organic PS (^{3}PS) followed by the formation of singlet oxygen, $^1O_2$ (or $^1\Delta_g$) may take place according to

$$^3\text{PS} + {}^3\text{O}_2 \rightarrow {}^1[{}^3\text{PS} \cdots {}^3\text{O}_2] \rightarrow {}^1[{}^1\text{PS}_0 \cdots {}^1\text{O}_2({}^1\Delta_g)] \rightarrow {}^1\text{PS}_0 + {}^1\Delta_g \tag{1.17}$$

The direct measurements of the $^1O_2$ emission (spectra and kinetics) in air-saturated toluene have shown that singlet oxygen generation by CdSe/ZnS QDs (d_{CdSe} = 3.0 nm) occurs with a quantum efficiency γ_Δ = 1.5%, and $^1O_2$ emission decay $\tau(^1\Delta_g)$ = 30.0 μs, while for larger CdSe/ZnS QDs with d_{CdSe} = 5.2 nm no singlet oxygen was observed. These results coincide with the corresponding data obtained for CdSe QDs (d_{CdSe} = 5 nm) in oxygen-saturated toluene at 293 K ($\gamma_\Delta \sim$ 1.5%, $\tau(^1\Delta_g)$ = 28.5 μs (Samia et al., 2003)). As discussed in details in (Zenkevich et al., 2011), the low quantum efficiency of $^1O_2$ generation by QDs in liquid solutions at ambient temperature may be explained by few reasons (Efros and Rosen, 2000; Klimov, 2000, 2010): (i) at ambient temperature the energy gap $\Delta E = E_{\text{"bright"}} - E_{\text{"dark"}} \sim$ 240 cm^{-1} between "bright" and "dark" states is comparable with $kT \sim$ 210 cm^{-1}, and "dark" states (often referred as the "triplet" states (Samia et al., 2003)) are shortened down to ~15 ns due to a effective thermal distribution; (ii) both for "bright" exciton (a total angular momentum N = 1) and for "dark" exciton (N = 2), the reaction between molecular oxygen and QD in these states does not completely follow Wigner spin conservation rule. It means that at ambient temperature most of the QDs are hardly perspective for direct generation of singlet oxygen.

In contrast to this, upon laser excitation of QD-porphyrin nanoassemblies, singlet oxygen is produced with a higher efficiency in comparison with only QDs under the same experimental conditions in air-saturated toluene. In Fig. 1.35C comparative results are presented on singlet oxygen $^1O_2$ luminescence (decays and intensities) sensitized by CdSe/ZnS QDs, $H_2P(m\text{-}Pyr)_4$ molecules as well by QD-porphyrin nanoassemblies (x = 4) at the same laser excitation conditions.

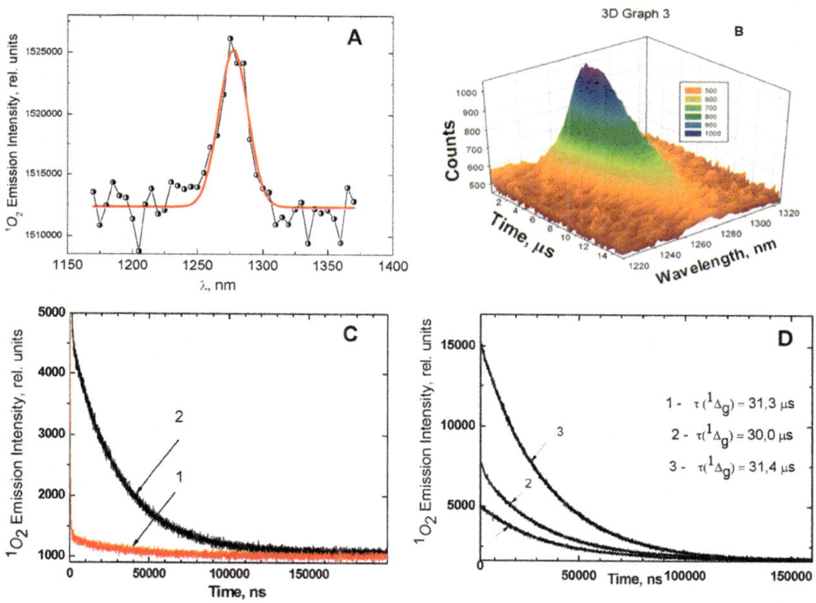

Figure 1.35 (A): Singlet oxygen $^1O_2$ luminescence band directly sensitized by CdSe/ZnS QDs (d_{CdSe} = 3.0 nm, two ZnS monolayers). (B): 3D presentation of $^1O_2$ emission. (C): $^1O_2$ luminescence decays sensitized by CdSe/ZnS QDs (1) and QD-porphyrin nanoassemblies (2) at the same molar concentrations of QDs and laser excitation conditions. (D): Decay kinetics of $^1O_2$ luminescence detected for $H_2P(m\text{-}Pyr)_4$ molecules (1), QD-porphyrin nanoassemblies (2) at the same molar concentrations of porphyrin and Pd-mesoporphyrin as a standard (3) at fixed laser excitation conditions. The molar ratio x = 4 was constant in air-saturated toluene, laser excitation at λ_{exc} = 532 nm, pulse duration $\Delta t_{1/2}$ = 0.7 ns, pulse energy E = 4 μJ, observation at λ_{reg} = 1270 nm. Adapted partly from Zenkevich et al. (2011).

For the quantitative analysis of FRET QD→porphyrin participation in the two-step process of $^1O_2$ generation by QD–porphyrin nanoassemblies two main aspects have been taken into account (Zenkevich et al., 2011; Zenkevich and von Borczyskowski, 2012, 2015). (i) Typically, PL measurements for QDs are carried out at essentially lower excitation energies with respect to laser experiments being used for $^1O_2$ generation. Indeed, a non-linear dependence of emission intensities for QDs and porphyrin molecules in QD–porphyrin nanoassemblies (x = 4) have been observed upon variation of the laser intensity (at λ_{exc} = 532 nm into the QD absorption band) (Zenkevich et al., 2011; Zenkevich and von Borczyskowski, 2015). This finding reflects the well-known pump-intensity dependence of the amplitudes of the single-exciton caused by non-radiative intraband Auger processes (Klimov, 2000, 2010). These results may be considered as a direct proof of the FRET QD → porphyrin realization in nanoassemblies competing with both radiative electron–hole recombination and non-radiative Auger processes. (ii) Therefore, using low-intensity laser excitation and the results of the comparative measurements of $^1O_2$ emission for QD–porphyrin nanoassemblies, $H_2P(m\text{-}Pyr)_4$ molecules and Pd-mesoporphyrin as a standard (in solution of low optical density) one may qualitatively derive the efficiency of FRET QD → porphyrin process in nanoassemblies (Φ^{Δ}_{FRET}) via this independent way (Zenkevich et al., 2011) according to

$$\Phi^{\Delta}_{FRET} = \frac{(I^{\Sigma}_{\Delta} - I^{A}_{\Delta})}{(I^{DA}_{\Delta})_{max}} \tag{1.18}$$

$$(I^{DA}_{\Delta})_{max} = \frac{I^{0}_{\Delta} \times \beta_{QD}}{\beta_0} \times \gamma^{A}_{\Delta} \tag{1.19}$$

where I^{Σ}_{Δ} is the total intensity of singlet oxygen emission; I^{A}_{Δ} is the $^1O_2$ emission intensity sensitized via direct excitation of acceptor; $(I^{DA}_{\Delta})_{max}$ is the expected increase of $^1O_2$ emission intensity sensitized via FRET in QD–porphyrin nanoassembly with the efficiency Φ_{FRET} = 1.0; I^{0}_{Δ} is the $^1O_2$ luminescence intensity sensitized by standard Pd-mesoporphyrin, γ^{A}_{Δ} = 0.6 is the quantum efficiency of singlet oxygen generation by $H_2P(m\text{-}Pyr)_4$ molecules in toluene (Sagun et al., 2002). $\beta_{QD} = (1\text{–}10^{-D_{QD}})$ and $\beta_0 = (1\text{–}10^{-D_0})$

are fractions of exciting light absorbed by QDs in a nanoassembly and a QD only. D_{QD} and D_0 are the corresponding optical densities of the sample at λ_{exc} = 532 nm.

It has been quantitatively evaluated that at low laser excitation intensity ($E \leq 4.5$ μJ/pulse) the efficiency of FRET QD → porphyrin found via comparative measurements of $^1O_2$ emission corresponds to Φ_{Δ}^{FRET} = 0.12–0.15. The most important result is that for the same QD–porphyrin nanoassemblies (x = 4), this value is in a good agreement with a FRET efficiency Φ_{FRET} = 0.12–0.14 found from the direct sensitization experiments discussed above.

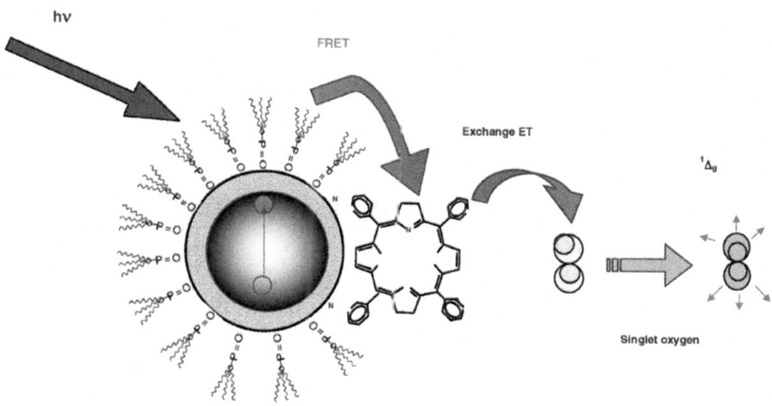

$$\text{(A)}\quad [QD-H_2P] + h\nu \rightarrow [QD^{\otimes} - H_2P] \xrightarrow{FRET} [QD-H_2P^1] \qquad (1.20)$$

$$\text{(B)}\quad [QD-H_2P]^1 \xrightarrow{ISC} [QD-H_2P^3] \qquad (1.21)$$

$$\text{(C)}\quad [QD-H_2P^3] + {}^3O_2 \rightarrow {}^1([QD\text{-}H_2P^3] \ldots {}^3O_2) \xrightarrow{Exc\ ET} {}^1([QD\text{-}H_2P] \ldots {}^3O_2({}^1\Delta_g))$$
$$\rightarrow [QD-H_2P] + {}^1\Delta_g \qquad (1.22)$$

Figure 1.36 Schematic presentation of the main processes leading to singlet oxygen generation by QD–porphyrin nanoassemblies in liquid solution. (A) QD excitation and FRET QD → porphyrin. (B) Non-radiative intersystem crossing S_1~~>T_1 in porphyrin molecule. (C) Formation of collision complex of the nanoassembly (having triplet-excited porphyrin) with molecular oxygen followed by the formation of singlet oxygen $^1O_2$ ($^1\Delta_g$) via exchange energy transfer from porphyrin.

This good quantitative coincidence of experimental values Φ_{FRET} and Φ_{Δ}^{FRET} (evaluated via independent experimental approaches) reveals that this FRET process QD→porphyrin (which is not the dominant one compared with non-FRET quenching) is the only reason of the increase of singlet oxygen generation by nanoassemblies in comparison with QDs only. The full scheme of the processes leading to singlet oxygen generation by QD–porphyrin nanoassemblies is presented in Fig. 1.36.

To the best of our knowledge, such an approach presents the first example showing quantitatively the role of FRET effects in the singlet oxygen generation by QD–dye nanoassemblies. Recently, our ideas have been proven upon studying the interaction of molecular oxygen with self-assembled conjugates based on CdSe core-shell QDs and pyridyl substituted Pd-porphyrins under both one- and two-photon excitation (Lemon et al., 2013).

1.4.2 Size-Dependent Non-FRET QD PL Quenching in QD–Dye Nanoassemblies

As was discussed in Section 1.1 due to quantum confinement (Klimov, 2000, 2010; Rogach, 2008; Gaponenko, 2010) the exciton in a QD is highly sensitive to local charges or distortions of the charge distribution. The excitonic wave function probes these local charges effectively because of its spatial extension even beyond the core of the QD. Correspondingly, the quantum efficiency of QD PL is sensitive to the involved interfaces including inorganic shell structures (Reiss et al., 2009; Kimura et al., 2004), surfactants (Wuister et al., 2004; Potapova et al., 2003; Querner et al., 2004; Willard et al., 2006) and solvent molecules (Leatherdale and Bawendi, 2001) as long as the exciton wave function extends beyond the QD core and—if present—ZnS capping layers. It is further evident that the time scale of these processes is related to that local volume (core, surface, and environment), which is responsible for the corresponding interaction processes.

It is evident that one of the demanding tasks is to probe the interactions locally by the directed or controlled modification of interfacial properties with a local probe. Such an approach may unravel microscopic details of either non-radiative processes occurring at the QD interfaces or the influence of chemical bonds at the QD surface. Clearly, organic dye molecules attached to the surface might act as such probes. Several experiments in this direction have been reported (Potapova et al., 2003; Querner et al., 2004; Willard et al., 2006). Following these ideas, we will describe here that the surface attachment of colloidal semiconductor QD by only a few pyridyl-functionalized dye molecules (such as porphyrins and perylene diimides discussed above) reveals the nature of the only recently identified non-FRET QD PL quenching mechanism for QD–porphyrin and QD–PDI nanoassemblies which (as was shown above) is the dominant PL quenching process compared to the additional FRET processes (Blaudeck et al., 2008, 2012; Kowerko et al., 2010). It should be noted that this non-FRET quenching mechanism has not been considered in most of the related publications and is connected with the extension of the wave function of the exciton beyond the QD core.

The main idea of corresponding titration experiments is based on the comparative study and analysis of PL quenching for QDs of different sizes (with and without ZnS monolayers) in QD–porphyrin nanoassemblies having one type of porphyrin molecule, namely $H_2P(m\text{-}Pyr)_4$ as a best quencher among other porphyrins (Blaudeck et al., 2008). Typical QD PL quenching curves are shown in Fig. 1.37A.

For the analysis of the QD PL quenching as a function of the number of H_2P molecules per QD for various QD sizes, the well-known Stern-Volmer formalism (Lakowicz, 2006), was used (Fig. 1.37B) and modified (Blaudeck et al., 2008). In our more generalized approach, QD PL quenching data can be presented in the following form

$$\frac{I_0}{I} = 1 + \int_0^{\infty} K(x)\cdot dx, \tag{1.23}$$

where $I(x)$ and I_0 are the QD PL emission intensities in the presence and absence of quenchers, respectively. In this case, the Stern-Volmer function $K(x)$ depends explicitly on the molar ratio x and is expressed as the first derivative of the experimental data plotted in a Stern-Volmer representation (Fig. 1.37B). In its turn, the Stern-Volmer "constant" $K(x)$ can be written as

$$K(x) = k_q(x)\cdot \tau_0 \tag{1.24}$$

where k_q corresponds to the total quenching rate induced by the number of attached quencher molecules and τ_0 is the intrinsic PL lifetime of the QD in absence of quencher molecules (known from independent measurements (Blaudeck et al., 2008) or taken from literature (Javier et al., 2003)).

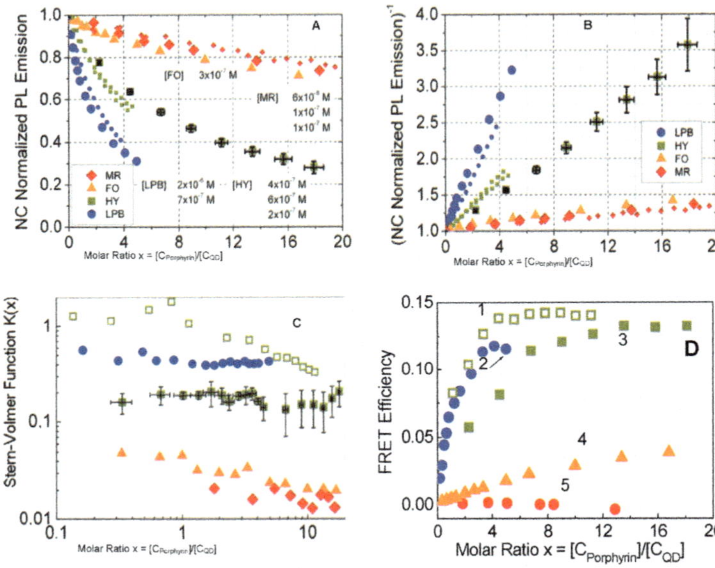

Figure 1.37 QD PL quenching (A), Stern-Volmer plots $I_0/I(x)$ (B), calculated Stern–Volmer function $K(x)$ (C) and FRET efficiencies calculated on the basis of approach described in (Zenkevich et al., 1986; Clapp et al., 2004; Knowles et al., 2010; Zenkevich et al., 2005) (D) for an uncapped CdSe QD (open markers) and CdSe/ZnS-capped QDs (solid markers) of various sizes upon titration by $H_2P(m\text{-}Pyr)_4$ as function of the molar ratio $x = [C_{Porphyrin}]/[C_{QD}]$ in toluene at 295 K: 1 (□) - d_{CdSe} = 3.5 nm, n_{ZnS} = 0; 2 (●) - d_{CdSe} = 2.1 nm, n_{ZnS} = 2; 3 (■) - d_{CdSe} = 3.0 nm, n_{ZnS} = 2; 4 (▲) - d_{CdSe} = 4.1 nm, n_{ZnS} = 2; 5 (◆) - d_{CdSe} = 5.2 nm, n_{ZnS} = 2. In part (C) $K(x)$ values have been corrected for the FRET contribution in PL quenching at every titration step for all QDs under study (presented in part (D)). Adapted from Blaudeck et al. (2008).

Figure 1.37B shows that the dependence $I_o/I(x)$ is not a linear correlation with x over the total molar ratio range. Moreover, from Fig. 1.37C where a double logarithmic plot is presented for

$K(x)$, it follows evidently that at the same molar ratio x, the QD PL quenching in nanoassemblies is dependent on the QD size. It can be seen also that $K(x)$ is initially constant but becomes smaller around a critical molar ratio (which we termed x_c (Blaudeck et al., 2008, 2012)). Additionally, this critical molar ratio x_c increases systematically with the diameter of the QD. A non-constant $K(x)$ indicates a deviation from the normal Stern-Volmer relationship in case of a bimolecular reaction.

It was shown in Section 1.4.1 that the detailed nature of the electronic structure of the various porphyrins does not influence QD PL quenching (Zenkevich et al. 2005, 2006, 2007a, 2007b). Thus we discuss only those aspects that are related directly to the outer interface of a QD as a function of size and shell thickness. Experimental data for mean $<K_{SV}(x)>$ values and PL quenching rate constants k_q being obtained from data shown in Figs. 1.37C,D are collected in Table 1.1.

Table 1.1 Mean $<K_{SV}(x)>$ values and PL quenching rate constants k_q for CdSe/ZnS-capped QDs of various core diameters d_{CdSe} (two ZnS monolayers) upon titration by $H_2P(m\text{-}Pyr)_4$ in toluene at 295 K

d_{CdSe}, nm	4.3	5.2	6.3	7.3
$<K_{SV}(x)>$	0.65 ± 0.10	0.115 ± 0.015	0.055 ± 0.007	0.020 ± 0.003
k_q, ns^{-1}	0.041 ± 0.016	0.0057 ± 0.0023	0.0027 ± 0.0016	0.0015 ± 0.0009

Note: In all cases QD PL quenching in QD–porphyrin nanoassemblies is caused by to at least two reasons (small FRET and dominant non-FRET). Thus, $<K_{SV}(x)>$ and k_q values have been corrected for the FRET contribution (shown in Fig. 1.37D).

Based on a detailed analysis of the above results described recently in (Blaudeck et al., 2012; von Borczyskowski and Zenkevich, 2014) the basic interpretation of the above findings can be presented in the following way. Obviously, QD PL quenching upon titration by porphyrin molecules occurs in two steps. (i) Immediately after titration, QD–porphyrin nanoassemblies are effectively formed, which is manifested in both QD PL quenching and low-effective FRET QD→porphyrin in nanoassemblies (Blaudeck et al., 2012). (ii) Then, during the following observation time, more porphyrin molecules become attached to QD surface in an exchange competition with capping TOPO molecules.

However, increased QD PL quenching and FRET do not follow the same dependence on the molar ratio *x*. In fact, the titration (followed by nanoassembly formation) induces predominantly other quenching mechanisms than those related to FRET. Figure 1.37D shows that the maximal FRET efficiency does not exceed ~12–15% over the total titration range, and becomes sequentially lower with increasing QD size. This makes us to believe that the exciton dynamics initiated by a single titration step are not only caused by the intrinsic properties of the porphyrin macrocycle itself, but also by a local change in the ligand shell and the QD surface states upon nanoassembly formation.

It follows from the data collected in Table 1.1 that for CdSe/ZnS QDs having two ZnS monolayers, k_q values decay considerably upon QD core diameter increase. In addition, for CdSe QDs without a ZnS shell a much stronger QD PL quenching is found compared to that obtained for the respective core-shell analogue. As far as the electronic structure and properties of attached porphyrin were the same in these experiments the observed QD PL quenching in QD–porphyrin nanoassemblies with various QDs has to be related with interface phenomena. In this respect, at the outer interface of the QD–porphyrin nanoassembly two contributions are important: (i) the presence of "binding" orbitals such as the lone pair of a pyridyl ring of the porphyrin or pyridine itself and (ii) the evanescent wave function of the confined exciton "leaking" out of the core (and through the ZnS shell) of the QD. From a physico-chemical point of view it means that upon interaction of the attached H_2P(m-Pyr)$_4$ molecule with the QD surface, the electron wave function may be locally modified (e.g., via inductive and/or mesomeric effects (Zenkevich et al., 2007a)) forming a surface local state capable to trap the electron of the photogenerated electron–hole pair. As in any related quantum mechanical problems the wave function of the delocalized electron may extend beyond an imposed barrier or, in other words, the corresponding particle may tunnel through the barrier. With respect to QDs, quantum confinement of the excitonic wave function in its most simple version confines the wave function of the exciton (electron–hole pair) to a spherical potential box (Dabbousi et al., 1997).

Following concepts discussed in (Dabbousi et al., 1997), the general conception of our model may be presented as follows.

Figures 1.38A,B show a general scheme of the QD–porphyrin interaction and PL quenching model under the conditions of quantum confinement where the electron wave function at the outer interface of the QD is presented schematically. This implies that the electron becomes partly localized in a volume element in the vicinity of the attachment site.

In its turn, upon the porphyrin attachment on QD surface, specific trap states are formed giving rise to a trapping of the delocalized electron. Correspondingly, all observed size dependence of the PL quenching (as expressed by the quenching rate k_q) should be directly related to the probability density $\Psi^2(r)$ calculated from the quantum-confined *s*-type electronic wave function for a 1*s* electron. Thus, the calculation of changes of the electron densities is therefore reduced to an evaluation of the radial probability function $r^2\psi^2(r)$ for the electron (normalized with respect to r^2, that is, with respect to the total QD surface area) as a function of the CdSe core diameter and ZnS shell thickness. The single-carrier envelope wave functions ψ in a spherical core/shell QD are calculated using the Schrödinger equation (the simplified particle-in-a-box model in effective-mass approximation (Haus et al., 1993) by

$$\left(\frac{\hbar^2}{2}\frac{\partial^2}{\partial r^2}\left(\frac{1}{m_a^*(r)} + V_a(r)\right)\right)\psi_a = E_a\psi_a. \qquad (1.25)$$

where the index $a = e, h$ represents electron and hole, respectively. Solutions were obtained by applying continuity relations for the single-carrier wave functions at the interfaces between the CdSe core ($i = 1$), ZnS shell ($i, j = 2$), and matrix ($j = 2$), respectively, by assuming that

$$\psi_i(r_{ji}) = \psi_j(r_{ji})$$

$$\frac{1}{m_i^*}\frac{d}{dr}\psi_i(r)\bigg|_{r=r_{ji}} = \frac{1}{m_j^*}\frac{d}{dr}\psi_j(r)\bigg|_{r=r_{ji}}. \qquad (1.26)$$

These calculations for four CdSe/ZnS QDs with two ZnS monolayers and one CdSe QD without a ZnS monolayer are presented in Fig. 1.38C (Blaudeck et al., 2008, 2012). It can be seen that $\psi^2(r)$ becomes smaller at the outer interface (marked by a circle) upon increasing QD diameter. The corresponding value is largest for the uncapped CdSe QD ($R = 1.73$ nm).

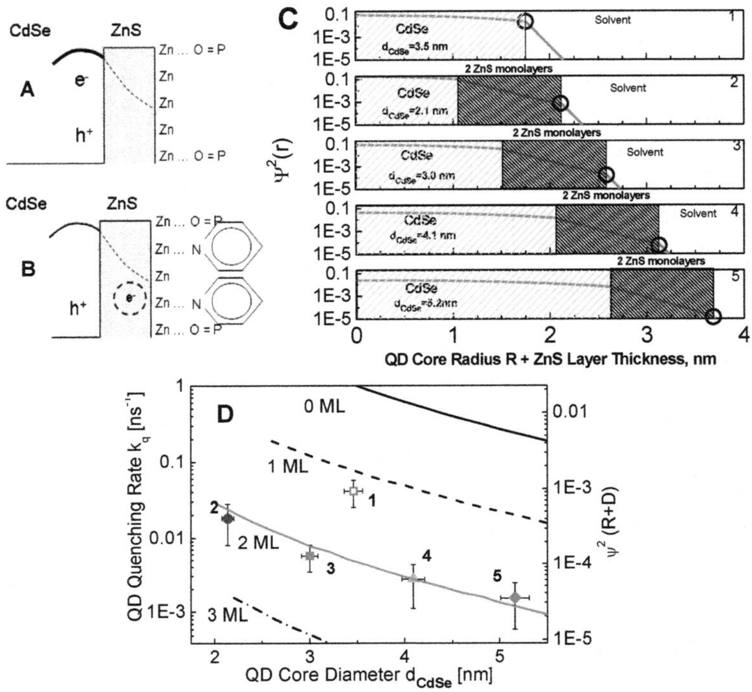

Figure 1.38 Basic scheme of the QD PL quenching model and comparison of experimental results and theoretical calculations for QD-porphyrin nanoassemblies. (A, B): Scheme of the QD–porphyrin interaction and PL quenching model: (A) Upon excitation, an electron–hole pair (exciton) is created in the CdSe core of the QD. In the case of finite barriers, e^- and h^+ are subject to tunneling from the core through the ZnS shell into the environment. (B) Because of the presence of a pyridyl-coordinated porphyrin molecule, the electron becomes partly localized in a volume element in the vicinity of the attachment site. On the QD surface, specific trap states are formed giving rise to the trapping of the delocalized electron. Charged QD remains dark. (C) Logarithmic presentation of the probability density $\Psi^2(r)$ calculated from the radial part of the quantum-confined *s*-type electronic wave function (simplified by a particle-in-a-box model in effective-mass approximation) for a 1*s* electron in a core-shell spherical potential for five different sizes of QDs with n_{ZnS} = 0 or 2. (D) Comparison of the experimentally determined quenching rate constants k_q (symbols with error bars, left axis) and calculated probability density functions $\psi^2(r = R + D)$ (lines, right axis) of a 1*s* electron

at the outer interface ($r = R + D$) between the ZnS shell and the environment as a function of core diameter d and ZnS shell thickness D. The constant C in Eq. (1.26) has been adjusted with respect to $\psi^2(R+D)$ to fit the experimental value at d_{CdSe} = 4.1 nm, n_{ZnS} = 2. All theoretical lines correspond to calculation of $\psi^2(R)$ with the same proportionality constant C. QD parameters are the same as described in caption to Fig. 1.37. Adapted from Blaudeck et al. (2008).

Now the QD PL quenching rate for various CdSe or CdSe/ZnS QDs may be written as

$$k_q(r) = C\psi^2(r), \tag{1.27}$$

where the constant C may be once adjusted with respect to $\psi^2(R + D)$ in order to fit the experimental value for a given QD (CdSe/ZnS with d_{CdSe} = 4.1 nm and two ZnS monolayers, in our case (Blaudeck et al., 2008)). Figure 1.38D shows a comparison of the experimentally determined quenching rate constants k_q and calculated (and scaled) values of probability densities $\psi^2(r = R + D)$. A remarkable coincidence is seen between experimental values k_q (symbols) and theoretical calculations $\psi^2(r)$ (lines).

On the basis of these results we conclude that one point-like charge density perturbation caused by an organic linker group or chromophore at the QD interface forces the electron of the delocalized exciton of the QD to become localized. Qualitatively the same physico-chemical picture is characteristic also for QD–PDI nanoassemblies (Blaudeck et al., 2012; Kowerko et al., 2009, 2010). In fact, the dependence of QD PL quenching induced by even only one single molecule on the QD surface clearly resembles the tunneling of an electron (through the ZnS barrier) to the outer interface of the QD. Such tunneling is followed by the (self-) localization of the electron. This corresponds to the creation or modification of trap states in the semiconductor band gap. Such traps might be subject to non-radiative channels, e.g., via enhanced electron-phonon coupling. Thus, changes in local charge densities result in changes in the quantum efficiency of the QD which have to be related to PL lifetime changes and modification of blinking in case of single QDs (Nirmal et al., 1996; Schlegel et al., 2002; Cichos et al., 2007; Zenkevich et al., 2015). These results reveal also that single functionalized porphyrin molecules can be considered as a probe for the complex interface physics and dynamics of colloidal semiconductor quantum dots (Blaudeck et al., 2012; Kowerko et al., 2009, 2010; Zenkevich et al., 2012, 2015). It is tempting to conclude that the quenching is merely due to the influence of the pyridyl anchor since the electronic properties of the porphyrins are obviously negligible (see Fig. 1.34) (Zenkevich et al., 2006, 2007a). Additionally, the PL quenching caused by PDI molecules is similar to the one observed for porphyrins as long as assemblies are formed in a toluene solvent. However, obviously the anchoring group itself is not the main source for quenching, since titration by orders of magnitude larger amounts of pyridine as compared to porphyrins does not result in noticeable quenching (see Fig. 1.11A).

Remember, as shown in Fig. 1.37B, that the Stern-Volmer quenching constant $K(x)$ is initially constant but becomes smaller close a "critical" molar ratio x_c while the x_c value increases systematically with the diameter of the QD (Blaudeck et al., 2008, 2012). These results can be connected to the fact that QD–porphyrin nanoassemblies are formed in a dynamic equilibrium of attachment and detachment of porphyrin molecules to the QD surface. As was discussed in Section 1.2.2.2, the time to reach the initial equilibrium is faster than our experimental observation time of about 60 s. This implies that there have to be some free or very easily accessible sites (Kalyuzhny and Murray, 2005; Munro et al., 2007, 2008; Ji et al., 2008; Koole et al., 2008; Morris-Cohen et al., 2010; Evans et al., 2010), or the dye attachment on the QD surface, since ligand replacement occurs typically on time scales of minutes or even longer. Taking into account these arguments and obtained results we conclude that assembly formation can be tentatively divided into a low-molar-ratio and high-molar-ratio range which is separated by a "critical" molar ratio x_c. Above the critical molar ratio x_c, the probability of attaching dye molecules to the QD surfaces is considerably reduced. The linear behavior of x_c as a function of the QD surface area for the same absolute QD concentration (Blaudeck et al., 2012) reveals that the number of freely accessible sites is proportional to the surface area of the respective QD. The experiments on QD–PDI nanoassemblies (Kowerko et al., 2009, 2010) show the same tendency. Thus, we conclude that upon detailed analysis of QD PL quenching in QD–dye nanoassemblies it is possible to probe the interface dynamics of QDs with dye molecules in solution even from ensemble experiments.

Summarizing all arguments discussed in this Section, we conclude that non-FRET quenching in QD–dye nanoassemblies is related to exchange and/or depletion of ligands by the respective dye molecules. In other words, non-FRET quenching may be caused by the removal of ligands at specific surface sites accompanied by the formation of intra-band states and possibly the creation of new (PL quenching) trap states. Hence, for tuning the PL properties, it does not need a mayor exchange of the ligand shell, but already very few dye molecules can induce very effectively PL quenching. This conclusion is supported by results (Gomez et al., 2006b) showing that the PL quantum yields of QDs and the photobrightening are determined primarily by the surface stoichiometry and by the resultant selective surface adsorption of passivants. Correspondingly, it follows from long-time behavior of PL quenching and its dependence on the amplitude of the electronic wavefunction at the outer interface, that both processes play an essential role in PL quenching.

1.4.3 Competition of FRET and Non-FRET Processes

Experimental findings have shown (see discussion in previous Sections) that besides FRET also non-FRET processes are induced upon attachment of dye molecules to a QD surface. In the following, we will show how the competition between these processes can be mechanistically explained. At the beginning we like to stress that at a molar ratio $x = 1$ not all constituents are forming assemblies (even less than expected according to a binomial or Poisson distribution). All our experiments at $x = 1$ show that the ratio $R_{nc/c}$ of non-complexed to complexed QDs is $2 < R_{nc/c} < 4$ (at least below $x < 4$) (Blaudeck et al., 2012; Kowerko et al., 2009, 2010). However, this will have only impact on the absolute efficiencies of PL quenching (for QDs) and fluorescence enhancement (for dyes) but not on the relative ones as we experimentally observe.

We approach our explanation by inspecting the PL of a single QD. As we have outlined in Chapters 4 and 6 of Von Borczyskowski and Zenkevich (2016), QD PL is subject to blinking. We and others have also shown (Hohng and Ha, 2004; Fomenko and Nesbitt,

2008; Gomez et al., 2006b; Hammer et al., 2006; Odoi et al., 2007; Cichos et al., 2007; Schmidt et al., 2014) that various PL intensities during a blinking time trace are inherently related to different PL decay times (see, e.g., Chapter 4 in Von Borczyskowski and Zenkevich (2016)). This implies that the PL decay is multi-exponential even for a single QD. Since a multi-exponential decay is also observed for an ensemble of QDs, we claim that the multi-exponential PL is (not only) due to ensemble heterogeneity but intrinsically related to the nature of the PL of QDs probably due to charging-discharging processes (see Chapter 4 in Von Borczyskowski and Zenkevich (2016)). Strictly speaking, a time average (single QD experiments) and an ensemble average are not completely identical because of weak ergodicity breaking. However, several experiments have shown that blinking of single QDs can at least qualitatively be compared with (reversible bleaching) of ensembles of QDs (Cichos et al., 2007; Chung and Bawendi, 2004). Figure 1.39 shows schematically part of a single QD blinking time trace in which we have tentatively assigned different PL intensities to PL decay times experimentally determined (see Table 1.2) in bulk experiments on QDs and H_2P-QD nanoassemblies (in brackets). The relative amplitudes of the ensemble PL decay are also given. We are aware, that taking only three decay times is a simplification, but recent experiments have shown, that this is a qualitatively sufficient approximation both for QD ensembles (Petrov et al., 2006) and single QDs (Schmidt et al., 2014) (see also Chapter 4 in this book).

Following Section 1.4.1, FRET transfer times between QDs and porphyrin molecules are between 10 and 30 ns, depending on the size of the QD. If FRET would be the only QD PL quenching process all PL intensities during a (hypothetical) blinking time trace would be reduced according to nearly the same FRET transfer time as long as the radiative decay times remain similar. This implies that the high (integrated) PL intensity (of the long PL decay time) will be affected more than the low intensity components (of the short PL decays). Additionally, if a detailed analysis beyond a three-component deconvolution would be possible three new decay components should show up in case that not all QDs are in an assembly. We will discuss this case later in this section (see Fig. 1.40). Restricting to a three-component deconvolution merely a shortening of the respective decay times will show up depending on the respective FRET efficiency with respect to the specific state. Inspection of Table 1.2 (see Section 1.4.4) and Fig. 4.13 (from Chapter 4 of this book) shows that decay times are in fact shortened. Moreover, the (normalized) amplitudes change considerably in favor of the shorter ones (see also Fig. 4.13 in Chapter 4 of this book). This is an important observation, since the relative (initial) population probabilities (amplitudes) of the individual three decay components should not be changed in the presence of merely FRET but only the decay times (and thus the integrated PL intensities).

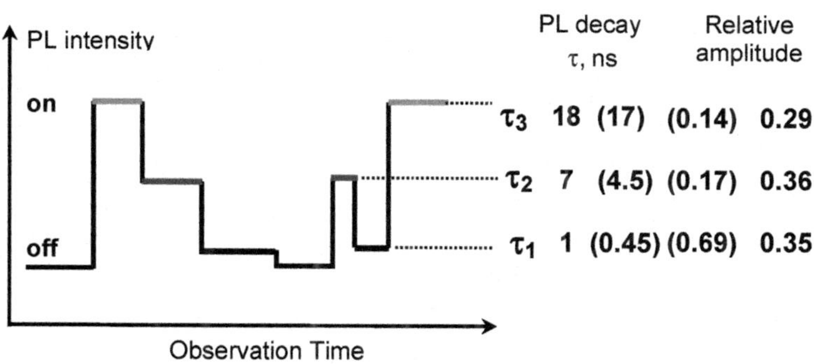

Figure 1.39 Schematic presentation of a hypothetical single QD blinking time trace. On the right side we relate PL decay times and amplitudes of CdSe/ZnS to the respective (hypothetical) intensities. Values in brackets correspond to the ones in CdSe/ZnS-H_2P nanoassemblies PL decay data are taken from Table 1.2.

Making use again of the single QD blinking model, these observations imply that PL intensities become smaller (since PL decay time decrease and, more importantly, the QD remains more often in one of the "off-" or "dim-" intensities (states). This is empirically exactly what we have termed non-FRET PL quenching. Since lowered PL intensities during a blinking time trace are related to charging-type processes (depending strongly on surface state properties) this proves that attachment of dye molecules results in an increased efficiency of such charge-related quenching processes depending strongly on the embedding environment (see Sections 1.2.2.2, 1.2.3.1, and 1.2.4 and Zenkevich et al. (2005)).

The question might arise why such non-FRET processes have not been observed in all QD–dye nanoassemblies reported in literature. We believe that this is due to the fact that in those cases the surface has been modified during synthesis procedure instead of during a titration experiment which allows monitoring the reorganization of surface state properties. We also like to emphasize that experiments discussed Sections in 1.2.2.2, 1.2.3.1, and 1.2.4 provide strong evidence that charge-related processes are involved which are in strong competition with the intended FRET processes.

In the following we will demonstrate how under the above conditions the FRET-related PL quenching is identified in the time resolved QD PL decay in parallel to porphyrin fluorescence kinetics for the examples of CdSe/ZnS QDs (three ZnS monolayers (Zenkevich et al., 2005)) and CdSe QDs (see Chapter 4 of this book and Blaudeck (2007)).

Figure 1.40 shows the deconvoluted PL decay of TOPO-capped CdSe/ZnS QDs (d_{CdSe} = 3.2 nm, three ZnS monolayers) for which we have calculated a FRET transfer time of about 37 ns (Zenkevich et al., 2005). Molar ratios have been changed between $0 \leq x \leq 3$. We have restricted to this range since the previous discussion (see Section 1.4.2) has shown that above this critical concentration more complex PL quenching takes place. Figure 1.40 reveals four broadly distributed decay times for $x = 0$. The longest very weak one longer than 50 ns belongs to trap states (discussed briefly in Section 1.2.2, see Fig. 1.13). The other three are typically in the range of three-exponential data fitting to PL decay with times centered at 20, 7 and 0.6 ns (compare also Table 1.2 for CdSe/ZnS QD of different size). Upon increasing the molar ratio x the shortest component becomes still shorter and new components appear at about 2 and 12 ns (the latter as a kind of shoulder) while the 7 and 20 ns components decrease in amplitude.

These experimental findings are in agreement with the above discussion. We like to point out that if restricting (for sensitivity reasons) a fit to only three components (as has been done for the CdSe QD to be discussed further on) this finding is in agreement with a shortening of all decay times, a decrease of the amplitude related to the long decay time and a strong increase of the amplitude for the two shortest PL decay components. We tentatively assign the ingrowth of the shoulder at ≈12 ns as to be due to FRET and obtain according to $k_{FRET} \approx (1/(12\text{ ns})-1/(20\text{ ns}))$ a FRET QD→porphyrin time of ≈ 30 ns in rough agreement with the calculated one of 37 ns (Zenkevich et al., 2005). This time is longer than the intrinsic PL decay of 20 ns. Next we inspect the fluorescence dynamics of the porphyrin molecule (H_2P).

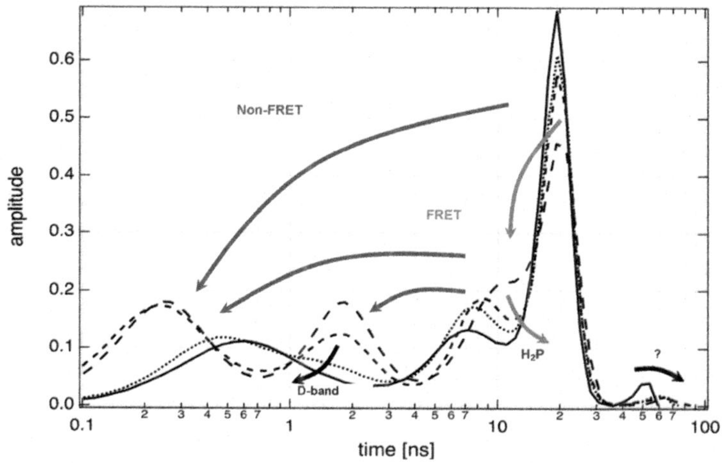

Figure 1.40 Deconvolution of the PL decay of TOPO capped CdSe/ZnS (d_{CdSe} = 3.2 nm; three ZnS monolayers, toluene solvent) as a function of molar ratio $x = [H_2P]/[QD]$ (x = 0: full line; x = 0.57: dotted line; x = 1.14: short dashed line; x = 2.3: long dashed line. Arrows mark the tentative relations among the different decay components following assembly formation. Arrows assigned to D-band (deep traps) and H_2P correspond to energy transfer out of PL of the QD. Red arrows correspond to FRET and blue arrows to non-FRET processes.

Figure 1.8C reveals at $x < 4$ a build-up component of about 2–3 ns and a fluorescence decay of 11 ns for H_2P which is by about 2 ns longer than the intrinsic H_2P decay under otherwise identical conditions. Above $x = 3$, both the build-up component and the lengthening of the H_2P decay disappear which can be tentatively explained by a more complex assembly formation (see discussion above). To become more qualitative Fig. 1.8C shows that the distribution of H_2P decay times becomes slightly broader and is on average shifted to longer times. Previously we have shown (Rempel et al., 1998) that if a donor decay time is longer than the one of the acceptor there will be still energy transfer lasting longer than the acceptors intrinsic decay time resulting in an apparent lengthening of the fluorescence decay. In this case we expect a bi-exponential decay with k_{H2P} and an additional "effective" energy transfer rate according to $(k_{FRET} + k_{QD})$. Taking the PL decay of the QD of 20 ns and the FRET transfer time of 30 ns (see previous discussion) we obtain 12 ns for the second component of the bi-exponential H_2P decay which is qualitatively in agreement with the experimental finding of an average decay time of H_2P of 12 ns.

The short build-up component of 3 ns is much too fast to be responsible for FRET QD→porphyrin for this kind of QD (d_{CdSe} = 3.2 nm, three ZnS monolayers). Therefore we assign this component to the population time of those intra-band trap states induced by H_2P or effectively populated in the presence of H_2P molecules on a QD surface. It is well known that a broad spectral distribution of intra-band trap states (called D-bands) exists (see, e.g., Blaudeck (2007)), depending on the concrete conditions of surface passivation. Such states decay slowly (e.g., 800 ns in Fig. 1.41)—if populated—with only weakly radiative decay times to the ground state. We will discuss the example of a CdSe QD without a ZnS layer in the following, since uncapped QDs show an increased density of such trap states.

We first present in Fig. 1.41 the PL decay of CdSe QDs without a ZnS layer (d_{CdSe} = 3.2 nm, TOPO capped) as a function of the detection wavelength. The decay has been fitted by four decay times which are in the range of 30 ns (close to the radiative lifetime of this QD (Blaudeck et al., 2008), a medium component close to 6 ns, a short component of about 1 ns and a very long component of 800 ns (D-band emission). Results are shown in Fig. 1.41 also for the relative amplitudes. The figure also contains the PL across the investigated wavelength range. Similar to what is shown in Table 1.2, the PL decay at the maximum of the PL at about 578 nm consists of three strong components. The contribution of the very long component dominates at long wavelengths. It is worthwhile to note that the mayor components at the wavelength of H_2P emission are the 6 and 800 ns components. We like to note that the very long lifetime might be strongly influenced by the presence of background signals. This implies that the absolute values might have a large systematic error.

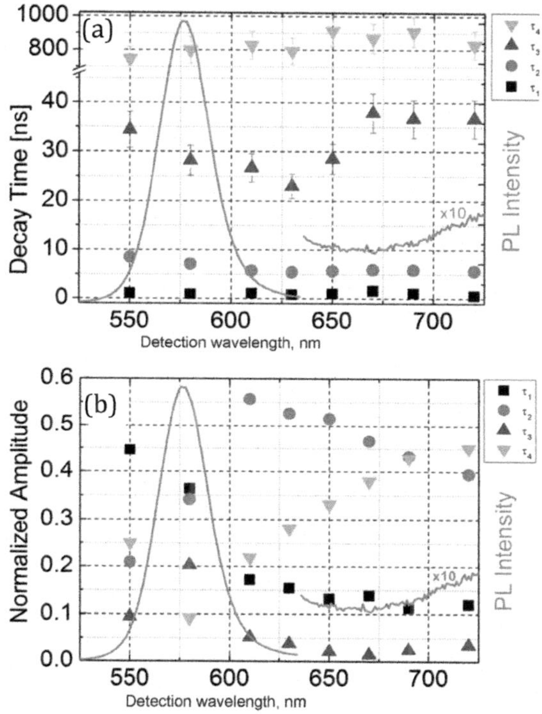

Figure 1.41 Detection wavelength dependence of PL decay dynamics fitted by 4 exponential decay times (a) and their respective amplitudes (b) for CdSe QDs (d_{CdSe} = 3.2 nm, TOPO capped) in toluene at 293 K. Adapted from Blaudeck (private communication).

Figure 4.13 in Chapter 4 of this book shows the influence of H_2P titration on the PL decay of the CdSe QD. The PL decay times become shorter and become closer in favor of the shorter components with increasing molar ratio, very similar to the observations for the earlier discussed CdSe/ZnS QD. In the range of H_2P fluorescence (at about 653 nm) we also observe as in the previous case a build-up component of a few ns and a lengthened H_2P decay time (from 7.8 ns to 11.5 ns). Also these two effects fade away with increasing molar ratio. A closer inspection of the H_2P fluorescence decay shows that the fluorescence decay is at least bi-exponential with decay times of 8 ns and ≈12 ns. The first one can be related to the intrinsic decay whereas the second one is probably due to FRET from the long lived QD to the

H_2P acceptor as discussed before. From the second component we approximate the FRET transfer rate to be $k_{\mathrm{FRET}} \approx (1/(12\ \mathrm{ns}) - 1/(29\ \mathrm{ns})) = 1/(20.4\ \mathrm{ns})$. Depending on the typical quantum yield for this type of QD we expect FRET transfer times between 10–20 ns (Blaudeck, 2007) which is in reasonable agreement with experimental findings.

The investigations of the two exemplary QDs show that FRET itself is effective as it is expected neglecting non-FRET processes. This confirms the applicability of Foerster type models for energy transfer in QD–dye nanoassemblies (Zenkevich et al., 2005; Clapp et al., 2004). However, attachment of various dye molecules both via functional groups and intercalation results during the replacement of ligands in alternative quenching channels which could in principle be described as a lowering of the QD PL quantum yield. In the language of single nanoassemblies, this results in a higher probability of off- and dim-states. This change of quantum yield is "permanent" and is not taking place from the excited state such as FRET or charge transfer. This behavior is the base to explain the large difference between static and dynamic PL quenching discussed in detail in Chapter 4.

The microscopic nature of these processes can be even followed spectroscopically, at least at reduced temperatures (Zenkevich et al., 2015). Discussion throughout this chapter has shown that attached dye molecules most probably induce a re-arrangement of (photo-induced) charges at the QD surface which is supported by the strong dependence of QD PL quenching in QD–dye nanoassemblies on solvent polarity (Zenkevich et al., 2005). We also suggest according to the above presented findings that certain dye-induced (dim) states populate subsequently traps in the band gap. Correspondingly, it is not excluded that these (shallow and deep) states may participate in energy transfer processes realized in QD–dye nanoassemblies (Lüa et al., 2006).

1.4.4 Tuning Quantum Dot Electronic States and Exciton Relaxation Dynamics by One Attached Dye Molecule

Results and discussion presented above have shown that the sensitivity of the QD PL intensity and decay upon dye attachment provides an opportunity to study and control the dynamics and fate of the exciton by controlling the properties of these molecules. With respect to CdSe/ZnS QDs as discussed in Chapter 2 of Von Borczyskowski and Zenkevich (2016), several principal aspects have been outlined which should be taken into account upon analysis of PL characteristics for QDs, especially to unravel as many as possible subtle optical properties of core and surface-related electronic states which gave new insights in several optical observables with respect to a basic understanding of surface properties. In addition, investigations of temperature dependent PL and absorption spectra for CdSe/ZnS (TOPO- or AM-capped) QDs, described in Chapter 3 in Von Borczyskowski and Zenkevich (2016), revealed that few mechanisms might be responsible for the formation of QD PL properties: exciton–phonon coupling depending on the kind of the capping ligand and the ligand-related phase transition, distribution and activation of surface trap states.

As we underlined earlier in this chapter, a significant progress has been achieved recently in quantum-chemical/quantum-mechanical calculations of QD structures and their electronic properties. What is most important with respect to QD–dye nanoassemblies discussed here, theoretical results give a profound background for the understanding of QD surfaces, interface phenomena and the impact of capping ligands, solvents and attached dye molecules (Puzder et al., 2004; Rempel et al., 2006; Frenzel et al., 2007; Inerbaev et al., 2009; Gomez-Campos and Califano, 2012; Kilina et al., 2009a, 2009b, 2012; Albert et al., 2011). In fact, calculations have shown that ligands cause (depending on their concrete number) surface morphology, modify even on a single ligand base electronic states or electron-phonon coupling and hot carrier relaxation (Kilin et al., 2007; Kilina et al., 2012; Fischer et al., 2012; Califano et al., 2013). Moreover, pronounced dependencies on structure-related specific positions of surface atoms have also been identified (Fischer et al., 2012) including the mobility of individual surface atoms or ligands (Voznyy and Sargent, 2014; Voznyy, 2011; Frantsuzov et al., 2013). On the basis of X-ray absorption spectroscopy and ab initio modeling of the experimental spectra, the influence of surface passivation on the unoccupied electronic states of CdSe QDs have been studied (Lee et al., 2012). It has been shown that significant differences are observed in the unoccupied electronic structure of CdSe QDs, which have been connected with variations in specific ligand surface bonding interactions. Thus, not wishing to analyze the whole set of interesting results being obtained in this direction, we will present some examples which are close to the processes and events experimentally studied and discussed here.

For $Cd_{33}Se_{33}$ QDs, DFT calculations predict significant surface reorganization for both QD and QD capped with amine or phosphine oxide ligands (Kilina et al., 2009a). In addition, strong surface-ligand interactions lead to substantial charge redistribution and polarization effects on the QD surface, resulting in the development of hybridized states, for which the electronic density is spread over the QD and the ligands. Calculations show also that the loss of one of the passivating ligands leads to either optically dark or bright additional states inside the band gap, depending on the position of the leaving ligand on the QD surface (Figs. 1.42 and 1.43). In its turn, being mostly optically dark, dense hybridized states open new relaxation channels for high-energy photoexcitations. In the result, a different manifestation of ligand binding may potentially lead to faster electron relaxation in QDs passivated by phosphine oxide than by amine ligands.

Using DFT and time-dependent DFT (TDDFT) quantum-chemical methodologies (Kilina et al., 2012), the impact of passivating ligands (i.e., amines, phosphines, phosphine oxides and pyridines) on the electronic and optical spectra of $Cd_{33}Se_{33}$ QDs (optimized geometry is presented in Fig. 1.43) has been examined in details. It was found that most ligand orbitals are deep in the valence and conduction bands of the QD, with pyridine being an exception by introducing new states close to the conduction band edge. Moreover, all ligands lead to the appearance of states which are highly delocalized over both the QD surface and ligands, forming hybridized orbitals rather than ligand-localized trap states. On the other hand, the states close to the band gap are delocalized over the QD atoms only and define the lower energy absorption spectra. In its turn, a random detachment of one of ligands from the QD surface results in the appearance of a highly localized unoccupied state inside the energy gap of the QD (Fig. 1.44). It was found that the degree to which the electronic structure is perturbed appears to be related to both the binding energy of the ligand and the degree of steric ligand-ligand interactions. It is interesting to note that a polar solvent significantly diminishes both effects leading to delocalization and stabilization of the surface states.

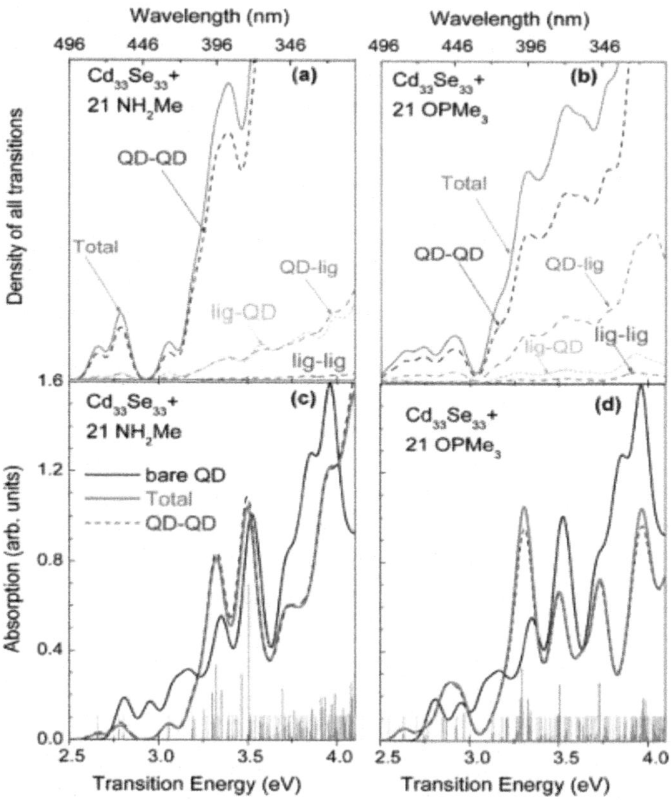

Figure 1.42 Comparison of the calculated density of states (DOS) and absorption spectra of $Cd_{33}Se_{33}$ QD capped by 21 NH_2Me (a, c) and $OPMe_3$ (b, d) molecules. The red solid line (a, b) shows the DOS spectra of all possible (including optically forbidden) transitions. Dashed lines show calculated transitions between valence and conduction bands. According to calculations, near the energy gap, transitions are dominated by QD transitions and are optically bright, while for higher energies, the fraction of transitions between hybridized states increases; however, most of such transitions are dark. Adapted from Kilina et al. (2009a).

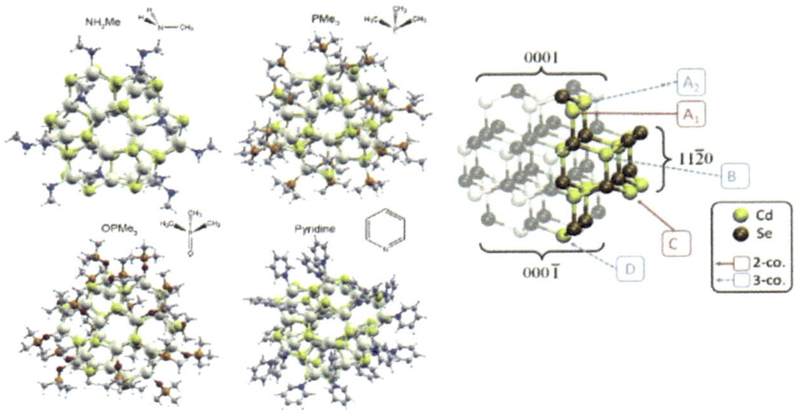

Figure 1.43 Optimized geometries of four differently ligated $Cd_{33}Se_{33}$ QDs. Chemical structures of the ligands [methylamine (NH_2Me), trimethylphosphine (PMe_3), trimethylphosphine oxide ($OPMe_3$), and pyridine] are shown also. Some of them have been used in our experiments. The conclusion is that geometry optimization of the $Cd_{33}Se_{33}$ QDs leads to a reconstruction of the surface so that all surface atoms become at least three-coordinated. Adapted from Kilina et al. (2009a).

Finally, we conclude that our experimental findings and qualitative description of the properties of QD–dye nanoassemblies can be closely related to these calculations of ligand-induced modifications of surface states of QDs. However, in order to achieve dye-mediated control of the energy relaxation processes and even modification of QD energies (note the influence of non-FRET processes) in QD–dye nanoassemblies we need to understand the specific ways and mechanisms by which attached dye molecules affect the PL spectral properties of QDs. Thus, we apt to believe that found non-FRET quenching of QD PL in QD–dye nanoassemblies may be a result of different binding strengths and/or different luminescent properties and/or charge localization features of surface and/or intra-band states. In case that there are various attachments sites with varying binding strengths it is reasonable to assume that these sites are in a different way sensitive to dye attachment either due to a formation of (dye specific) trap states, which quench the QD PL effectively, or are more easily exchanged by dye molecules.

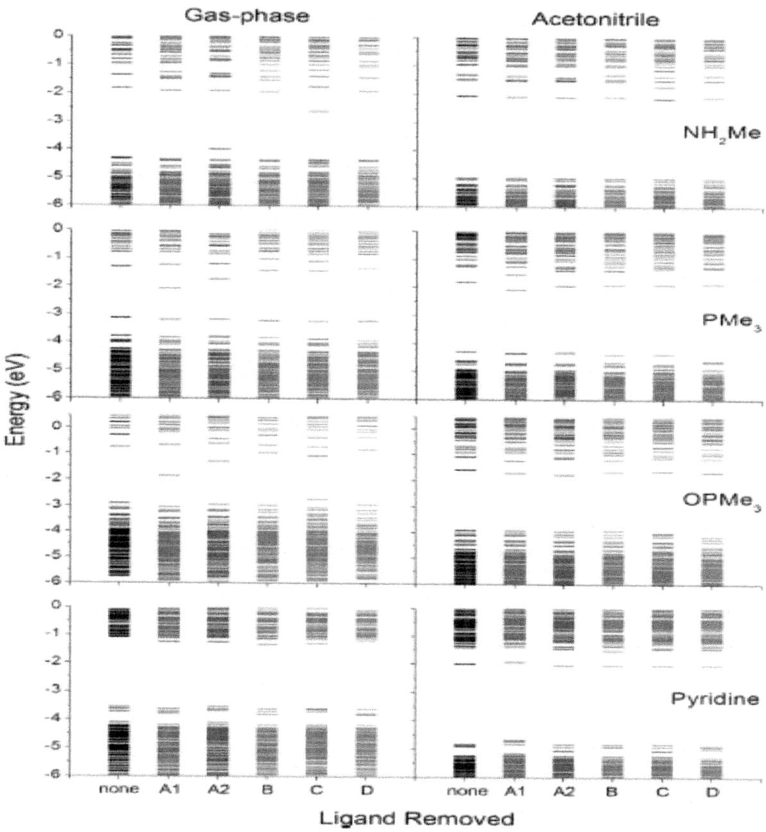

Figure 1.44 Calculated energy level diagram for the fully passivated $Cd_{33}Se_{33}$ with 21 ligands and $Cd_{33}Se_{33}$ with 20 ligands (i.e., one ligand has been removed from the specific position as indicated underneath the diagram; for definitions see Fig. 1.43). Positions A1 and C are two-coordinated Cd atoms while the rest are three-coordinated Cd atoms. Solvent stabilizes the energy levels that are disrupted by a ligand loss. Adapted from Kilina et al. (2009a).

In this respect, the basic understanding is on a one-molecule-to-one-QD base of special interest. It follows from the analysis of literature data in Section 1.1.1, that theoretical considerations predict a specific influence of attached dye molecules on QD surface and PL properties followed by formation of related trap states of various nature.

The results on systematic investigations of QD–porphyrin nanoassemblies (based on TOPO-capped CdSe/ZnS QDs) and QD–PDI nanoassemblies (based on AM-capped CdSe/ZnS QDs) with respect to the replacement of a few ligands by only one dye molecule (using a combination of steady-state, time-resolved experiments on both bulk and single nanoassembly detection level) have been published in part by us recently (Zenkevich et al., 2012, 2015). Those investigations have shown that the combination of bulk and single nanoassembly experiments allows for a detailed and up to now not yet reported complex analysis of the QD PL in CdSe/ZnS-Dye nanoassemblies. In both cases, namely ensemble and time averaged single QD detection, electronic states of different nature with varying PL energies and decay dynamic are subsequently explored on slow time scales typical for blinking phenomena which are buried but nevertheless present in ensemble experiments. Moreover, on the basis of these results it becomes possible to analyze (in some cases even quantitatively) what the reasons for non-FRET quenching processes are. In other words, for QD–dye nanoassemblies, we are in a position to connect electron tunneling under the conditions of quantum confinement with the formation of new surface states having various energies and decay times. Some of these recently obtained and published results will be discussed in the following.

Let us summarize the main ideas evaluated from the analysis of the results of comprehensive temperature experiments (77–295 K) carried out for QD–porphyrin and QD–PDI nanoassemblies (at molar ratio $x = 1$) (Zenkevich et al., 2015) in combination with newly designed and complementary (time and spectrally resolved) spectroscopy on single QDs (Schmidt et al., 2014) based on change point analysis of QD PL blinking (Watkins and Jang, 2005; Schmidt et al., 2014). Remember that Figs. 1.13 and 1.19 show the temperature dependence of spectral properties for TOPO capped CdSe/ZnS QDs and amine-capped CdSe/ZnS-AM QDs, respectively. As was discussed earlier for both dependencies the manifestation of a low-temperature phase transition of the capping ligand layer is a characteristic feature. The attachment of functionalized dye molecule does not influence the phase transition temperature T_{crit} effectively, but QD PL changes at the phase transition temperature become much more pronounced upon nanoassembly formation. Upon temperature lowering PL spectra of QDs and those in nanoassemblies are shifted to the blue (and narrowed in spectral width) which is due to a decrease of electron-phonon coupling (see Chapter 3 in Von Borczyskowski and Zenkevich (2016)). In addition, comparing PL bands at 77–95 K of CdSe/ZnS QDs with those of QDs in nanoassemblies an additional small blue-shift is observed in the case of H_2P-CdSe/ZnS nanoassemblies, while the PL band is more strongly blue-shifted and considerably broadened for QDs in CuP-CdSe/ZnS nanoassemblies (Zenkevich et al., 2015). These observations are quite remarkable since they are caused by on average at most 1 porphyrin molecule per QD.

The observed variations of PL parameters for QDs and QD–porphyrin nanoassemblies may be considered as a manifestation of the presence of several assembly dependent spectral components (states) constituting a PL band. We have shown (supported by accuracy statistics analysis and confidence interval determination (Zenkevich et al., 2015)) that an improved approximation of the QD PL band characteristics is achieved by assuming at least two Gaussian PL components with an energy separation >5–10 meV. Figure 1.45 shows as a function of temperature (A) the maximum $(E_{max})_i$, (B) the related spectral line width $(FWHM)_i$, and (C) the integrated intensity I_i for the two spectral components obtained for CdSe/ZnS QDs (left) and CdSe/ZnS in H_2P-CdSe/ZnS (middle) or in CuP-CdSe/ZnS (right) nanoassemblies. In all cases a broad (B) and a narrow (N) spectral components are present. It is evidently seen, that all temperature dependencies differ strongly among each other. For CdSe/ZnS QDs, above $T_{crit} \approx 220$ K $(E_{max})_i$ values are nearly identical. Deviations from a continuous change with temperature are observed in the vicinity of the phase transition temperature $T_{crit} \approx 220$ K, and this difference becomes most obvious when inspecting the PL intensity ratio in Fig. 1.45D. In the case of QDs in H_2P-CdSe/ZnS nanoassemblies, an increase of the energy separation δE between B and N Gaussian spectral components as well as a large change of the intensity ratio are observed, while the temperature dependence of $(E_{max})_i$ and $(FWHM)_i$ is similar with respect to that found for QDs. Assigning PL energies according to the spectral widths N and B crossing of energies is observed at T_{crit} for CdSe/ZnS QDs and CdSe/ZnS in H_2P-CdSe/ZnS nanoassemblies (Fig. 1.45A). The situation is much more complex for QDs in CuP-CdSe/ZnS nanoassemblies.

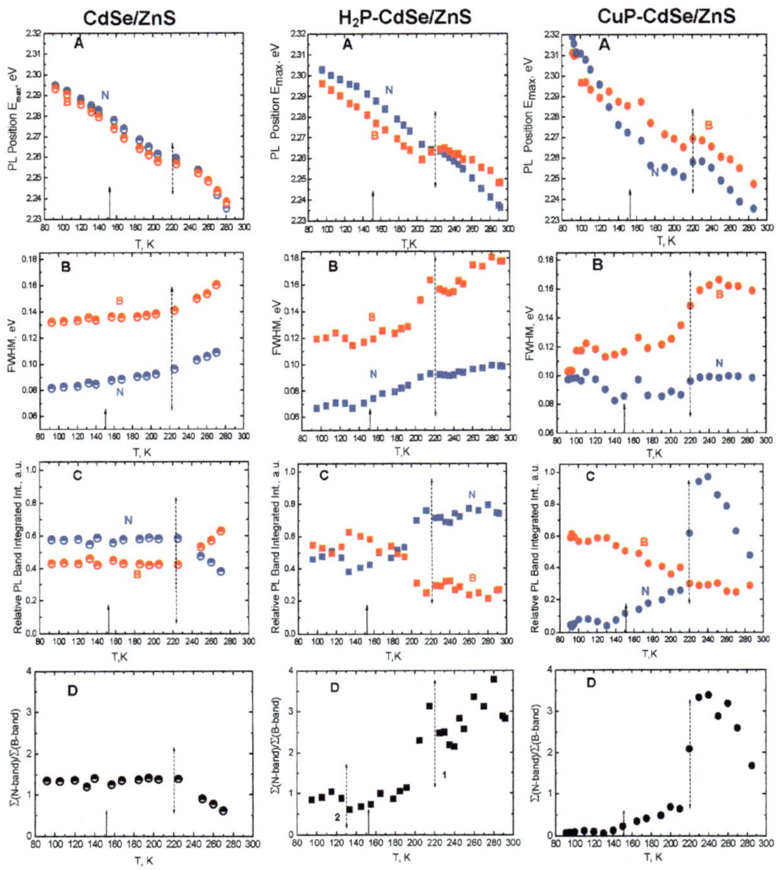

Figure 1.45 Temperature dependence of PL properties for (left) TOPO-capped CdSe/ZnS QDs (d_{CdSe} = 3.0 nm, two ZnS monolayers) as well as for QDs in H_2P-CdSe/ZnS (middle) and CuP-CdSe/ZnS (right) nanoassemblies in methylcyclohexane-toluene (6:1) mixture at λ_{exc} = 450 nm. (Fits have been performed with two Gaussian components, termed B (broad) and N (narrow)). The temperature T_{crit} (≈220 K) of a phase transition of the capping TOPO layer is indicated by the dashed arrow. The glass transition temperatures: 146.7 K (methylcyclohexane), 180 K (toluene), and 151.6 K (methylcyclohexane/toluene (6:1) mixture, shown by a solid arrow. Properties for N and B components are shown as A: PL band energies E_{max}; B: Full spectral widths at half maximum (FWHM); C: Relative integrated PL band intensities; D: Ratio of integrated PL band intensities. Adapted from Zenkevich et al. (2015).

While the energy variation of the broad component B is similar to the one of CdSe/ZnS QDs, the corresponding variation of the narrow component N deviates from the corresponding one of QDs in H_2P-CdSe/ZnS. The separation of PL energies is larger for QDs in nanoassemblies as compared to CdSe/ZnS. A remarkable finding is that at low temperatures the energy separation δE between the B and N components is increasing from CdSe/ZnS to QDs in H_2P-CdSe/ZnS and further to CuP-CdSe/ZnS nanoassemblies. A close analysis of these results permits the conclusion that a narrow component N emerges for CuP-CdSe/ZnS nanoassemblies which is 20 meV above the energies of QDs in H_2P-CdSe/ZnS nanoassemblies and which depends much stronger on temperature than the broad component B.

It is interestingly to note that using the above deconvolution procedure for AM-capped CdSe/ZnS QDs and the respective QD–DPP (1:1) nanoassemblies we found very similar temperature dependencies for N and B components of the QD PL (Fig. 1.46). These findings show that the existence of several assembly-induced spectral components of the PL band is an intrinsic property of CdSe/ZnS QDs not strongly dependent on the nature of capping ligands.

Comparative time-resolved PL decay data obtained for TOPO-capped CdSe/ZnS QDs and QD–porphyrin nanoassemblies at 77 and 290 K at various detection wavelengths across the PL bands of QDs, provides additional information indicating energetic inhomogeneities of QD emission beyond the one caused by the size distribution. At the moment, it is well documented that at ambient conditions for bulk solutions of CdSe/ZnS QDs (Kapitonov et al., 1999; Javier et al., 2003; Zenkevich et al., 2005; Petrov et al., 2005, 2006) as well for single QDs (Schmidt et al., 2014; Kowerko et al., 2010; Watkins and Jang, 2005; Schlegel et al., 2002; Fischer et al., 2004), PL decay times are broadly distributed and are dependent on detection wavelength, PL intensity and temperature (Petrov et al., 2006; Watkins and Jang, 2005; Schlegel et al., 2002; Fischer et al., 2004). Here, we found that at both investigated temperatures (77 and 290 K) the PL decays for CdSe/ZnS QDs as well as for H_2P-CdSe/ZnS and CuP-CdSe/ZnS nanoassemblies are also multi-exponential. At least three exponential functions (with decay times $\tau_1 \leq 1$ ns, $\tau_2 \approx 4$–7 ns and $\tau_3 \approx 10$–18 ns) are needed within our time resolution of 0.2 ns to approximate the PL decay.

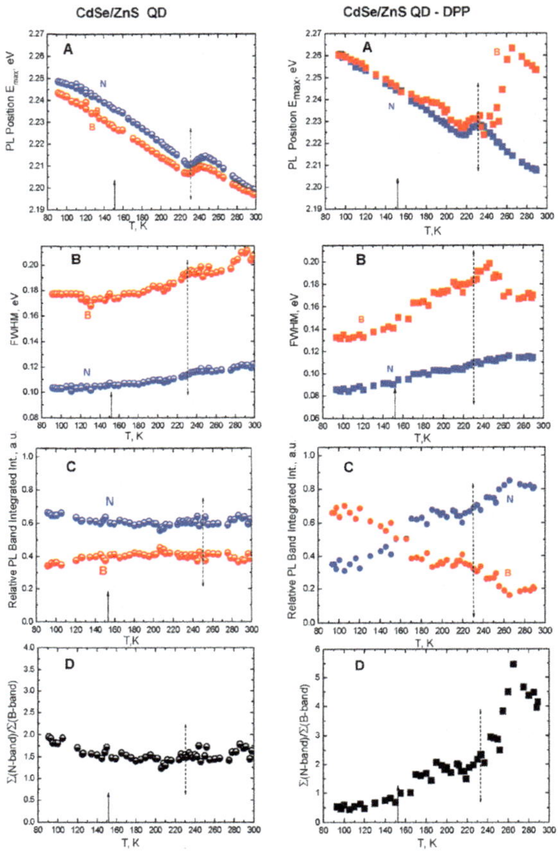

Figure 1.46 Temperature dependence of PL properties for AM-capped CdSe/ZnS QDs (d_{CdSe} = 3.2 nm, 2 ZnS monolayers) and for QDs in QD–DPP nanoassemblies (1:1) in methylcyclohexane-toluene (6:1) mixture at λ_{exc} = 430 nm. Fits have been obtained by two Gaussian components, named B (broad) and N (narrow). Properties for N and B components are shown as (A) PL band energies E_{max}; (B) Full spectral widths at half maximum (FWHM); (C): Relative integrated PL band intensities; (D): Ratio of integrated PL band intensities. The temperature T_{crit} (≈230 K) of a phase transition of the capping AM layer is indicated by the dashed arrow. The glass transition temperatures: 146.7 K (methylcyclohexane), 180 K (toluene), and 151.6 K (methylcyclohexane/toluene (6:1) mixture, shown by a solid arrow. Data for a broad B-component of QD–AM PL band in DPP-CdSe/ZnS-AM (1:1) nanoassemblies at T > 250 K (A, right) are overestimated because of the overlap with DPP emission in this spectral range.

Details of the experimental three-exponential fit of the PL decay for these objects as well the comparison of the correlations between time-resolved PL data for CdSe/ZnS QDs in bulk solutions and single QDs (spin coated onto a quartz substrate, change point analysis of QD blinking events (Schmidt et al., 2014)) are presented in our recent publication (Zenkevich et al., 2015). Some findings and main results are collected in Fig. 1.47 and Table 1.2.

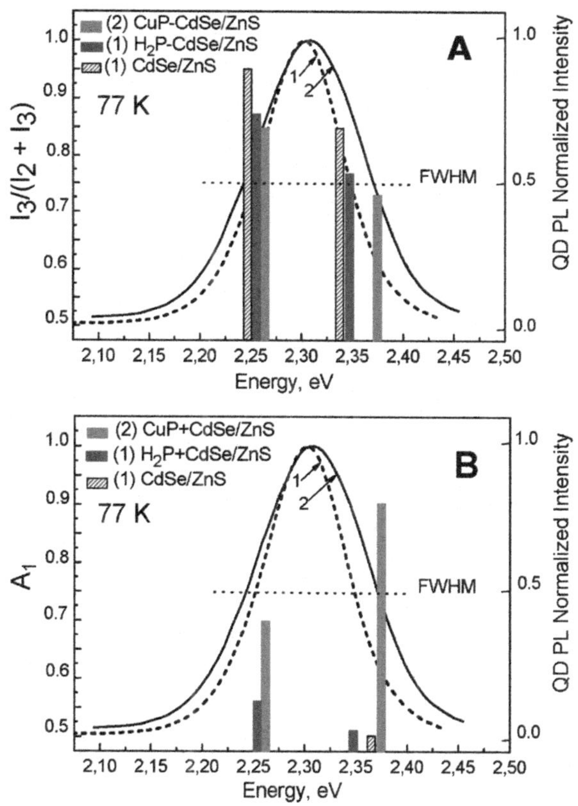

Figure 1.47 Relative PL intensities $I_3/(I_2 + I_3)$ (A) and normalized amplitudes A_1 (B) for TOPO capped CdSe/ZnS QDs and QDs in H_2P-CdSe/ZnS or CuP-CdSe/ZnS nanoassemblies at 77 K. For assignment of I_i and A_i see Table 1.2. Normalized PL bands are shown as broken lines for CdSe/ZnS (1) or QDs in H_2P-CdSe/ZnS (1) and as full lines for CuP-CdSe/ZnS (2) nanoassemblies at 77 K. I_3 corresponds to the longest decay component relevant in spectral deconvolution while A_1 is related to the fastest decaying PL component and for most cases not detected in PL spectra.

Table 1.2 QD PL decay times τ_i, normalized amplitudes A_i and intensities I_i for TOPO-capped CdSe/ZnS QDs (d_{CdSe} = 3.0 nm, two ZnS monolayers) and QDs in H_2P-CdSe/ZnS or CuP-CdSe/ZnS nanoassemblies in methylcyclohexane-toluene (6:1) mixture at T = 77 and T = 290 K[a]

	τ_i/ns	Single QD[b]		Ensemble of QDs[a]					
	I_i/%	CdSe/ZnS		CdSe/ZnS		H_2P-CdSe/ZnS		CuP-CdSe/ZnS	
T/K	$E_{i,FWHM}$	**High**	**Low**	**High**	**Low**	**High**	**Low**	**high**	**low**
77	τ_1			0.5		0.3		0.2	0.4
	A_1			0.2		0.51	0.56	0.9	0.7
	I_1			1		3	3	21	5
	τ_2			4	3.5	3.8	4	2.8	3.3
	A_2			0.28	0.2	0.27	0.21	0.07	0.2
	I_2			14	5	22	16	21	15
	τ_3			14.5	16	11.5	12.5	9.8	10.3
	A_3			0.52	0.8	0.31	0.35	0.03	0.1
	I_3			85	95	75	82	59	80
	$<\tau>^c$			14.5	17.5	9.5	10.5	9.0	8.5
290	τ_1	2	0.8	0.8	0.8	0.3	0.6	0.5	0.5
	A_1	0.13	0.44	0.37	0.33	0.72	0.65	0.63	0.64
	I_1			4	3	7	10	7	8
	τ_2	9	8	7.7	7	4	5	6	5
	A_2	0.31	0.27	0.36	0.36	0.15	0.2	0.19	0.17
	I_2			35	30	20	25	25	21
	τ_3	21	18	18	18	16.5	17.5	17	16.5
	A_3	0.56	0.29	0.27	0.31	0.13	0.15	0.18	0.17
	I_3			61	67	72	65	68	71
	$<\tau>^c$	13.9	14.4	14.5	14.5	12.5	12.0	12.5	11

[a]Parameter ranges (high, low) correspond to the variation of the QD PL detected at high and low PL energy at the corresponding FWHM values of the QD PL band. Typical errors for τ_i are 0.2–0.5 ns. Typical errors for A_i are 5%. (λ_{exc} = 410 nm, solvent is methylcyclohexane-toluene mixture 6:1).

[b]Single QD data according to (Zenkevich et al., 2015). Data have been taken for the peak values of the decay corresponding distributions.

[c]QD PL multi-exponential decay curves $A(t)$ were fitted by three components A_i according to

$$A(t) = \sum A_i \exp(-t/\tau_i). \tag{1.28}$$

Mean decay times $<\tau>$ were calculated according to

$$<\tau> = (\sum A_i\tau_i^2)/(\sum A_i\,\tau_i). \tag{1.29}$$

Adapted from Zenkevich et al. (2015).

In Table 1.2 fitting parameters at two PL energies correspond to the respective high and low FWHM position of the total PL line width. Data collected in Table 1.2 show that both the absolute decay times τ_i, related amplitudes and their relation to the spectral range depend noticeably on the kind of nanoassembly, which is in accordance with the observation that the composition of PL spectra depends strongly on the kind of nanoassembly. In addition, at both 77 and 295 K, the decrease of decay times upon nanoassembly formation reflects the presence of QD PL quenching processes following attachment of only one porphyrin molecule as was discussed by us earlier in Sections 1.2.2.1, 1.2.2.2, 1.4.1, and 1.4.2. It should be noted that typically electron-phonon coupling will result in faster relaxation processes upon temperature increase (see Chapter 3 in Von Borczyskowski and Zenkevich (2016)). However, the opposite temperature dependence is observed inspecting Table 1.2. This is a clear indication that the nature of PL states is different below and above T_{crit}.

Summarizing, the whole set of experimental data presented above shows that the QD PL decay dynamics depend strongly on detection wavelength, temperature and the kind of nanoassembly but not on excitation power under the present conditions. These findings prove again that we are dealing, besides spectral broadening related to the size distribution of QDs, with an inhomogeneity of PL energies caused by the presence of various electronic states as they can be characterized according to the large variation of PL decay times τ_i. It was found (Zenkevich et al., 2015) that mean decay times $<\tau>$ are almost identical over the total spectral range for CuP-CdSe/ZnS nanoassemblies at 77 K but decrease with energy at 290 K. While mean decay times $<\tau>$ for both CdSe/ZnS and QDs in H_2P-CdSe/ZnS nanoassemblies are shorter at high PL energies at 77 K (see Fig. 1.47 and Table 1.2) as compared to those at low PL energies, they remain more equal over the total spectral range at 290 K. The small variation of $<\tau>$ values for CdSe/ZnS at 290 K is in complete agreement with recent findings for a somewhat larger CdSe/ZnS QD at room temperature (Petrov et al., 2006).

From the spectral distribution of PL decay times (see Fig. 1.47A) it follows that at 90 K the component with the decay time τ_2 has to be related to the energetically higher narrow N component identified in Fig. 1.39A, while the decay time τ_3 has to be assigned to the energetically lower broad spectral component B. The very short decay component τ_1 in the wing of the PL spectral line is obviously at still higher energy than the energies related to A_3 and A_2 (see Table 1.2 and Fig. 1.47B) but cannot be detected readily due to the corresponding low intensities I_i. A noticeably detectable contribution of the fast decaying component τ_1 is, however, found in case of CuP-CdSe/ZnS nanoassemblies at 77 K which amounts up to 21% of the total integrated intensity (see Table 1.2). Hence we assign τ_1 to the energetically highest component N of the CuP-CdSe/ZnS nanoassembly. This spectroscopically non-negligible short component also explains the broad FWHM for QD PL in case of CuP-CdSe/ZnS (see a comparison of normalized PL spectra for alone QDs and those in nanoassemblies presented in Fig. 1.47).

Based on these conclusions, Scheme 1.1 shows a qualitative correlation of PL energies and PL decay dynamics in ensembles of QDs as determined from deconvolution of PL energies upon temperature changes according to results presented in Table 1.2 and shown in Figs. 1.45 and 1.47. Thus, taken together all experimental temperature dependent spectral and kinetic data on ensembles of QDs and QD–porphyrin nanoassemblies as well as results on study of blinking statistics for single QDs (Zenkevich et al., 2015; Schmidt et al., 2014), one can explain the origin of at least two different spectral PL components for CdSe/ZnS QDs and QDs in the nanoassemblies. An attached porphyrin free base molecule, H_2P, induces at least one new emissive CdSe/ZnS state, which is at all temperatures higher in energy than the one of CdSe/ZnS QD, while the attachment of CuP molecule to the same QD leads to the formation of new states: one state is similar to that caused by H_2P attachment ($\tau_2 \approx$ 3–4 ns), and the other state is a short-lived state with $\tau_1 <$ 0.5 ns. A compilation of all these findings with respect to PL energies is that all PL stems from states below the band-edge energy and has to be assigned to intrinsic or induced surface or trap states.

Concluding, we have shown experimentally for the first time, that only one dye molecule is capable to change the surface-related distribution of near-band-edge or surface state energies. Correspondingly, both the formation of these states (either band-edge states, surface states or shallow traps) in QD–dye nanoassemblies and the extension of the exciton wave function beyond the core/shell structure of the respective QD (discussed in Section 1.4.2 (Blaudeck et al., 2008; Dabbousi et al., 1997)) may be considered as physico-chemical background of the realization of a QD PL non-FRET quenching.

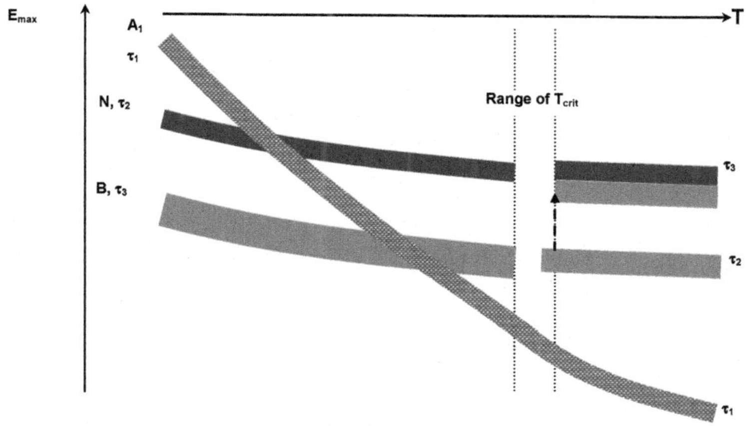

Scheme 1.1 Qualitative temperature dependence of up to three PL energies responsible for QD PL emission of alone CdSe/ZnS QDs and those in QD–porphyrin nanoassemblies derived from the combined analysis of experimental QD PL decays (τ_1, τ_2, τ_3) and deconvoluted Gaussian components (N, B) for the QD PL band. Time resolved single QD data show (Zenkevich et al., 2015) that a short-lived component of type τ_1 is predominantly detected at the red edge of the QD PL at 290 K.

As we mentioned above our findings are confirmed by theoretical calculations (time-dependent DFT, (Kilina et al., 2009a, 2012; Fischer et al., 2012)) showing a few interesting effects which are directly related to our main conclusions: (i) organic ligands (such as amine- or pyridine-like) may introduce new states to the QD band edge states of both the conduction and valence band, and (ii) the removal of only one ligand out of an initially symmetric set of ligands will shift the lowest absorption noticeably to higher or lower energies strongly depending on the type of coordination of the removed ligand.

In the light of these considerations we present below the comparison of photoluminescence excitation (PLE) spectra detected for TOPO-capped CdSe/ZnS QDs and the corresponding

CuP-CdSe/ZnS (1:1) nanoassemblies at various detection wavelengths at ambient temperature (Fig. 1.48).

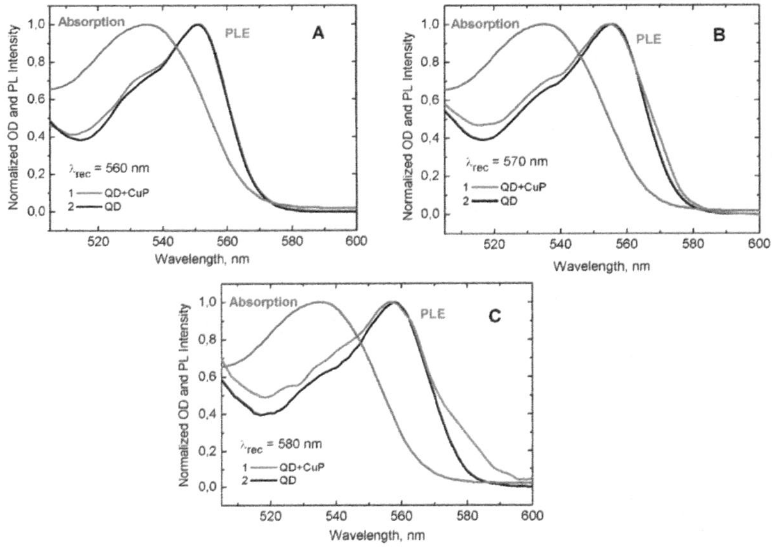

Figure 1.48 Photoluminescence excitation (PLE) spectra detected for TOPO-capped CdSe/ZnS QDs and the corresponding CuP-CdSe/ZnS (1:1) nanoassemblies (methylcyclohexane-toluene mixture 6:1, ambient temperature) at various detection wavelengths: λ_{rec} = 560 nm (A); λ_{rec} = 570 nm (B); λ_{rec} = 570 nm (C).

First, at all registration wavelengths, PLE spectral bands for both QDs and CuP-CdSe/ZnS nanoassemblies do not coincide with the absorption band corresponding to the first excitonic transition of the QD. Three reasons account for this: (i) Stokes shifts (see Chapters 2 and 3 in Von Borczyskowski and Zenkevich (2016)); (ii) effective size selection and (iii) PL from shallow traps below the band edge. In addition, at the same detection wavelengths, PLE excitation spectra for CuP-CdSe/ZnS nanoassemblies are slightly different from those measured for QDs. Moreover, this difference becomes more evident upon shifting the registration to the red. Additionally, the appearance and rise of new bands (especially at long wavelength side) for CuP-CdSe/ZnS nanoassemblies in comparison with QD are noticeable. These findings are considered as an experimental proof of the formation of new surface states in the band gap induced by only one attached Cu-porphyrin molecule. In Section 1.2.2.2 we presented and discussed also an appearance of still deeper trap states for CuP-CdSe/ZnS nanoassemblies upon temperature lowering (see Fig. 1.13). As was described in (Zenkevich et al., 2015) new PL bands show up upon assembly formation (see also Figs. 1.45 and 1.46). It follows from PLE experiments that also modified absorption bands show up even for highly excited exciton states.

Thus, increasing knowledge of specific interface phenomena at the QD surface will permit to realize the driven control of the complex exciton dynamics in QD–dye nanoassembly, and, in the result, will allow for a guided tailoring functional nanostructures for possible applications in photovoltaics and nanosensorics.

1.5 Conclusions

Basically, in this chapter we have shown that the strategy based on "bottom-up" self-assembly approach may be successfully used for the formation of heterogeneous organic-inorganic nanoassemblies containing colloidal TOPO- or amine-capped semiconductor CdSe QDs of various sizes and dye molecules with and/or without functionalized anchoring substituents (like meso-pyridyl substituted porphyrins and perylene diimides). Using a combination of ensemble (steady-state and time-resolved technique in a temperature range 77–295 K) and single molecule spectroscopy of nanoassemblies, we have shown that single functionalized dye molecules can be considered as extremely sensitive probes for studying the complex interface physics and exciton relaxation processes in QDs. Herein we focused on a quantitative experimental analysis of QD PL properties as well as the observed QD PL quenching in QD–dye nanoassemblies. Our findings have shown that surface-mediated processes dictate the probability of several of the most interesting and potentially useful photophysical phenomena observed for colloidal QDs. The principal aspects are the following:

(i) It has been quantitatively clarified that the major part of the observed QD PL quenching in QD–dye nanoassemblies, namely non-FRET processes can be understood on one hand, in terms of electron tunneling beyond the CdSe core under conditions of quantum confinement, and on the other hand, by the influence of ligand dynamics. FRET QD→Dye often plays a minor role in the QD PL overall quenching. The competition between FRET and non-FRET quenching processes drastically depends on the solvent and ligand properties. Quenching by non-FRET in QD–dye nano-assemblies is effectively enhanced already in slightly polar solvents which is often overlooked in literature data.

(ii) Nevertheless, we have shown quantitatively that even low FRET efficiencies (experimentally found for QD–porphyrin nanoassemblies in toluene at ambient temperature) may increase significantly the photochemical activity of these complexes. Moreover, a good quantitative coincidence of experimental FRET efficiencies Φ_{FRET} (found from the acceptor fluorescence sensitization) and Φ_{Δ}^{FRET} values (evaluated via independent measurements of singlet oxygen generation) reveals that namely FRET process QD→porphyrin is the only reason of the essential increase of singlet oxygen generation by nanoassemblies in comparison to QDs without assembly formation.

(iii) The observed non-FRET quenching of QD PL induced by attachment of only one dye molecule (or incorporation in the ligand shell) can be explained by the presence of a limited number of empty or easily accessible sites on the QD surface followed by ligand competitive exchange (TOPO, amines and dyes) on various time scales.

The efficiency of non-FRET quenching depends on various factors, including QD size (electron tunneling in conditions of quantum confinement), the number of attachment sites, the microscopic nature of these sites, which might be additionally influenced by solvent properties, crystal structure (facets) and QD surface geometry. We argue that non-FRET quenching is related to depletion of capping ligands by the respective dye molecules followed-up by creating more and even new quenching states according to the amplitude of the tunneling ("leaking") excitonic wave function. Deep intraband traps are effectively populated.

(iv) Temperature variation (77–295 K) and related changes in QD absorption and emission reveal drastic changes of the ligand shell structure in a narrow temperature range for organic (TOPO and amine) ligands (so-called "phase transition" of capping layer). The effects on QD PL intensities, energies and dynamics at this transition become considerably pronounced upon attachment of only one or a few dye molecules (porphyrin or perylene diimide) to the QD surface. It follows from these experiments, that the ligand phase transition has impact even on the QD core structure and exciton–phonon coupling.

(v) The comparison of averaged and deconvoluted spectral PL properties (such as band energies, spectral widths and relative integrated band intensities) of CdSe/ZnS QDs and QDs in QD–dye nanoassemblies in a wide temperature range proofs that already one attached dye molecule causes not only PL quenching but also changes the energy landscape of the QD PL noticeably. Temperature controls the energetic ordering of electronic states. Especially below the "phase transition" of TOPO ligands, PL energies depend critically on the type of the surface-attached porphyrin molecule nearly independent on the type of ligand.

(vi) Based on QD PL time-resolved experiments (including those for 77 and 295 K as well as change point analysis of single QD blinking events) we find at least three basically different types of emissive states. Although we identified via spectral deconvolution only two PL energies this is easily understood since the short PL decay component will in most cases not contribute significantly to the time averaged PL spectrum.

To conclude, this chapter shows that surface-mediated processes dictate the probability of several of the most interesting and potentially useful photophysical phenomena observed for colloidal QDs. This study underscores the dramatic sensitivity of the optical properties of QDs to their surface chemistry, and the great potential that exists to tune these properties by rationally designing organic molecules. The specific selectivity of QD PL to the surface-attached dye molecule provides new and not yet reported experimental insights into QD surface properties. In this respect, a qualitative comparison with theoretical calculations (time-dependent DFT, (Kilina et al., 2009a, 2012; Fischer et al., 2012)), which takes the number, position and chemical nature of ligands and their specific removal into account, explains at least qualitatively the observed modifications of PL energies and dynamics. Scheme 1.2 shows qualitatively how the lowest band edge states respond to depletion of the ligand shell due to either geometric constraints or ligand replacement following dye attachment. The basic conclusion is that a dye molecule removes ligands from (specific) surface sites thus acting as a new "ligand" creating a modified set of new surface states. This explains specifically why non-FRET processes of QD PL quenching upon QD–dye nanoassembly formation are often not exclusively related to energy (FRET) or charge transfer. In a certain way we use dye molecules as single molecular surface probes. In addition, we have shown experimentally for the first time, that only one dye molecule is capable to change the surface-related distribution of near-band-edge or surface state energies.

Moreover, theoretical analysis of the formation and properties of QD–organic molecule nanoassemblies as well as a detailed understanding of the ligand–QD interactions and dynamics may give benefits to experimental studies as well as to exploit by controllable manner such nanostructures in a wide field of possible applications.

Finally, an additional pathway to form still more complex QD–dye nanostructures is now open in combining both self-aggregation of dye molecules to QDs and self-aggregation among functionalized dyes among themselves. We show in Scheme 1.3 the principle approach of self-aggregation of porphyrin aggregates to the surface of a QD.

We have previously shown that such a scheme can be realized: porphyrin being anchored to the QD surface via two (p-Pyr) rings, the porphyrin molecule $H_2P(p\text{-}Pyr)_4$ may be additionally attached to the dimer $(ZnOEP)_2Ph$ via twofold coordination interactions (Zenkevich and Von Borczyskowski, 2009). Titration of a solution of QDs with porphyrin triads and pentads results in QD PL quenching which, however, is reduced due to the competition of aggregation among porphyrins themselves. Nevertheless, quenching is in the same order of magnitude as for quenching with only one type of functionalized porphyrin (Zenkevich and Von Borczyskowski, 2009). Realization of such complex nanoaggregates allows for architectures combing alternative routes for energy and charge transfer and opens pathway to nanoaggregates with a manifold of functions.

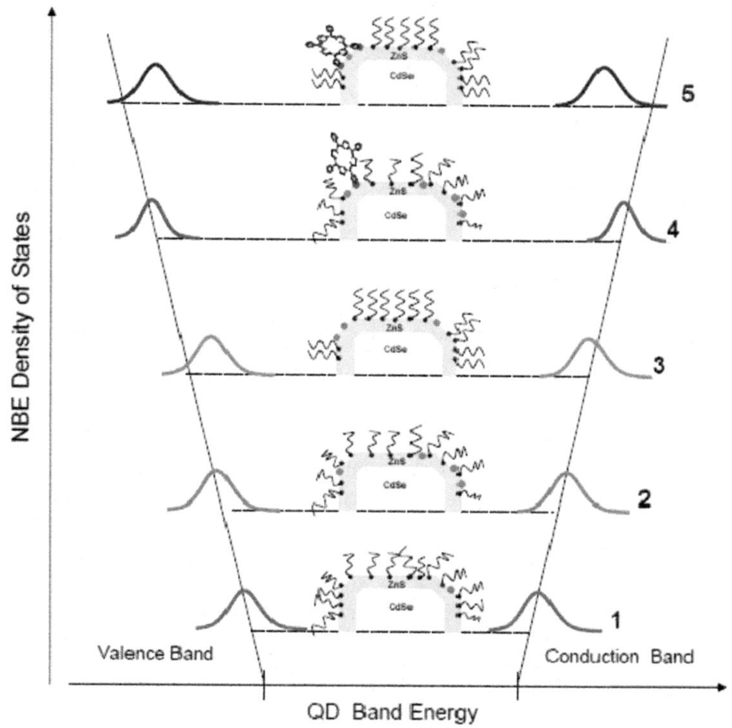

Scheme 1.2 Schematic qualitative presentation for the density of near band energy (NBE) states as a function of CdSe/ZnS QD surface organization (based on considerations described in (Kilina et al., 2009a, 2012)): (1) complete coverage of QD surface by capping ligand molecules; (2) partial coverage of QD surface by capping ligand molecules caused by steric constraints; (3) "phase transition" of capping layer (e.g., crystallization and ordering) followed by the formation of additional surface states; (4) competing attachment of a porphyrin molecule on QD surface before "phase transition" leading to the removal of few capping ligand molecules due to steric interactions; (5) competing two-point attachment of a porphyrin molecule on QD surface after "phase transition" in the vicinity of surface defects. Increase of the band gap upon ligand depletion follows recent calculations by (Kilina et al., 2009a, 2012). The phase transition is accompanied by creation of surface traps due to strain formation and increased ligand coverage. The transition from 4 to 5 is tentatively accompanied by a transition from a onefold coordination of tetra-pyridyl substituted porphyrin to a twofold one. Adapted from Zenkevich et al. (2015).

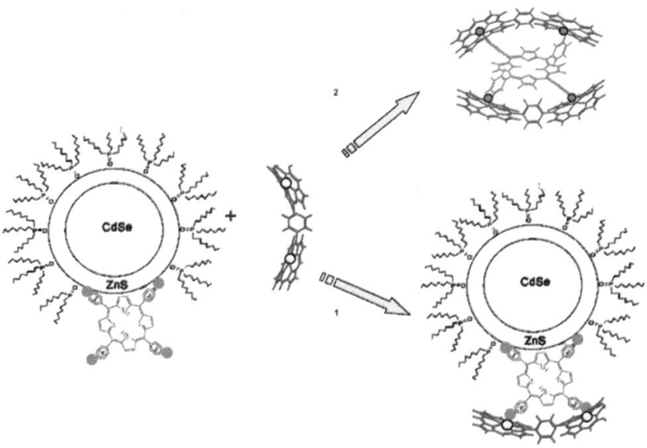

Scheme 1.3 Competitive self-aggregation of porphyrin molecules H_2P (p-Pyr)$_4$ to the QD surface followed by triad formation among porphyrins (Route 1) as well as with two $(ZnOEP)_2Ph$ dimers followed by formation of a porphyrin pentad $2(ZnOEP)_2Ph \otimes H_2P$ in solution (Route 2). Pentads (or triads) are formed via coordination interactions of central Zn ions in both porphyrin macrocycles of the dimer $(ZnOEP)_2Ph$ with N atoms of pyridyl (p-Pyr)$_4$ groups of the porphyrin monomer H_2P(p-Pyr)$_4$. Adapted from Zenkevich and Von Borczyskowski (2009).

References

Agranovich, V. M., and Galanin, M. D. (1982). *Electronic Excitation Energy Transfer in Condensed Matter* (North-Holland Pub. Co., Amsterdam, New York).

Albert, V. V., Ivanov, S. A., Tretiak, S., and Kilina, S. **(2011).** Electronic structure of ligated CdSe clusters: Dependence on DFT methodology, *J. Phys. Chem. C*, **115**, p. 15793–15800.

Anni, M., Manna, L., Cigolani, R., Valerini, D., Creti, A., and Lomascolo, M. (2004). Forster energy transfer from blue-emitting polymers to colloidal CdSe/ZnS core shell quantum dots, *Appl. Phys. Lett.*, **85**, pp. 4169–4171.

Atkins, P., and de Paula, J. (2002). *Atkin's Physical Chemistry*, 7th ed. (Oxford University Press, Oxford) p. 577.

Baker, D. R., and Kamat, P. V. (2010). Tuning the emission of CdSe quantum dots by controlled trap enhancement, *Langmuir*, **26**, pp. 11272–11276.

Baldenebro-Lopez, J., Flores-Holguin, N., Castorena-Gonzalez, J., Glossman-Mitnik, D., Adegoke, O., and Nyokong, T. (2013). Molecular design of copper complexes as sensitizers for efficient dye-sensitized solar cells,*J. Photochem. Photobiol. A: Chem.*, **267**, pp. 58–66.

Basko, D., La Rocca, J. C., Bassani, F., and Agranovich, V. M. (1999). Foerster energy transfer from a semiconductor quantum well to an organic material overlayer, *Eur. Phys. J. B*, **8**, pp. 353–362.

Beane, G., Boldt, K., Kirkwood, N., and Mulvaney, P. (2014). Energy transfer between quantum dots and conjugated dye molecules, *J. Phys. Chem. C*, **118**, pp. 18079–18086.

Bischof, T., Ivanda, M., Lermann, G., Materny, A., Kiefer, W., and Kalus, J. (1996). Linear and nonlinear Raman studies on CdS_xSe_{-x} doped glasses, *J. Raman Spectrosc.*, **27**, pp. 297–302.

Blaudeck, T. (2007). Self-assembly of functionalized porphyrin molecules on semiconductor nanocrystal surfaces, Technische Universität Chemnitz, Germany.

Blaudeck, T., Zenkevich, E., Abdel-Mottaleb, M., Szwaykowska, K., Kowerko, D., Cichos, F., and von Borczyskowski, C. (2012). Formation principles and ligand dynamics for nanoassemblies of CdSe quantum dots and functionalized dye molecules, *ChemPhysChem.*, **13**, pp. 959–972.

Blaudeck, T., Zenkevich, E. I., Cichos, F., and von Borczyskowski, C. (2008). Probing wave functions at semiconductor quantum-dot surfaces by non-FRET photoluminescence quenching. *J. Phys. Chem. C*, **112**, pp. 20251–20257.

Boldt, K., Jander, S., Hoppe, K., and Weller, H. (2011). Characterization of the organic ligand shell of semiconductor quantum dots by fluorescence quenching experiments, *ACS Nano*, **5**, pp. 8115–8123.

Brown, P. R., Kim, D., Lunt, R. R., Zhao, N., Bawendi, M. G., Grossman, J. C., and Bulovic, V. (2014). Energy level modification in lead sulfide quantum dot thin films through ligand Exchange, *ACS Nano*, **8**, pp. 5863–5872.

Bullen, C., and Mulvaney, P. (2006). The effects of chemisorption on the luminescence of CdSe quantum dots, *Langmuir*, **22**, pp. 3007–3013.

Burda, C., Green, T. C., Link, S., and El-Sayed, M. A. (1999). Electron shuttling across the interface of CdSe nanoparticles monitored by femtosecond laser spectroscopy, *J. Phys. Chem. B*, **103**, pp. 1783–1788.

Califano, M., Franceschetti, A., and Zunger, A. (2005). *Temperature dependence of excitonic radiative decay in CdSe quantum dots: The role of surface hole traps, Nano Lett.*, **5**, pp. 2360–2364.

Califano, M., and Gómez-Campos, F. M. (2013). Universal trapping mechanism in semiconductor nanocrystals, *Nano Lett.*, **13**, pp. 2047–2052.

Chernook, A. V., Rempel, U., von Borczyskowski, C., Zenkevich, E. I., and Shulga, A. M. (1996b). Formation and optical properties of self-organized pentameric porphyrin arrays, *Chem. Phys. Lett.*, **254**, pp. 229–241.

Chernook, A. V., Shulga, A. M., Zenkevich, E. I., Rempel, U., and von Borczyskowski, C. (1996a). Complexation and interchromophoric interactions in self-organized porphyrin and chlorin triads, *J. Phys. Chem.*, **100**, pp. 1918–1926.

Chung, I., and Bawendi, M. G. (2004). Relationship between single quantum-dot intermittency and fluorescence intensity decays from collections of dots, *Phys. Rev. B*, **70**, p. 165304.

Cichos, F., Martin, J., and von Borczyskowski, C. (2004). Emission intermittency in silicon nanocrystals, *Phys. Rev. B*, **70**, p. 1153141.

Cichos, F., von Borczyskowski, C., and Orrit, M. (2007). Power-law intermittency of single emitters, *Curr. Opin. Colloid Interface Sci.*, **12**, pp. 272–284

Cina, J. A., Kilin, D. S., and Humble, T. S. (2003). Wave packet interferometry for short-time electronic energy transfer: Multidimensional optical spectroscopy in the time domain, *J. Chem. Phys.*, **118**, pp. 46–61.

Clapp, A., Medintz, I., Mauro, J., Fisher, B., Bawendi, M., and Mattoussi, H. (2004). Fluorescence resonance energy transfer between quantum dot donors and dye-labeled protein acceptors, *J. Am. Chem. Soc.*, **126**, pp. 301–310.

Clapp, A. R., Medintz, I. L., and Mattoussi, H. (2006). Foerster resonance energy transfer investigations using quantum-dot fluorophores, *Chem. Phys. Chem.*, **7**, pp. 47–57.

Clapp, A. R., Medintz, I. L., Uyeda, H., Brent, T., Fisher, R., Goldman, E. R., Bawendi, M. G., and Mattoussi, H. (2005). Quantum dot-based multiplexed fluorescence resonance energy transfer, *J. Am. Chem. Soc.*, **127**, pp. 18212–18221.

Coe-Sullivan, S., Woo, W.-K., Steckel, J. S., Bawendi, M., and Bulovic, V. (2003). Tuning the performance of hybrid organic/inorganic quantum dot light-emitting devices, *Org. Electron.*, **4**, pp. 123–130.

Crooker, S. A., Barrick, T., Hollinthworth, J. A., and Klimov, V. I. (2003). Multiple temperature regimes of radiative decay in CdSe nanocrystal quantum dots: Intrinsic limits to the dark-exciton lifetime, *Appl. Phys. Lett.*, **82**, pp. 2793–2795.

Cui, J., Beyler, A. P., Bischof, T. S., Wilson, M. W. B., and Bawendi, M. G. (2014). Deconstructing the photon stream from single nanocrystals: From binning to correlation, *Chem. Soc. Rev.*, **43**, pp. 1287–1310.

Cui, S.-C., Tachikawa, T., Fujitsuka, M., and Majima, T. (2011). Photoinduced electron transfer in a quantum dot–cucurbituril supramolecular complex, *J. Phys. Chem. C*, **115**, pp. 1824–1830.

Cui, S.–C., Tachikawa, T., Fujitsuka, M., and Tetsuro, M. (2010). Solvent-polarity dependence of electron-transfer kinetics in a CdSe/ZnS quantum dot-pyromellitimide conjugate, *J. Phys. Chem. C*, **114**, pp. 1217–1225.

D'Souza, S., Antunes, E., Litwinski, Ch., and Nyokong, T. (2011). Photophysical effects of zinc monoamino phthalocyanines linked to mercaptopropionic-capped CdTe quantum dots, *J. Photochem. Photobiol. A: Chem.*, **220**, pp. 011–019.

Dabbousi, B. O., Redriguez-Vejo, J., Mikulec, F. V., Heine, J. R., Mattousi, H., Ober, R., Jensen, K. F., and Bawendi, M. G. (1997). (CdSe)ZnS core-shell quantum dots: Synthesis and optical and structural characterization of a size series of highly luminescent materials, *J. Phys. Chem. B*, **101**, pp. 9463–9475.

Dayal, S., and Burda, C. (2007). Surface effects on quantum dot-based energy transfer, *J. Am. Chem. Soc.*, **129**, pp. 7977–7981.

Dayal, S., Lou, Y., Samia, A. C. S., Berlin, J. C., Kenney, M. E., and Burda, C. (2006). Observation of non-Foerster-type energy-transfer behavior in quantum dot-phthalocyanine conjugates, *J. Am. Chem. Soc.*, **128**, pp. 13974–13975.

De Mello Donegá, C. (2014). *Nanoparticles* (Springer-Verlag, Berlin, Heidelberg).

De Mello Donega, C., Bode, M., and Meijerink, A. (2006). Size- and temperature-dependence of exciton lifetimes in CdSe quantum dots, *Phys. Rev. B*, **74**, p. 085320

Dworak, L., Matylitzky, V. V., Ren, T., Basche, T., and Wachtveitl, J. (2014). Acceptor concentration dependence of Förster resonance energy transfer dynamics in dye-quantum dot complexes, *J. Phys. Chem. C*, **118**, pp. 4396–4402.

Dzhagarov, B. M., Sagun, E. I., Ganzha, V. A., and Gurinovich, G. P. (1987). Mechanism for triplet state quenching of chlorophyll and related substances with molecular oxygen, *Khimicheskaya Fizika*, **6**, pp. 919–928.

Efros, A. L., and Rosen, M. (2000). The electronic structure of semiconductor nanocrystals, *Ann. Rev. Mater. Sci.*, **30**, pp. 475–521.

Empedocles, S. A., and Bawendi, M. G. (1999). Influence of spectral diffusion on the line shapes of single CdSe nanocrystallite quantum dots, *J. Phys. Chem. B*, **103**, pp. 1826–1830.

Evans, C. M., Evans, M. E., and Krauss, T. D. (2010). Mysteries of TOPSe revealed: Insights into quantum dot nucleation, *J. Am. Chem. Soc.*, **132**, pp. 10973–10975.

Fernee, M. J., Littleton, B. N., Cooper, S., Rubinsztein-Dunlop, H., Gomez, D. E., and Mulvaney, P. (2008). Acoustic phonon contributions to the emission spectrum of single CdSe nanocrystals, *J. Phys. Chem. C*, **112**, pp. 1878–1886.

Fernee, M. J., Plakhotnik, T., Louyer, Y., Littleton, B. N., Potzner, C., Tamarat, P., Mulvaney, P., and Lounis, B. (2012). Spontaneous spectral diffusion in CdSe quantum dots, *J. Phys. Chem. Lett.*, **3**, pp. 1716–1720.

Fischer, S. A., Crotty, A. M. M., Kilina, S. V., Ivanov, S. A., and Tretiak, S. (2012). Passivating ligand and solvent contributions to the electronic properties of semiconductor nanocrystals, *Nanoscale*, **4**, pp. 904–914.

Fisher, B. R., Eisler, H.-J., Stott, N. E., and Bawendi, M. G. (2004). Emission intensity dependence and single-exponential behavior in single colloidal quantum dot fluorescence lifetimes, *J. Phys. Chem. B*, **108**, pp. 143–148.

Foerster, T. (1965). *Modern Quantum Chemistry* (Sinanoglu, O., ed.), Academic Press, New York.

Fomenko, V., and Nesbitt, D. J. (2008). Solution control of radiative and nonradiative lifetimes: A novel contribution to quantum dot blinking suppression, *Nano Lett.*, **8**, pp. 287–293.

Frantsuzov, P. A., Volkán-Kacscó, S., and Jankó, B. (2013). Universality of the fluorescence intermittency in nanoscale systems: Experiment and theory, *Nano Lett.*, **13**, pp. 402–408.

Frasco, M. F. and Chaniotakis N. (2009). Semiconductor quantum dots in chemical sensors and biosensors, *Sensors*, **9**, pp. 7266–7286.

Frasco, M. F., Vamvakaki, V., and Chaniotakis, N. (2010). Porphyrin decorated CdSe quantum dots for direct fluorescent sensing of metal ions, *J. Nanopart. Res.*, **12**, pp. 1449–1458.

Frenzel, J., Joswig, J. O., and Seifert, G. (2007). Optical excitations in cadmium sulfide nanoparticles, *J. Phys. Chem. C*, **111**, pp. 10761–10770.

Fron, E., Pilot, R., Schweitzer, G., Qu, J., Herrmann, A., Muellen, K., Hofkens, J., der Auweraer, M. V. F., and Schryver, C. D. (2008). Photoinduced electron-transfer in perylenediimide triphenylamine-based dendrimers: Single photon timing and femtosecond transient absorption spectroscopy, *Photochem. Photobiol. Sci.*, **7**, pp. 597–604.

Gaponenko, S. V. (2010). *Introduction to Nanophotonics* (Cambridge University Press, Cambridge).

Gerlach, F., Täuber, D., and von Borczyskowski, C. (2013). Correlated blinking via time dependent energy transfer in single CdSe quantum dot-dye nanoassemblies, *Chem. Phys. Lett.*, **572**, pp. 90–99.

Ginger, D. S., Greenham, N. C. (1999). Photoinduced electron transfer from conjugated polymers to CdSe nanocrystals. *Phys. Rev. B*, **59**, pp. 10622–10629.

Gomez, D. E., van Embden, J., Jasieniak, J., Smith, T. A., and Mulvaney, P. (2006b). Blinking and surface chemistry of single CdSe nanocrystals, *Small*, **2**, pp. 204–208.

Gomez, D. E., van Embden, J., and Mulvaney, P. (2006a). Spectral diffusion of single semiconductor nanocrystals: The influence of the dielectric environment, *Appl. Phys. Lett.*, **88**, p. 154106.

Gomez-Campos, F. M., and Califano, M. (2012). Hole surface trapping in CdSe Nanocrystals: Dynamics, rate fluctuations, and implications for blinking, *Nano Lett.*, **12**, pp. 4508–4517.

Guo, W., Sun, X., Jacobson, O., Yan, X., Min, K., Srivatsan, A., Niu, G., Kiesewetter, D. O., Chang, Jin., and Chen, X. (2015). Intrinsically radioactive [64Cu]CuInS/ZnS quantum dots for PET and optical imaging: Improved radiochemical stability and controllable cerenkov luminescence, *ACS Nano*, **9**, pp. 488–495.

Gur, I., Fromer, N. A., Chen, C.-P., Kanaras, A. G., and Alivisatos, A. P. (2007). Hybrid solar cells with prescribed nanoscale morphologies based on hyperbranched semiconductor nanocrystals, *Nano Lett.*, **7**, pp. 409–414.

Guyot-Sionnest, P., Shim, M., Matranga, C., and Hines, M. (1999). Intraband relaxation in CdSe quantum dots, *Phys. Rev. B Condens. Matter Mater. Phys.*, **60**, pp. R2181(R).

Hadar, I., Halivni, S., Even-Dar, N., Faust, A., and Banin, U. (2015). Dimensionality effects on fluorescence resonance energy transfer between single semiconductor nanocrystals and multiple dye acceptors, *J. Phys. Chem. C*, **119**, pp. 3849–3856.

Haghi, A. K., Sabu, T., Pourhashemi, A., Hamrang, A., and Klodzinska, E. (2015). *Nanomaterials and Nanotechnology for Composites: Design, Simulation and Applications*, (Apple Academic Press).

Halivni, S., Sitt, A., Hadar, I., and Banin, U. (2012). Effect of nanoparticle dimensionality on fluorescence resonance energy transfer in nanoparticle-dye conjugated systems, *ACS Nano*, **6**, pp. 2758–2765.

Halpert, J. E., Tischler, J. R., Nair, G., Walker, B. J., Liu, W., Bulovic, V., and Bawendi, M. G. (2009). Electrostatic formation of quantum dot/J-aggregate FRET pairs in solution, *J. Phys. Chem. C*, **113**, pp. 9986–9992.

Hammer, N. I., Early, K. T., Sill, K., Odoi, M. Y., Emrick, T., and Barnes, M. D. (2006). Coverage-mediated suppression of blinking in solid state quantum dot conjugated organic composite nanostructures, *J. Phys. Chem. B*, **110**, pp. 14167–14171.

Haus, J. W., Zhou, H. S., Homma, I., Komiyama, H. (1993). Quantum confinement in semiconductor heterostructure nanometer-size particles, *Phys. Rev. B* 1993, **47**, pp. 1359–1365.

He, X., and Ma, N. (2014). An overview of recent advance of quantum dots for biomedical applications, *Colloids Surf. B: Biointerfaces*, **124**, pp. 118–131.

Hillhouse, H. W., and Beard, M. C. (2009). Solar cells from colloidal nanocrystals: Fundamentals, materials, devices, and economics, *Curr. Opin. Colloid Interface Sci.*, **14**, pp. 245–259.

Hohlbein, J., and Huebner, C. G. (2005). Simple scheme for rapid three-dimensional orientation determination of the emission dipole of single molecules, *Appl. Phys. Lett.*, **86**, pp. 121104–121107.

Hohng, S., and Ha, T. (2004). Near-complete suppression of quantum dot blinking in ambient conditions, *J. Am. Chem. Soc.*, **126**, pp. 1324–1325.

Holt, B., Kudera, S., Weiss, A., Schrader, T. E., Manna, L., Parak, W. J., and Braun, M. (2008). Ligand exchange of CdSe nanocrystals probed by optical spectroscopy in the visible and mid-ir, *J. Mater. Chem.*, **18**, pp. 2728–2732.

Inerbaev, T. M., Masunov, A. E., Khondaker, S. I., Dobrinescu, A., Plamadă, A.-V., and Kawazoe, Y. (2009). Quantum chemistry of quantum dots: Effects of ligands and oxidation, *J. Chem. Phys.*, **131**, p. 044106.

Issac, A., Jin, S., Lian, T. (2008). Intermittent electron transfer activity from single CdSe/ZnS quantum dots, *J. Am. Chem. Soc.*, **130**, pp. 11280–11281.

Issac, A., von Borczyskowski, C., and Cichos, F. (2005). Correlation between photoluminescence intermittency of CdSe quantum dots and self-trapped states in dielectric media, *Phys. Rev. B*, **71**, p. 161302(R).

Javier, A., Magana, D., Jennings, T., and Strouse, J. F. (2003). Nanosecond exciton recombination dynamics in colloidal CdSe quantum dots under ambient conditions, *Appl. Phys. Lett.*, **83**, pp. 1423–1425.

Jhonsi, M. A., and Rengnathan, R. (2010). Investigations on the photoinduced interaction of water soluble thioglycolic acid (TGA) capped CdTe quantum dots with certain porphyrins, *J. Colloid Interface Sci.*, **344**, pp. 596–602.

Ji, X., Copenhaver, D., Sichmeller, C., and Peng, X. (2008). Ligand bonding and dynamics on colloidal nanocrystal at room temperature: The case of alkylamines on CdSe nanocrystals, *J. Am. Chem. Soc.*, **130**, pp. 5726–5735.

Ji, X., Makarov, N. S., Wang, W., Palui, G., Robel, I., and Mattoussi, H. (2015a). Tuning the redox coupling between quantum dots and dopamine in hybrid nanoscale assemblies, *J. Phys. Chem. C*, **119**, pp. 3388–3399.

Ji, X., Wang, W., and Mattoussi, H. (2015b). Effects of separation distance on the charge transfer interactions in quantum dot-dopamine assemblies, *Phys. Chem. Chem. Phys.*, **17**, pp. 10108–10117.

Jin, Z., and Hildebrandt, N. (2012). Quantum dot nanoparticles for in vitro sensing. In *Frontiers of Nanoscience*, vol. 4: *Nanobiotechnology—Inorganic Nanoparticles vs Organic Nanoparticles*, pp. 291–306.

Kalyuzhny, G., and Murray, R. W. (2005). Ligand effects on optical properties of CdSe nanocrystals, *J. Phys. Chem., C*, **109**, pp. 7012–7021.

Kang, S., Yasuda, M., Miyasaka, H., Umeyama, T., Matano, Y., Yoshida, K., Isoda, S., and Imahori, H. (2008). Multiporphyrin-modified CdSe nanoparticles as efficient light-harvesting materials, *ChemSusChem.*, **1**, pp. 254–255.

Kapitonov, A. M., Stupak, A. P., Gaponenko, S. V., Petrov, E. P., Rogach, A. L., and Eychmueller, A. (1999). Luminescence properties of thiol-stabilized CdTe nanocrystals, *J. Phys. Chem. B*, **103**, pp. 10109–10113.

Kilin, D. S., Tsemekham, K., Zenkevich, E. I., Prezhdo, O. V., and von Borczyskowski, C. (2007). Ab initio study of transfer dynamics from a Core-Shell semiconductor quantum dot to a porphyrin sensitizer, *J. Photochem. Photobiol. A*, **190**, pp. 342–351.

Kilin, D., Zenkevich, E., and von Borczyskowski, C. (2015). Dynamics of photoinduced electron transfer in multiporphyrin nanoassemblies, *Proceeding of the International Conference "Nanomeeting-2015". Physics, Chemistry and Applications of Nanostructures. Reviews and Short Notes* (Borisenko, V. E., Gaponenko, S. V., Gurin, V. S., and Kam, C. H., eds.), World Scientific Publishing Co., New Jersey, London, Singapore, Bejing, Shanghai, Hong-Kong, Taipei, Chenna, pp. 14–17.

Kilina, S. V., Ivanov, S., and Tretiak, S. (2009a). Effect of surface ligands on optical and electronic spectra of semiconductor nanoclusters, *J. Am. Chem. Soc.*, **131**, pp. 7717–7726.

Kilina, S. V., Kilin, D. S., and Prezhdo, O. V. (2009b). Breaking the phonon bottleneck in PbSe and CdSe quantum dots: Time-domain density functional theory of charge carrier relaxation, *ACS Nano*, **3**, pp. 93–99.

Kilina, S., Velizhanin, K. A., Ivanov, S., Prezhdo, O. V., and Tretiak, S. (2012). Surface ligands increase photoexcitation rates in CdSe quantum dots, *ACS Nano*, **6**, pp. 6515–6524.

Kim, D., Holten, D., and Gouterman, M. (1984). *Evidence for picosecond transient absorption studies of charge transfer states in Cu^{II} Porphyrins*, *J. Am. Chem. Soc.*, **106**, pp. 2793–2798.

Kimura, J., Uematsu, T., Maenosono, S., Yamaguchi, Y. (2004). Photoinduced fluorescence enhancement in CdSe/ZnS quantum dot submonolayers sandwiched between insulating layers: influence of dot proximity, *J. Phys. Chem. B*, **108**, pp. 13258–13264.

Klimov, V. (2000). In *Handbook of Nanostructured Materials and Nanotechnology* (Nalwa, H. S., ed.), Acad. Press, USA, vol. **4**, pp. 451–527.

Klimov, V. (2010). *Nanocrystal Quantum Dots.* (CRC Press LLC, Washington).

Klimov, V. I., Mikhailovsky, A. A., McBranch, D. W., Leatherdale, C. A., and Bawendi, M. G. (2000). Mechanisms for intraband energy relaxation in semiconductor quantum dots: The role of electron-hole interactions, *Phys. Rev. B*, **61**, p. 13349(R).

Knowles, K. E., Frederick, M. T., Tice, D. B., Morris-Cohen, A. J., and Weiss, E. A. (2012). Colloidal quantum dots: Think outside the (particle-in-a-)box, *J. Phys. Chem. Lett.*, **3**, pp. 18–26.

Knowles, K. E., Tice, D. B., McArthur, E. A., Solomon, G. C., and Weiss, E. A. (2010). Chemical control of the photoluminescence of CdSe quantum dot-organic complexes with a series of para-substituted aniline ligands, *J. Am. Chem. Soc.*, **132**, pp. 1041–1050.

Ko, H. C., Yuan, C. T., Lin, S. H., and Tang, J. (2011). Observation of inverted regime electron transfer in CdSe/ZnS QDs from pH-sensitive single-particle and ensemble fluorescence measurements, *J. Phys. Chem. C*, **115**, pp. 13977–13984.

Koole, R., Schapotschnikow, P., de Mello Donega, C., Vlugt, T. J. H., and Meijerink, A. (2008). Time-dependent photoluminescence spectroscopy as a tool to measure the ligand exchange kinetics on a quantum dot surface, *ACS Nano*, **2**, pp. 1703–1714.

Koposov, A. Y., Szymanski, P., Cardolaccia, T., Meyer, T. J., Klimov, V. I., and Sykora, M. (2011). Electronic properties and structure of assemblies of CdSe nanocrystal quantum dots and Ru-polypyridine complexes probed by steady state and time-resolved photoluminescence, *Adv. Funct. Mater.*, **21**, pp. 3159–3168.

Kovalenko, M. V., Manna, L., Cabot, A., Hens, Z., Talapin, D. V., Kagan, C. R., Klimov, V. I., Rogach, A. L., Reiss, P., Milliron, D. J., Guyot-Sionnnest, P., Konstantatos, G., Parak, W. J., Hyeon, T., Korgel, B. A., Murray, C. B., and Heiss, W. (2015). Prospects of nanoscience with nanocrystals, *ACS Nano*, **9**, pp. 1012–1057.

Kowerko, D. (2010). Dynamic processes in functionalised perylene bisimide molecules, semiconductor nanocrystals and assemblies. PhD Thesis. Technische Universität Chemnitz, Germany.

Kowerko, D., Krause, S., Amecke, N., Abdel-Mottaleb, M., Schuster, J., and von Borczyskowski, C. (2009a). Identification of different donor-acceptor structures via Förster resonance energy transfer (FRET) in quantum-dot-perylene bisimide assemblies, *Int. J. Mol. Sci.*, **10**, pp. 5239–5256.

Kowerko, D., Schuster, J., Amecke, N., Abdel-Mottaleb, M., Dobrawa, R., Wuerthner, F., and von Borczyskowski, C. (2010). FRET and ligand related non-FRET processes in single quantum dot-perylene bisimide assemblies, *Phys. Chem. Chem. Phys.*, **12**, pp. 4112–4123.

Kowerko, D., Schuster, J., and von Borczyskowski, C. (2009b). Restricted conformation dynamics of single functionalized perylene bisimide molecules on SiO_2 surfaces and in thin polymer films, *Mol. Phys.*, **107**, pp. 1911–1921.

Krause, S., Aramendía, P. F., Täuber, D., and von Borczyskowski, C. (2011a). Freezing single dye dynamics on interfaces and in polymers, *Phys. Chem. Chem. Phys.*, **13**, pp. 1754–1761.

Krause, S., Kowerko, D., Börner, R., Hübner, C. G., and von Borczyskowski, C. (2011b). Spectral diffusion of single molecules in a hierarchical energy landscape, *Chem. Phys. Chem.*, **12**, pp. 303–312.

Labeau, O., Tamarat, P., and Lounis, B. (2003). Temperature dependence of the luminescence lifetime of single CdSe/ZnS quantum dots, *Phys. Rev. Lett.*, **90**, p. 257404.

Lakowicz, J. (2006). *Principles of Fluorescence Spectroscopy* (Springer, New York).

Lang, E., Würthner, F., and Köhler, J. (2005). Photophysical properties of a tetraphenoxy-substituted perylene bisimide derivative characterized by single-molecule spectroscopy, *ChemPhysChem.*, **6**, pp. 935–941.

Leatherdale C. A., Bawendi M. G. (2001). Observation of solvatochromism in CdSe colloidal quantum dots, *Phys. Rev. B.*, **63**, pp. 165315. http://dx.doi.org/10.1103/PhysRevB.63.165315.

Lee, J. Kim, H.-J., Cheng, T., Lee, K., Kim, K.-S., Glotzer, S. C., Kim, J., and Kotov, N. A. (2009). Control of energy transfer to CdTe nanowires via conjugated polymer orientation, *J. Phys. Chem. C*, **113**, pp. 109–116.

Lee, W. Z., Shu, G. W., Wang, J. S., Shen, J. L., Lin, C. A., Chang, W. H., Ruaan, R. C., Chou, W. C., Lu, C. H., and Lee, Y. C. (2005). Recombination dynamics of luminescence in colloidal CdSe/ZrS quantum dots, *Nanotechnology*, **16**, pp. 1517–1521.

Lee, J. R. I., Whitley, H. D., Meulenberg, R. W., Wolcott, A., Zhang, J. Z., Prendergast, D., Lovingood, D. D., Strouse, G. F., Ogitsu, T., Schwegler, E., Terminello, L. J., and van Buuren, T. (2012). Ligand-mediated modification of the electronic structure of CdSe quantum dots, *Nano Lett.*, **12**, pp. 2763–2767.

Lehn, J.-M. (1990). Perspectives in supramolecular chemistry—from molecular recognition towards molecular information processing and self-organization, *Angew. Chem., Int. Ed. Engl.*, **29**, pp. 1304–1319.

Lemon, C. M., Karnas, E., Bawendi, M. G., and Nocera, D. G. (2013). Oxygen sensing with quantum dot-porphyrin conjugates, *Inorg. Chem.*, **52**, pp. 10394–10406.

Liang, G.-X., Li, L.-L., Liu, H.-Y., Zhang, J.-R., Burda, C., and Zhu, J.-J. (2010). Fabrication of near-infrared-emitting CdSeTe/ZnS core/shell quantum dots and their electrogenerated chemiluminescence, *Chem. Commun.*, **46**, pp. 2974–2976.

Lim, T. C., Bailey, V. J., Ho, Y.-P., and Wang, T.-H. (2008). Intercalating dye as an acceptor in quantum-dot-mediated FRET, *Nanotechnology*, **19**, p. 075701.

Liptay, T. J., Marshall, L. F., Rao, P. S., Ram, R. J., and Bawendi, M. G. (2007). Anomalous Stokes shift in CdSe nanocrystals, *Phys. Rev. B*, **76**, p. 155314.

Liptay, T. J., and Ram, R. J. (2006). Temperature dependence of the exciton transition in semiconductor quantum dots, *Appl. Phys. Lett.*, **89**, p. 223132.

Liu, L., Hensel, J., Fitzmorris, R. C., Li, Y., and Zhang, J. Z. (2010). Preparation and photoelectrochemical properties of CdSe/TiO_2 hybrid mesoporous structures, *J. Phys. Chem. Lett.*, **1**, pp. 155–160.

Lu, L., Xuhong, G., Yan, L., and Xinhua, Z. (2010). Bifunctional multidentate ligand modified highly stable water-soluble quantum dots, *Inorg. Chem.*, **49**, pp. 3768–3775.

Lüa, W., Tokuhiro, Y., Umezu, I., and Sugimura, A. (2006). Resonance energy transfer based on shallow and deep energy levels of biotin-polyethylene glycol/polyamine stabilized CdS quantum dots, *Appl. Phys. Lett.*, **89**, p. 143901.

Mansur, H. S., and Mansur, A. A. P. (2011). CdSe quantum dots stabilized by carboxylic-functionalized PVA: Synthesis and UV–vis spectroscopy characterization, *Mater. Chem. Phys.*, **125**, pp. 709–717.

Marcus, R. A., (1993). Electron transfer reactions in chemistry. Theory and experiment, *Rev. Modern Phys.*, **65**, pp. 599–610.

Marcus, Y. (2002). *Solvent Mixtures* (Marcel Dekker, New York).

Mattoussi, H., Palui, G., and Na, H. B. (2012). Luminescent quantum dots as platforms for probing in vitro and in vivo biological processes, *Adv. Drug Deliv. Rev.*, Biological Interactions of Nanoparticles, **64**, pp. 138–166.

McArthur, E. A., Godbe, J. M., Tice, D. B., and Weiss, E. A. (2012). A study of the binding of cyanine dyes to colloidal quantum dots using spectral signatures of dye aggregation, *J. Phys. Chem. C*, **116**, pp. 6136–6142.

McBride, J. R., Pennycook, T. J., Pennycook, S. J., and Rosenthal, S. J. (2013). The possibility and implications of dynamic nanoparticle surfaces, *ACS Nano*, **7**, pp. 8358–8365.

McNaught, A. D., and Wilkinson, A. (1997). *IUPAC Compendium of Chemical Terminology*, 2nd ed. (the "Gold Book"). (Blackwell Scientific Publications, Oxford).

Michler, P., ed. (2009) *Single Semiconductor Quantum Dots* (Springer-Verlag, Berlin).

Mocatta, D., Cohen, G., Schattner, J., Millo, O., Rabani, E., Banin, U. (2011). Heavily doped semiconductor nanocrystal quantum dots, *Science*, **332**, pp. 77–81.

Moreels, I., Martins, J. C., and Hens, Z. (2007). Solution NMR techniques for investigating colloidal nanocrystal ligands: A case study on trioctylphosphine oxide at InP quantum dots, *Sens. Actuators B*, **126**, pp. 283–288.

Morello, G., De Giorgi, M., Kudera, S., Manna, L., Cingolani, R., and Anni, M. (2007). Temperature and size dependence of nonradiative relaxation and exciton-phonon coupling in colloidal CdTe quantum dots, *J. Phys. Chem. C*, **111**, pp. 5846–5849.

Morris-Cohen, A. J., Donakowski, M. D., Knowles, K. E., and Weiss, E. A. (2010). The effect of a common purification procedure on the chemical composition of surfaces of CdSe quantum dots synthesized with trioctylphosphine oxide (TOPO), *J. Phys. Chem. C*, **114**, pp. 897–906.

Morris-Cohen, A. J., Vasilenko, V., Amin, V. A., Reuter, M. G., and Weiss, E. A. (2012). Model for adsorption of ligands to colloidal quantum dots with concentration-dependent surface structure, *ACS Nano*, **6**, pp. 557–565.

Moyano, D. F., Goldsmith, M., Solfiell, D. J., Landesman-Milo, D., Miranda, O. R., Peer, D., and Rotello, V. M. (2012). Nanoparticle hydrophobicity dictates immune response, *J. Am. Chem. Soc.*, **134**, pp. 3965–3967.

Müller, J., Lupton, J. M., Rogach, A. L., Feldmann, J., Talapin, D. V., and Weller, H. (2005). Monitoring surface charge migration in the spectral dynamics of single CdSe/CdS nanodot/nanorod heterostructures, *Phys. Rev. B*, **72**, p. 205339.

Munro, A. M., and Ginger, D. S. (2008). Photoluminescence quenching of single CdSe nanocrystals by ligand adsorption, *Nano Lett.*, **8**, pp. 2585–2590.

Munro, A. M., Plante, I. J.-L., Ng, M. S., and Ginger, D. S. (2007). Quantitative study of the effect of surface ligand concentration on CdS nanocrystal photoluminescence, *J. Phys. Chem. C*, **111**, pp. 6220–6227.

Murray, C. B., Kagan, C. R., and Bawendi, M. G. (2000). Synthesis and characterization of monodisperse nanocrystals and close-packed nanocrystal essemblies, *Ann. Rev. Mater. Sci.*, **30**, pp. 545–610.

Nicolini, C. (2010). *Nanotechnology and Nanobiosciences*. Chapter 1, Nanoscale Materials, vol. 1 (Pan Stanford Series on Nanobiotechnology).

Nirmal M., Dabbousi B. O., Bawendi M. G., Macklin J. J., Trautman J. K., Harris T. D., Brus L. E. (1996). Fluorescence intermittency in single cadmium selenide nanocrystals, *Nature*, **383**, pp. 802–804.

Odoi, M. Y., Hammer, N. I., Early, K. T., McCarthy, K .D., Tangirala, R., Emrick, T., and Barnes, M. D. (2007). Fluorescence lifetimes and correlated photon statistics from single CdSe/oligo (phenylene vinylene) composite nanostructures, *Nano Lett.*, **7**, pp. 2769–2773.

Orlova, A., Gubanova, M., Maslov, V., Vinogradova, G., Baranov, A., Fedorov, A., and Gounko, L. (2010). Spectro-luminescent properties of the complexes formed by charged CdTe quantum dots and molecules of the tetrasulfophthalocyanin, *Optika i Spectroscopiya (in Russian)*, **108**, pp. 975–982.

Peng, X., Schlamp, M. C., Kadavanich, A. V., Alivisatos, A. P. (1997). Epitaxial growth of highly luminescent CdSe/CdS core/shell nanocrystals with photostability and electronic accessibility, *J. Am. Chem. Soc.*, **19**, pp. 7019–7029.

Petrov, E. P., Cichos, F., Zenkevich, E., Starukhin, D., and von Borczyskowski, C. (2005). Time resolved photoluminescence anisotropy of CdSe/ZnS nanoparticles in toluene at 300 K, *Chem. Phys. Lett.*, **402**, pp. 233–238.

Petrov, E. P., Cichos, F., von Borczyskowski, C. (2006). Intrinsic photophysics of semiconductor nanocrystals in dielectric media: formation of surface states. *J. Luminesc.*, **119–120**, pp. 412–417.

Pons, T., Medintz, I. L., Wang, X., English, D. S., and Mattoussi, H. (2006). Solution-phase single quantum dot fluorescence resonance energy transfer, *J. Am. Chem. Soc.*, **128**, pp. 15324–15331.

Potapova, I., Mruk, R., Hübner, C., Zentel, R., Basché, T., and Mews, A. (2005). CdSe/ZnS nanocrystals with dye-functionalized polymer ligands containing many anchor goups. *Angew. Chem.*, **117**, pp. 2490–2493.

Potapova, I., Mruk, F., Prehl, S., Zentel, R., Basché, T., and Mews, A. (2003). Semiconductor Nanocrystals with Multifunctional Polymer Ligands, *J. Am. Chem. Soc.*, **125**, p. 320.

Prezhdo, O. V. (1999). Mean field approximation for the stochastic Schrodinger equation, *J. Chem. Phys.*, **111**, pp. 8366–8377.

Puzder, A., Williamson, A. J., Zaitseva, N., Galli, G., Manna, L., and Alivisatos, A. P. (2004). The effect of organic ligand binding on the growth of CdSe nanoparticles probed by ab initio calculations, *Nano Lett.*, **4**, pp. 2361–2365.

Qu, L. H., and Peng, X. G. (2002). Control of photoluminescence properties of CdSe nanocrystals in grouth, *J. Am. Chem. Soc.*, **124**, pp. 2049–2055.

Querner, C., Reiss, P., Bleuse, J., and Pron, A. (2004). Chelating ligands for nanocrystals' surface functionalization, *J. Am. Chem. Soc.*, **126**, pp. 11574–11582.

Rakovich, A., Savateeva, D., Rakovich, T., Donegan, J. F., Rakovich, Y. P., Kelly, V., Lesnyak, V., and Eychmueller, A. (2010). CdTe quantum dot/dye hybrid system as photosensitizer for photodynamic therapy, *Nanoscale Res. Lett.*, **5**, pp. 753–760.

Reiss, P., Protiere, M., and Li, L. (2009). Core/Shell semiconductor nanocrystals, *Small*, **5**, pp. 154–168.

Rempel, U., Meyer, S., von Maltzan, B., and von Borczyskowski, C. (1998). Energy transfer and distance independent charge separation inself-organised porphyrin-quinone aggregates, *J. Luminesc.*, **78**, pp. 97–110.

Rempel, J. Y., Trout, B. L., Bawendi, M. G., and Jensen, K. F. (2006). Density functional theory of ligand binding on CdSe (0001), (000), and (110) single crystal relaxed and reconstructed surfaces: Implications for nanocrystalline growth, *J. Phys. Chem. B*, **110**, pp. 18007–18016.

Ren, T., Mandal, P. K., Erker, W., Liu, Z., Avlasevich, Y., Puhl, L., Muellen, K., and Basche, T. (2008). A simple and versatile route to stable quantum dot-dye hybrids in nonaqueous and aqueous solutions, *J. Am. Chem. Soc.*, **130**, pp. 17242–17243.

Rogach, A. L. (2008). Ed. *Semiconductor Nanocrystal Quantum Dots: Synthesis, Assembly, Spectroscopy and Applications* (Springer-Verlag, Wien).

Rogers, B., Adams, J., and Pennathur, S. (2014). *Nanotechnology: Understanding Small Systems*. Mechanical and Aerospace Engineering Series, CRC Press Textbook.

Roy, R., Hohng, S., and Ha, T. (2008). A practical guide to single-molecule FRET, *Nat. Methods*, **5**, pp. 507–516.

Sadhu, S., and Patra, A. (2013). A brief overview of some physical studies on the relaxation dynamics and Förster resonance energy transfer of semiconductor quantum dots, *ChemPhysChem*, **14**, pp. 2641–2653.

Sagun, E. I., Zenkevich, E. I., Knyukshto, V. N., Shulga, A. M., Starukhin, D. A., and von Borczyskowski, C. (2002). *Interaction of multiporphyrin systems with molecular oxygen in liquid solutions: Extra ligation and screening effects, Chem. Phys.*, **275**, pp. 211–237.

Samia, A. C. S., Chen, X., and Burda, C. (2003). Semiconductor quantum dots for photodynamic therapy, *J. Am. Chem. Soc.*, **123**, pp. 15736–15737.

Schapotschnikow, P., Hommersom, B., and Vlugt, T. J. H. (2009). Adsorption and binding of ligands to CdSe nanocrystals, *J. Phys. Chem. C*, **113**, pp. 12690–12698.

Schlegel, G., Bohnenberger, J., Potapova, I., and Mews, A. (2002). Fluorescence decay time of single semiconductor nanocrystals, *Phys. Rev. Lett.*, **88**, p. 137401.

Schmelz, O., Mews, A., Basché, T., Herrmann, A., and Müllen, K. (2001). Supramolecular complexes from CdSe nanocrystals and organic fluorophors, *Langmuir*, **17**, p. 2861.

Schmidt, R., Krasselt, C., Göhler, C., and von Borczyskowski, C. (2014). The fluorescence intermittency for quantum dot is not power-law distributed: A luminescence intensity resolved approach, *ACS Nano*, **8**, pp. 3506–3521.

Schmitt, F. J. (2010). Temperature induced conformational changes in hybrid complexes formed from CdSe/ZnS nanocrystals and the phycobiliprotein antenna of Acaryochloris marina, *J. Opt.*, **12**, p. 084008.

Schmitt, F.-J., Maksimov, E. G., Suedmeyer, H., Jeyasangar, V., Theiss, C., Paschenko, V. Z., Eichler, H. J., and Renger, G. (2011). Time resolved temperature switchable excitation energy transfer processes between CdSe/ZnS nanocrystals and phycobiliprotein antenna from Acaryochloris marina, *Photon. Nanostruct. Fundamentals Appl.*, **9**, pp. 190–195.

Schreiber, M., Kilin, D., and Kleinekathöfer, U. (1999). Comparison of two models for bridge-assisted charge_transfer, *J. Lumin.*, **83 & 84**, pp. 235–240.

Shatkin, J. A. (2012). *Nanotechnology: Health and Environmental Risks*, 2nd ed., Series: Perspectives in Nanotechnology (CRC Press).

Sitt, A., Even-Dar, N., Halivni, S., Faust, A., Yedidya, L., and Banin, U. (2013). Analysis of shape and dimensionality effects on fluorescence resonance energy transfer from nanocrystals to multiple acceptors, *J. Phys. Chem. C*, **117**, pp. 22186–22197.

Smith, A. M., Duan, H. W., Rhyner, M. N., Ruan, G., and Nie, S. M. (2006). A systematic examination of surface coatings on the optical and chemical properties of semiconductor quantum dots, *Phys. Chem. Chem. Phys.*, **8**, pp. 3895–3903.

Smith, A. M., and Nie, S. (2010). Semiconductor nanocrystals: Structure, properties, and band gap engineering, *Acc. Chem. Res.*, **43**, pp. 190–200.

Song, N., Zhu, H., Jin, S., Zhan, W., and Lian, T. (2011). Poisson-distributed electron-transfer dynamics from single quantum dots to C_{60} molecules, *ACS Nano*, **5**, pp. 613–621.

Stewart, M. H., Huston, A. L., Scott, A. M., Efros, A. L., Melinger, J. S., Gemmill, K. B., Trammell, S. A., Blanco-Canosa, J. B., Dawson, P. E., and Medintz, I. L. (2012). Complex Förster energy transfer interactions between semiconductor quantum dots and a redox-active osmium assembly, *ACS Nano*, **6**, pp. 5330–5347.

Tachiya, M. (1975). Application of a generating function to reaction-kinetics in micelles: Kinetics of quenching of luminescent probes in micelles, *Chem. Phys. Lett.*, **33**, pp. 289–292.

Tagliazucchi, M., Tice, D. B., Sweeney, C. M., Morris-Cohen, A. J., and Weiss, E. A. (2014). Ligand-controlled rates of photoinduced electron transfer in hybrid CdSe nanocrystal/poly(viologen) films, *Inorg. Chem.*, **53**, pp. 1900–1915.

Talapin, D. V., Lee, J.-S., Kovalenko, M. V., and Shevchenko, E. V. (2010). Prospects of nanocrystal solids as electronic and optoelectronic materials, *Chem. Rev.*, **110**, pp. 389–458.

Talapin, D. V., Rogach, A. L., Kornowski, A., Haase, M., and Weller, H. (2001). *Highly luminescent monodisperse CdSe and CdSe/ZnS nanocrystals synthesized in a hexadecylamine—trioctylphosphine oxide—trioctylphospine mixture, Nano Lett.*, **1**, pp. 207–211.

Tripathi, S. K., Ramneek, K., and Sharma, M. (2015). Synthesis and characterization of TGA-capped CdTe nanoparticles embedded in PVA matrix, *Appl. Phys. A.*, **118**, pp. 1287–1295.

Tsay, J. M., Trzoss, M., Shi, L., Kong, X., Selke, M., Jung, M. E., and Weiss, S. (2007). Singlet oxygen production by peptide-coated quantum dot-photosensitizer conjugates, *J. Am. Chem. Soc.*, **129**, pp. 6865–6871.

Underwood, D. F., Kippeny, T., and Rosenthal, S. J. (2001). Ultrafast carrier dynamics in CdSe nanocrystals determined by femtosecond fluorescence upconversion spectroscopy, *J. Phys. Chem. B*, **105**, pp. 436–443.

Vajtai, R. (2013). *Springer Handbook of Nanomaterials* (Springer-Verlag, Berlin, Heidelberg).

Van Driel, A. F, Allan, G., Delerue, C., Lodahl, P., Vos, W. L., and Vanmaekelbergh, D. (2005). Frequency-dependent spontaneous emission rate from CdSe and CdTe nanocrystals: Influence of dark states, *Phys. Rev. Lett.*, **95**, p. 236804.

Van Sark, W. G. J. H. M., Frederix, P. L. T. M., Bol, A. A., Gerritsen, H. C., and Meijerink, A. (2002). Blueing, bleaching, and blinking of single CdSe/ZnS quantum dots, *ChemPhysChem.*, **3**, pp. 871–879.

Vo-Dinh, T. (2003). ed. *Biomedical Photonics Handbook.* (CRC Press LLC: New York, Washington).

Von Borczyskowski, C., and Zenkevich, E. (2014). Formation principles and exciton relaxation in semiconductor quantum dot—dye nanoassemblies. In: *Quantum Dot Molecules, Lecture Notes in Nanoscale Science and Technology 14*. Springer Series in Materials Science, eds. Wu, J. and Wang, Z. M. (Springer Science + Business Media, New York). vol. 14, Quantum Dot Molecules, pp. 77–148.

Von Borczyskowski, C., and Zenkevich, E. (2016). Eds. *Tuning Semiconducting and Metallic Nanostructures: Spectroscopy, Dynamics, and Self-Assembly.* Copyright © 2016 Pan Stanford Publishing Pte. Ltd. ISBN 978-981-4745-24-6 (Hardcover), 978-981-4745-25-3 (eBook).

Von Holt, B., Kudera, S., Weiss, A., Schrader, T. E., Manna, L., Parak, W. J., and Braun, M. (2008). Ligand exchange of CdSe nanocrystals probed by optical spectroscopy in the visible and mid-IR, *J. Mater. Chem.*, **18**, pp. 2728–2732.

Voznyy, O. (2011). Mobile surface traps in CdSe nanocrystals with carboxylic acid ligands, *J. Phys. Chem. C*, **115**, pp. 15927–15932.

Voznyy, O., and Sargent, E. H. (2014). Atomistic model of fluorescence intermittency of colloidal quantum dots, *Phys. Rev. Lett.*, **112**, p. 157401.

Wang, M., Moon, S.-J., Xu, M., Chittibabu, K., Wang, P., Cevey-Ha, N.-L., Humphry-Baker, R., Zakeeruddin, S. M., and Graetzel, M. (2010). Efficient and stable solid-state dye-sensitized solar cells based on a high-molar-extinction-coefficient sensitizer, *Small*, **6**, pp. 319–324.

Wang, L., Long, R., and Prezhdo, O. V. (2015). Time-domain ab initio modeling of photoinduced dynamics at nanoscale interfaces, *Ann. Rev. Phys. Chem.*, **66**, pp. 549–579.

Watkins, L. P., Yang, H. (2005). Detection of intensity change points in time-resolved single molecule measurements, *J. Phys. Chem. B.*, **109**, pp. 617–628.

Whitesides, G. M., and Grzybowski, B. (2002). Self-assembly at all scales, *Science*, **295**, pp. 2418–2421.

Willard, D. M., Mutschler, T., Yu, M., Jung, J., van Orden, A. (2006). Directing energy flow through quantum dots towards nanoscale sensing, *Anal. and Bioanalyt. Chem.*, **384**, pp. 564–571.

Williard, D. M., Carillo, L. L., Jung, J., and Van Orden, A. (2001). CdSe-ZnS quantum dots as resonane energy transfer donors in a model protein-protein binding assay, *Nano Lett.*, **1**, pp. 469–474.

Wuerthner, F. (2004). Perylene bisimide dyes as versatile building blocks for functional supramolecular architectures, *Chem. Commun.*, **14**, pp. 1564–1579.

Wuister, S. F., De Mello Donega, C., Bode, M., and Meijerink, A. (2004). Luminescence temperature anti-quenching of water-soluble CdTe quantum dots: Role of the solvent, *J. Am. Chem. Soc.*, **126**, pp. 10397–10402.

Xie, R. G., Kolb, U., Li, J. B., Basche, T., and Mews, A. (2005). Synthesis and characterization of highly luminescent CdSe-core CdS/$Zn_{0.5}Cd_{0.5}S$/ZnS multishell nanocrystals, *J. Am. Chem. Soc.*, **127**, pp. 7480–7488.

Yu, W. W., Wang, Y. A., and Peng, X. (2003). Formation and stability of size-, shape-, and structure-controlled CdTe nanocrystals: Ligand effects on monomers and nanocrystals, *Chem. Mater.*, **15**, pp. 4300–4308.

Zen kevich, E. I., Blaudeck, T., Shulga, A. M., Cichos, F., and von Borczyskowski, C. (2007b). Identification and assignment of porphyrin-CdSe heteronanoassemblies, *J. Luminesc.*, **122–123**, pp. 784–788.

Zenkevich, E., Cichos, F., Shulga, A., Petrov, E., Blaudeck, T., and von Borczyskowski, C. von (2005). Nanoassemblies designed from semiconductor quantum dots and molecular arrays, *J. Phys. Chem. B*, **109**, pp. 8679–8692.

Zenkevich, E. I., Sagun, E. I., Knyukshto, V. N., Stasheuski, A. S., Galievsky, V. A., Stupak, A. P., Blaudeck, T., and von Borczyskowski, C. von (2011). Quantitative analysis of singlet oxygen ($^1O_2$) generation via energy transfer in bioconjugates based on semiconductor quantum dots and porphyrin ligands, *J. Phys. Chem. C*, **115**, pp. 21535–21545.

Zenkevich, E. I., Sagun, E. I., Knyukshto, V. N., Shulga, A. M., Mironov, A. F., Efremova, O. A., Bonnett, R., Songca, S. P., and Kassem, M. M. (1996). *J. Photochem. Photobiol. B: Biol.*, **33**, pp. 171–180.

Zenkevich, E. I., Sagun, E. I., Yarovoi, A. A., Shulga, A. M., Knyukshto, V. N., Stupak, A. P., and von Borczyskowski, C. von (2007a). Photoinduced relaxation processes in complexes based on semiconductor nanocrystals CdSe and organic molecules, *Optika i Spectroscopiya (in Russian)*, **103**, pp. 998–1009.

Zenkevich, E., Shulga, A., Chernook, A., and Gurinovich, G. (1986). Migration of energy of electronic excitation in the chemical dimers of the cyclopentaneporphyrins. *J. Appl. Spectr.*, **45**, pp. 984–991.

Zenkevich, E., Stupak, A., Göhler, C., Krasselt, C., and von Borczyskowski, C. (2015). Tuning electronic states of a CdSe/ZnS quantum dot by only one functional dye molecule, *ACS Nano*, **9**, pp. 2886–2903.

Zenkevich, E. I., Stupak, A. P., Kowerko, D., and von Borczyskowski, C. (2012). Influence of single dye molecules on temperature and time dependent optical properties of CdSe/ZnS quantum dots: Ensemble and single nanoassembly detection, *Chem. Phys.*, **406**, pp. 21–29.

Zenkevich, E. I., and von Borczyskowski, C. (2002). Multiporphyrin self-assembled arrays in solutions and films: Thermodynamics, spectroscopy and photochemistry. In: *Handbook of Polyelectrolytes and Their Applications* (Tripathy, S. K., Kumar, J., and Nalwa, H. S., eds.), American Scientific Publishers, USA, **2**, pp. 301–348.

Zenkevich, E. I., von Borczyskowski, C., Shulga, A. M. (2003). Structure and excited states properties of mutiporphyrin arrays formed by supramolecular design, *J. Porph. Phthalocyan*, **7**, pp. 731–754.

Zenkevich, E. I., and von Borczyskowski, C. von (2009). Photoinduced relaxation processes in self-assembled nanostructures: Multiporphyrin complexes and composites "CdSe/ZnS quantum dot-porphyrin," *Macroheterocycles*, **2**, pp. 206–220.

Zenkevich, E., and von Borczyskowski, C. (2012). Formation principles and excited states relaxation in self-assembled complexes: Multiporphyrin arrays and "semiconductor CdSe/ZnS quantum dot-porphyrin" nanocomposites. In: *Handbook of Porphyrin Science with Application to Chemistry, Physics, Materials Science, Engineering, Biology and Medicine. Volume 22 - Biophysical and Physicochemical Studies of Tetrapyrroles* (Kadish, K., Smith, K. M., and Guilard, R., eds.), World Scientific Publishing Co. Pte. Ltd., Singapore, Chapter 104, pp. 68–159.

Zenkevich, E. I., and von Borczyskowski, C. (2014). Self-organization principles in the formation of multiporphyrin complexes and "semiconductor quantum dot-porphyrin" nanoassemblies, *J. Porphyrins Phthalocyanines*, **18**, pp. 1–19.

Zenkevich, E., and von Borczyskowski, C. (2015). Surface photochemistry of quantum dot-porphyrin nanoassemblies for singlet oxygen generation. In: *Photoinduced Processes at Surfaces and in Nanomaterials* (Kilin, D. ed.), ACS Symposium Series, USA, v. **1196**, Chapter 12, pp. 235–272.

Zhang, K., Chang, H., Fu, A., Alivisatos, A. P., and Yang, H. (2006). Continuous distribution of emission states from single CdSe/ZnS quantum dots, *Nano Lett.*, **6**, pp. 843–847.

Zhu, H., Song, N., and Lian, T. (2010). Controlling charge separation and recombination rates in CdSe/ZnS type I core-shell quantum dots by shell thicknesses, *J. Am. Chem. Soc.*, **132**, pp. 15038–15045.

Chapter 2

Interrelation of Assembly Formation and Ligand Depletion in Colloidal Quantum Dots

Danny Kowerko

Faculty of Computer Science, Junior Professorship Media Computing, Technische Universität Chemnitz, Straße der Nationen 62, D-09107 Chemnitz, Germany

danny.kowerko@informatik.tu-chemnitz.de

Spectral and time-resolved ensemble and single-particle experiments on colloidal CdSe/ZnS semiconductor nanocrystals (quantum dots, QDs) are presented. Processes decreasing the photoluminescence (PL) lifetime and thus the PL intensity of QDs like ligand depletion, assembly formation with organic dye molecules, photo-oxidation and charge trapping/de-trapping induced blinking of single particles are analysed with special regard to the interrelation between PL intensity, lifetime and spectral fluctuations of the PL. Noticeable similarities of the respective PL lifetime–intensity relations (PLIR) independent of the method of surface modification by dyes, ligands or oxygen suggest a common mechanism behind all observed PL quenching effects. The PLIR for single QD intensity fluctuations will be

Self-Assembled Organic-Inorganic Nanostructures: Optics and Dynamics
Edited by Eduard Zenkevich and Christian von Borczyskowski

ISBN 978-981-4745-43-7 (Hardcover), 978-981-4745-44-4 (eBook)
www.panstanford.com

described analytically to quantify the contribution of strong and weak quenching states. In this way ensemble phenomena accompanied by PL intensity changes are related to averaged, modified single-particle blinking. Finally we discuss charge separation-induced Auger recombination and electron-phonon coupling as possible physical origins of PL quenching. Analysing spectral diffusion of single QDs, the existence of different charged states is substantiated by proving the occurrence of quantum-confined Stark effect-related spectral shifts of different magnitude.

2.1 Introduction

Quantum dots (QDs) are a class of inorganic fluorophores with variable photophysical properties resulting in a large variation of application fields. QDs provide high quantum yield, spectral tunability in absorption or luminescence spectrum and offer high extinction coefficients (Vosch et al., 2009; Yu et al., 2003). QD surfaces offer many degrees of freedom for functionalization with different surfactants. Potential applications are their use as fluorescent labels, chemical sensors and in electronic devices. While some photophysical properties of QDs are comparable to (ideal) fluorophores a deeper insight reveals that QDs are intrinsically quite heterogeneous with respect to morphology: multiple crystal structures, sizes, core/shell intermixing, surface facets, variable surface ligand coverage, and oxidation.

Theoretical models relate the existence of interband states directly to the surface saturation conditions of dangling bonds. Such states are interwoven with photophysical properties like the quantum yield (QY), PL lifetime and photostability. However, single-particle (molecule) spectroscopy (SPS or SMS) reveals that QDs are subject to distinct fluorescence intermittency, spectral diffusion, PL lifetime fluctuations. Beyond that they reveal a pronounced heterogeneity between QDs from the same batch and sample.

Both, SPS and bulk spectroscopy are consistent in terms of multi-exponential and spectral dependent PL lifetime distributions, inhomogeneities in spectrum and photo-degradation effects like photooxidation. Even though blinking and oxidation suppression strategies are known, it is still important to understand the blinking mechanism to utilize QDs in sensor applications or as donor (D) or acceptor (A) in Charge Tranfer (CT) and Foerster Resonant Energy Transfer (FRET) complexes.

Physical origins of single-particle blinking frequently discussed in literature are electron and/or hole trapping in intrinsic or external (trap) states (Saba et al., 2013; Voznyy et al., 2013), the Quantum Confined Stark Effect (QCSE) (Empedocles, 1997; Seufert et al., 2001), Auger recombination (Efros and Rosen, 1997; Peterson and Nesbitt, 2009), electron-phonon coupling (Martin et al., 2008; Prezhdo, 2008) and multi-exciton generation (Kambhampati, 2011; Peterson and Nesbitt, 2009). All effects are known to affect radiative and/or non-radiative decay rates.

Herein, the local environment plays a critical role as dielectricity and polarity have been proven to influence photophysical parameters, manifested for example in markedly changed power-law blinking in different dielectric environments (Issac et al., 2005). Beside these direct effects, the solvent polarity influences also the chemical affinity of ligands to the surface (Bullen et al., 2006).

The characterization of assemblies of QDs and dyes relies (among others) on PL intensity changes, in case of FRET on PL quenching of donor (D) emission and enhancement of acceptor (A) emission and in case of CT on PL quenching of donor and acceptor emission.

A Förster theory based on both a complete quantitative description of QD PL quenching and dye PL enhancement thereby depending on the molar ratio taking fractional labelling into account is rare (Clapp et al., 2005; Kowerko et al., 2010; Blaudeck et al., 2012). Some studies focus merely on PL quenching, others present only qualitative data showing PL QD quenching and dye fluorescence enhancement. Several publications indicate the limited applicability of Förster theory to such assemblies which originate from the complicated underlying photophysics of QDs themselves (Blaudeck et al., 2008; Kowerko et al., 2010; Blaudeck et al., 2012). Other studies discuss QD-dye assemblies in the context of CT by means of transient absorption spectroscopy (Issac et al., 2008). However, a complete quantitative study of both, donor and acceptor (fluorescence) in the context of charge transfer is still missing. The challenge obviously consists in disentangling intrinsic and extrinsic (surface related) effects in QDs' photophysics.

Due to the intrinsic heterogeneity of QDs time-resolved single-molecule spectroscopy has become an established method complementary to bulk counterparts. Throughout all studies SMS revealed the unique behaviour of each individual QD but also helped explaining observations related to ensemble type experiments. In this chapter, we will focus on three chemical surface phenomena and the related experimental approaches changing the overall quantum yields (QY) of QDs:

(i) ligand depletion (bulk),
(ii) photo-degradation (bulk and single QDs), and
(iii) assembly formation of QDs with perylene bisimide (PBI) molecules as potential FRET acceptors.

We will demonstrate the consequences of these effects on the single-particle level, quantify their heterogeneity and discuss them in the context of their complementary bulk experiments. As spectral fingerprint, we investigate the PL intensity, quantum yield, PL spectra (shape analysis) and PL lifetime–intensity distribution (PLID) resolved on sub-second to time scales of hours (Zhang et al., 2006).

2.2 Experimental Methods

2.2.1 Bulk Fluorimetry of QDS, PBI Molecules and QD-PBI Assemblies

Colloidal CdSe/ZnS core/shell quantum dots (QDs) passivated by hexadecylamine (HDA) ligands have been purchased from Evident Technologies (see Fig. 2.1) with an emission maximum at 565 nm that overlaps with the absorption spectrum of perylene bisimide (PBI) molecules. The latter have been functionalized with various pyridyl groups which are also shown in Fig. 2.1. Synthesis of such PBI derivatives is described elsewhere (Würthner et al., 2001). Ensemble experiments of QDs, PBIs and QD-PBI assemblies were carried out in spectroscopic grade toluene (Merck) in a "Cary Scan 100" UV/vis and a "Cary Eclipse" fluorescence spectrometer (Varian). Stock solutions were diluted down to 10^{-7} M and photoluminescence (PL) spectra are recorded immediately after dilution with 12 repetition cycles every 5 min and 10–20 cycles every 15 min to cover several hours observation time. The sample was kept in dark between each measurement.

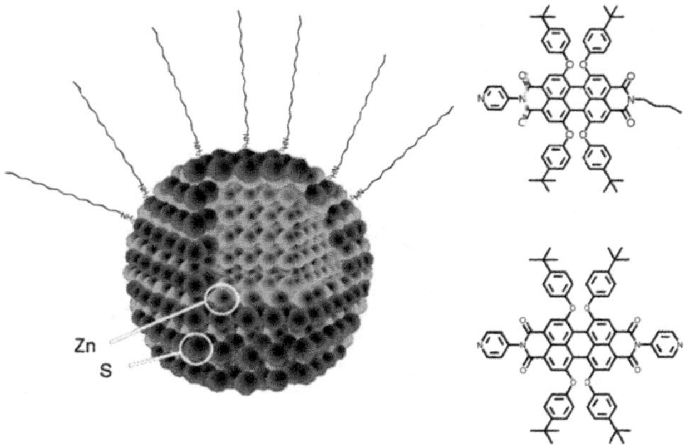

Figure 2.1 (left) Scheme of a CdSe/ZnS quantum dot (QD) with hexadecylamine (HDA) ligands. (right top) pyridyl perylene bisimide (PP), (right, bottom) di-pyridyl perylene bisimide (DPP).

The QD PL intensity decreases initially but remains constant after 30–40 min. After this period of equilibration the PBI dye molecules were added to the 100 nM QD solution to adjust molar ratios ($x = c_{Dye}/c_{QD}$) of 0.1 to 10. In contrast to standard titration experiments, each titration step was done in a separate cuvette to avoid convolution of multiple time-dependent processes caused by the titrated PBI dye molecules. In both, single and PBI assembled QD samples, samples were exposed to daylight for a couple of minutes after all experiments. Subsequently a PL spectrum was recorded to follow the PL intensity, lifetime and spectra.

2.2.2 Single-Molecule Spectroscopy

Single-molecule (particle) experiments were carried out in a home-built laser scanning confocal microscope that allows simultaneous recordings of PL spectra and PL lifetime of QDs and PBI dye molecules for long observation times. For details, we refer the reader to Chapter 4 in von Borczyskowski and Zenkevich (2016). All molecules, particles and assemblies discussed in Section 2.4.2 were excited continuously with a repetition rate of 10 MHz at 465 nm for several minutes with constant average excitation power of approximately 400 W/cm^2.

2.3 Data Analysis

For details on exponential fit models (single, multi, stretched exponential) and photon counting detection schemes we refer to Chapter 4 in von Borczyskowski and Zenkevich (2016).

2.3.1 Software and Gauss-Fit Based Analysis

Goal of a profound spectral data analysis is the characterization of the temporal evolution of spectral features such as the maximum, the shape/width and the overall intensity. Fitting Gaussian functions to PL spectra is a typical method to extract such information. Due to the large sets of data, a Matlab-based routine was written allowing

(i) import of comma separated value type files (as provided by the spectrometers);
(ii) fitting up to four Gaussians to a data set using the Matlab curve fitting toolbox and;
(iii) export the fit parameters a, b and c (amplitude, centre and width) together with the integrated intensity.

2.3.2 Extracting PL Decay Rates and Electronic States from PL Lifetime–Intensity Distributions (PLID)

PL quenching and blinking phenomena are directly related to electronic (excitonic) states with varying radiative rate k_r and/or non-radiative rates k_{nr}. Physical processes changing the radiative rates are, e.g. the (quantum confined) Stark effect (Empedocles, 1997), while non-radiative rates are changed upon electron-phonon coupling (Martin et al., 2008), multi-exciton generation (Klimov et al., 2007) and/or Auger recombination (Wang et al., 2009) and charge or energy transfer (Clapp et al., 2005; Lakowicz,

2006; Holman et al., 2003; Valeur, 2002). Both rates are directly related to the experimentally accessible PL lifetime τ and quantum yield QY. Knowing the relation between QY and τ provides access to k_r and k_{nr}. The respective formalism is given in Section 2.3.2.1. While the mathematical formalism for systems consisting of a single state is rather simple and summarized in Section 2.3.2.1, it becomes more sophisticated for systems undergoing transitions between 2, 3 or even a (quasi-) continuum of states. Section 2.3.2.2 will be dedicated to the latter more complex case.

2.3.2.1 Monoexponential PL lifetime distributions

In case of (organic) molecules conformational changes or isomerization are known to influence both radiative and non-radiative rates (Würthner, 2004). Both rates directly influence the PL lifetime τ which is defined as the reciprocal sum of the radiative and non-radiative depopulation rates (k_r and k_{nr}) of the excited state according to (Lakowicz, 2006):

$$\tau = \frac{1}{k_r + k_{nr}} \tag{2.1}$$

To separate which one of the rates changes the PL lifetime, a relation to the intensity is required. The PL intensity of an emitter depends on its absorption cross-section σ, the excitation rate k_{exc} and the quantum yield QY according to

$$I \propto k_{exc} \cdot \sigma \cdot \mathrm{QY}, \tag{2.2}$$

whereby the QY is defined as

$$\mathrm{QY} = \frac{k_r}{k_r + k_{nr}}. \tag{2.3}$$

In case photoluminescence originates from only one state as often assumed especially in ensemble experiments and if radiative and non-radiative rates are at hand, Eqs. (2.1) and (2.3) can be used to calculate k_r and k_{nr} from the experimentally accessible measured QY and τ. Figure 2.2 visualizes the dependence of PL lifetime τ and quantum yield QY on both rates and highlights in red some typical experimentally observed values for CdSe type

QDs. The latter is schematized on the left of Fig. 2.2 and denoted by <1>. State <2> represents a quenched state with shortened lifetime due to an increased non-radiative rate k_{nr2}, while <3> represents a dark state where $k_{\text{nr3}} >> k_{\text{nr1}}$.

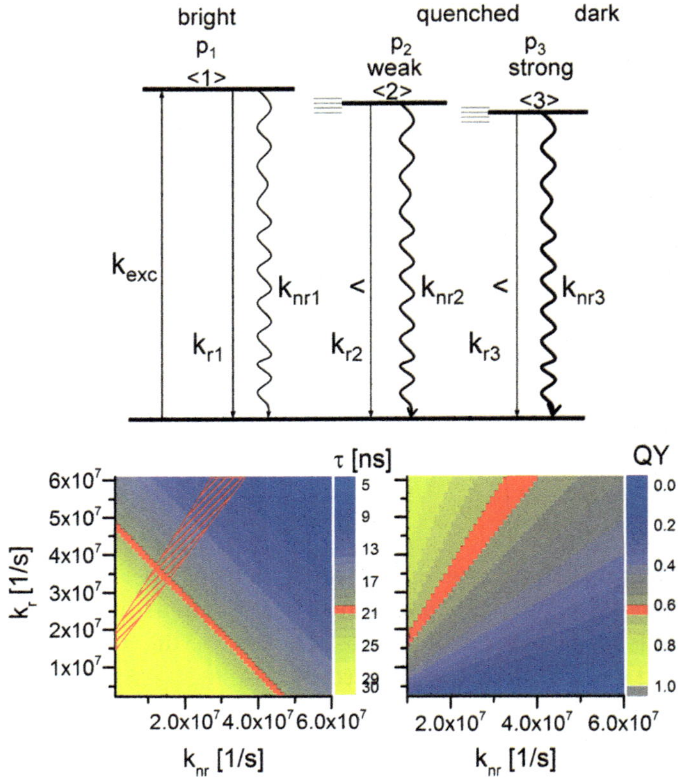

Figure 2.2 (Top) Schematic representation of energy levels of electronic states and interstate rates for emitters with different emissive states, e.g. a bright, an intermediate and a dark state with increasing non-radiative rates k_{nr} with $k_{\text{nr1}} < k_{\text{nr2}} << k_{\text{nr3}}$, respectively. (Bottom) Graphical representation of quantum yield QY and PL lifetime τ as a function of the radiative and non-radiative rates in a range typically observed for QDs. The red area marks characteristic values for bulk QD quantum yields and the dominant lifetime component associated with the bright state at 20 ns. Overlay of both red areas serves as graphical visualization to determine k_{r} and k_{nr}, e.g. from experimental QY-PL lifetime data.

The mono-exponential character of the PL lifetime histogram is preserved if the photoluminescence of all emitters is quenched at the same time and to the same extent. In such a homogeneous type of PL quenching the PL of the ensemble is described by only one state with changed rates represented by the (false) blue colour code in Fig. 2.2. Such a process generally changes the PL lifetime, while the sum of radiative and non-radiative rates remains constant.

Another scenario yielding mono-exponential PL lifetimes is the case where emitters have one or more additional states of type <3> and the forward/backward rates k_1 and k_{-1} between <1> and <3>, respectively, are on the time-scale of the experiment. This effect is schematized in the upper part of Fig. 2.2.

2.3.2.2 Multi-exponential PL lifetime distributions

Here, we discuss two cases where the PL lifetime distribution becomes multi-exponential:

(i) the existence of n subspecies or subpopulations with different PL lifetimes $\tau_1, \tau_2, \ldots, \tau_n$ and
(ii) individual emitters interconverting between two (or more) states of distinguishable PL lifetimes represented by <1> and <2> while the system is in equilibrium, i.e. the inverse sum of forward and backward rates $(k_1 + k_{-1})$ is distinctly smaller than the time-scale of the experiment.

In both cases the PL lifetime histograms would appear as illustrated in the lower part of Fig. 2.3 and the overall detected intensity and quantum yield turns into a linear combination of each individual intensity and $\mathrm{QY_i}$ as follows:

$$I \sim I_{\mathrm{abs}} \cdot \sum_{i=1}^{i_{\max}} p_i \cdot \mathrm{QY_i} \quad \text{with} \quad \sum_{i}^{i_{\max}} p_i = 1 \tag{2.4a}$$

$$\mathrm{QY} = \sum_{i}^{i_{\max}} p_i \cdot \mathrm{QY_i} \quad \text{with} \quad \sum_{i}^{i_{\max}} p_i = 1. \tag{2.4b}$$

Equation (2.4b) represents the effective QY, which is directly proportional to the observed intensity if the experimental detection efficiency is constant for each state <*i*>. In case (i), p_i represent the fractions of each subspecies while in case (ii), p_i is a function of the ratio of forward and backward rates as defined in Fig. 2.3. Here, for simplicity we assume I_{abs} to be constant for all states.

Single-molecule fluorescence experiments are capable to unravel whether case (i) or (ii) causes the multi-exponentiality in an ensemble PL lifetime histogram.

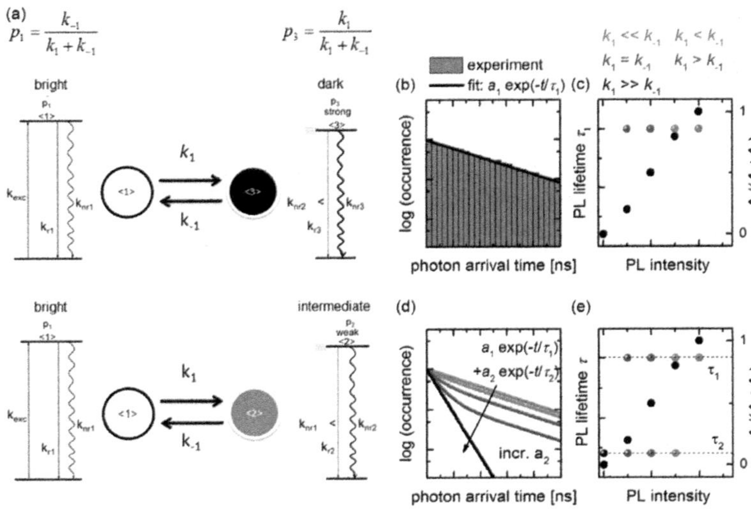

Figure 2.3 (a) Schematic two-state system: (top) consisting of a bright state <1> ($k_{\text{nr1}} < k_{\text{r1}}$) and a dark state <3> ($k_{\text{nr3}} >> k_{\text{r3}}$). (bottom) State <2> is darker than <1> ($k_{\text{nr2}} > k_{\text{nr1}}$). (b) Determination of PL lifetime τ_1 from a mono-experimental fit to the histogram of photon arrival times using TCSPC method. (c) Increased population of state <3> without evidence of a second lifetime component. (d) Bi-exponential decay upon increasing population of <2>. (e) Two PL lifetime components τ_1 and τ_2 resulting from the bi-exponential fit of decay data presented in (d).

Since <3> is defined here as a spectroscopically dark state ($QY_3 \to 0$) its probability of occupation p_3 can only be determined indirectly from the PLIR. The required analytic expression to determine p_3 is derived below from Eqs. (2.4a) to (2.4b).

Population probabilities of state <*i*> denoted here as p_1, p_2, ..., p_n are available from the overall time spent in each state <*i*>, which might be directly accessible in single-molecule experiments. In an ensemble experiments the photon arrival time histogram

$h(t)$ and the cumulative histogram $H(t)$ become multi-exponential according to

$$h(t) = \sum_i a_i \cdot \exp\left(-\frac{t}{\tau_i}\right) \tag{2.5a}$$

$$H(t) = \sum_i A_i \cdot \exp\left(\frac{t}{\tau_i}\right) + y_0 \tag{2.5b}$$

$$I_i = \int_0^{\tau_{end}} a_i \cdot \exp\left(-\frac{t}{\tau_i}\right) dt = a_i \cdot \tau_i = A_i. \tag{2.5c}$$

According to Eq. (2.5c), the number of detected photons I_i emitted from each state is either the product of amplitudes a_i and lifetimes τ_i from Eq. (2.5a) or the amplitudes A_i from Eq. (2.5b), where y_0 is the total number of photons in $H(t)$. The equality sign in (2.5c) holds only if τ_{end} is infinity but a good approximation is achieved already for values of τ_{end} = 5–10$\cdot \tau_i$, where τ_i should be the longest of the lifetime components.

For a three-state system, the relative population p_i of state <*i*> can be determined in three different ways:

(i) By determining the ratio of (dwell) times spent in a state <*i*> to the overall time spent in all three states according to

$$p_i = \frac{\sum_j t_{\text{dwell}<1>,j}}{\sum_j t_{\text{dwell}<1>,j} + \sum_k t_{\text{dwell}<2>,k} + \sum_l t_{\text{dwell}<3>,l}} \tag{2.6a}$$

with *j*, *k* and l being the number of dwell time events accessible, e.g. in single-molecule experiments.

(ii) From fitting the photon arrival time histogram with a multi-exponential function and utilizing the amplitudes A_i or a_i together with the quantum yields QY_i as

$$p_i = \frac{\frac{A_i}{\text{QY}_\text{i}}}{\sum_i \frac{A_i}{\text{QY}_\text{i}}} = \frac{\frac{a_i \tau_i}{\text{QY}_\text{i}}}{\sum_i \frac{a_i \tau_i}{\text{QY}_\text{i}}} = \frac{\frac{a_i}{k_{ri}}}{\sum_i \frac{a_i}{k_{ri}}} = \frac{a_i}{\sum_i a_i}. \tag{2.6b}$$

Note that using Eqs. (2.1) and (2.3) and assuming k_{ri} = const. for all states <i>, this equation simplifies to

$$p_i = \frac{\frac{a_i}{k_{ri}}}{\sum_i \frac{a_i}{k_{ri}}} = \frac{a_i}{\sum_i a_i}. \tag{2.6c}$$

Hence, the population of each state is directly available from a multi-exponential fit using Eq. (2.5a) to the photon arrival time histogram in case that each state differs only in the non-radiative recombination rate k_{nri}.

(iii) From the transition rate matrix k given by

$$k = \begin{pmatrix} 0 & k_{12} & k_{13} \\ k_{21} & 0 & k_{23} \\ k_{31} & k_{32} & 0 \end{pmatrix}, \tag{2.6d}$$

where k_{ij} are the forward and backward rates between each state in units of s^{-1} using i, j = 1, 2 and 3 for a three-state system as depicted in Fig. 2.4a. To derive p_i, it is mandatory to calculate the transition probability matrix M as obtained by taking recourse to a discrete time Markov chain (DTMC) illustrated in Fig. 2.4b (Krengel, 2005; McKinney et al., 2006; Messina et al., 2006) with p_{ij} being the probabilities of the respective transition from i to j per time step/bin as

$$M = \begin{pmatrix} p_{11} & p_{12} & p_{13} \\ p_{21} & p_{22} & p_{23} \\ p_{31} & p_{32} & p_{33} \end{pmatrix} = \begin{pmatrix} 1-\left(\frac{k_{12}}{f}+\frac{k_{13}}{f}\right) & \frac{k_{12}}{f} & \frac{k_{13}}{f} \\ \frac{k_{21}}{f} & 1-\left(\frac{k_{21}}{f}+\frac{k_{23}}{f}\right) & \frac{k_{23}}{f} \\ \frac{k_{31}}{f} & \frac{k_{32}}{f} & 1-\left(\frac{k_{31}}{f}+\frac{k_{32}}{f}\right) \end{pmatrix}, \tag{2.6e}$$

with

$$f = \frac{1}{t_{\text{bin}}}, \tag{2.6f}$$

where f is the frame rate in frames per second (fps) obtained from the experimental binning time t_{bin}. From M the values

p_i are at hand by either solving the following Eigenvalue/ Eigenvector problem:

$$\pi M = \pi \quad \text{with} \quad \pi = (p_1, p_2, p_3). \tag{2.6g}$$

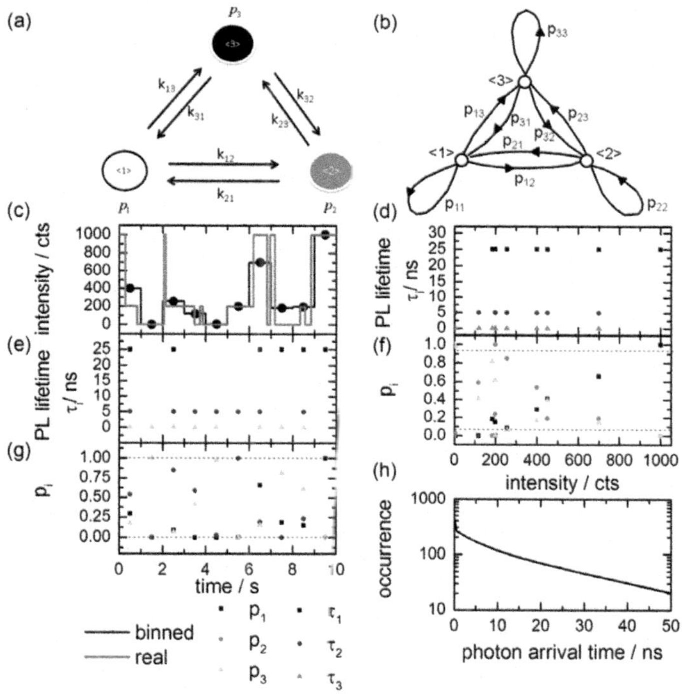

Figure 2.4 (a) Three-level system scheme with forward/backward rates $k_{12} = k_{21} = 1$, $k_{13} = k_{31} = 2$, $k_{23} = k_{32} = 1$. (b) Markov chain with three states and probabilities p_{ij}. The respective transition rate and probability matrices for (a) and (b) are presented in Eqs. 2.6d and 2.6e. The total detected intensities of state <1>, <2> and <3> were simulated with I_1 = 1000 cps, I_2 = 200 cps and I_3 = 1 cps, and PL lifetimes of τ_1 = 25 ns, τ_2 = 5 ns and τ_1 = 0.025 ns, respectively. (c) Simulated time trace (red) applying fixed time binning as mostly used in camera-based detection. (e) Potentially observable PL lifetimes and (g) their relative population time within a binned time unit t_{bin} = 1 s. (d, f) PL lifetime and relative population time vs. binned intensity for time trace shown in (a). (h) Theoretical photon arrival time histogram as obtained after 12 s. Note that only five photons were emitted from state <3> which was defined to have a QY_3 = 0.1% and thus appears almost dark.

Alternatively, we get p_i from calculating M to the power of a considerably high power of n as follows (Krengel, 2005):

$$\lim_{n\to\infty} M^n = \begin{pmatrix} p_1 & p_2 & p_3 \\ p_1 & p_2 & p_3 \\ p_1 & p_2 & p_3 \end{pmatrix}. \tag{2.6h}$$

To provide an example, if k = [0 1 5; 5 0 1; 1 1 0], setting f = 10 fps we get M = [0.4 0.1 0.5; 0.5 0.4 0.1; 0.1 0.1 0.8] and thus end up with p_1 = 22.4%, p_2 = 14.3%, p_3 = 63.3%. The example given in Fig. 2.4c yields $p_1 = p_2 = p_3$ = 33.3%.

In general, the PL amplitudes A_i and lifetimes τ_i are experimentally accessible as long as the quantum yield QY_i is large enough to give a sufficient number of photons. A detailed calculation of accuracy of τ_i determination as function of photon and background counts is given in (Köllner, 1991).

In cases where the rates k_1 and k_{-1} are slower than the inverse binning time of the experiment, Hidden Markov Models (HMM) can be applied to time traces (McKinney et al., 2006) and the respective rates k_{in} and k_{ni} can be derived from a dwell-time histogram of these transitions. p_i will finally be determined according to Eq. (2.6c). If k_1 or k_{-1} are equal or faster than the inverse binning time, intensity time traces are averaged over sub-binning blinking/interconversion effects, as exemplified in Fig. 2.4c. Here, a reasonable dwell-time analysis with HMM fails, while change point analysis is still capable to resolve such fast transitions if single photon arrival times are available (Schmidt et al., 2014, 2012; Watkins and Yang, 2005). However, while fixed time binning on the single-molecule level might not resolve the interconversion rates k_1 and k_{-1}, it is still possible to derive the relative populations from PL lifetime histogram analysis according to Eq. (2.6b). If dark states are present it is necessary to evaluate the amplitudes and lifetimes of the spectroscopically detectable components at given QY or intensity levels.

In the following we will assume a three-level system with a bright, intermediate and dark state referred to as <1>, <2> and <3>, respectively. For single CdSe/ZnS QDs, typically one often finds at intermediate time resolution only two bright PL lifetime regimes around τ_1 = 20 ± 5 ns and τ_2 = 5 ± 2 ns with a single QD-specific amplitude ratio, as exemplarily shown in Fig. 2.5.

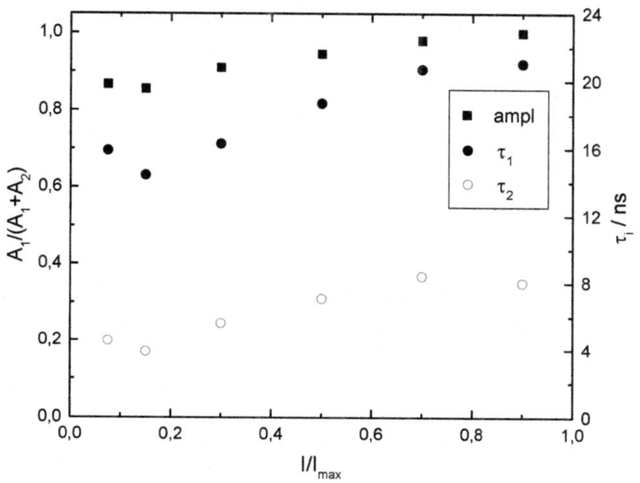

Figure 2.5 Typical amplitude ratio as obtained from a single CdSe/ZnS QD at room temperature within different intensity intervals (0.1 represents the interval 0–0.2, and so forth) with respect to the maximal intensity for a single QD PL intensity time trace.

The aim of the following section is to describe a formalism to determine the relative populations at different absolute intensities to quantify how strong the dark state contributes to the overall quenching at different quenching intensities. In other words we search for p_i as a function of $A_1/(A_1 + A_2)$.

Experimentally we determine the ratio of $A_1/(A_1 + A_2)$ which is equivalent to $I_1/(I_1 + I_2)$ as a function of the intensity or more precisely as a function of the effective quantum yield. Per definition, $\mathrm{QY}_1(p_1)$ and $\mathrm{QY}_2(p_2)$ refer to the states with the PL lifetimes τ_1 and τ_2. For a three-state system the intensity ratio depends on the probability of occupation (p_i, i = 1, 2, 3) and can be inferred from (2.4a) resulting in

$$\frac{I_1}{I_1+I_2}(p)=\frac{p_1\cdot I_{\mathrm{abs}}\cdot \mathrm{QY}_1}{p_1\cdot I_{\mathrm{abs}}\cdot \mathrm{QY}_1+(1-p_1-p_3)\cdot I_{\mathrm{abs}}\mathrm{QY}_2}$$

$$=\frac{p_1}{p_1+(1-p_1-p_3)\cdot \mathrm{QY}_2/\mathrm{QY}_1}. \qquad (2.7\mathrm{a})$$

If $p_3 = 0$, Eq. (2.7a) simplifies to

$$\frac{I_1}{I_1+I_2}(p_1) = \frac{p_1}{p_1+(1-p_1)\cdot \mathrm{QY}_2/\mathrm{QY}_1} = \frac{1}{\left(1+\frac{(1-p_1)}{p_1}\cdot\frac{\mathrm{QY}_2}{\mathrm{QY}_1}\right)}$$

$$= \frac{1}{\left(1+\left(\frac{1}{p_1}-1\right)\cdot\frac{\mathrm{QY}_2}{\mathrm{QY}_1}\right)}. \tag{2.7b}$$

In a two-level system, the overall QY depends only on the population of one state due to the normalization condition in (2.4b) and simplifies to

$$\mathrm{QY} \approx p_1\mathrm{QY}_1 + p_2\cdot\mathrm{QY}_2 = p_1\mathrm{QY}_1 + (1-p_1)\cdot\mathrm{QY}_2. \tag{2.8}$$

Solving (2.8) to obtain p_1 yields (2.9)

$$p_1 = \frac{\mathrm{QY}-\mathrm{QY}_2}{\mathrm{QY}_1-\mathrm{QY}_2}. \tag{2.9}$$

By combining (2.9) and (2.7b) we get

$$\frac{I_1}{I_1+I_2}(\mathrm{QY}) = \frac{A_1}{A_1+A_2}(\mathrm{QY}) = \frac{1}{\left(1+\left(\frac{\mathrm{QY}_1-\mathrm{QY}_2}{\mathrm{QY}-\mathrm{QY}_2}-1\right)\cdot\frac{\mathrm{QY}_2}{\mathrm{QY}_1}\right)}$$

$$= \frac{1}{1+\frac{(\mathrm{QY}_1-\mathrm{QY}_2)\mathrm{QY}_2}{(\mathrm{QY}-\mathrm{QY}_2)\mathrm{QY}_1}-\frac{\mathrm{QY}_2}{\mathrm{QY}_1}}. \tag{2.10}$$

For a three-state system, we now have an equation at hand which relates the experimentally accessible amplitude ratio as a function the quantum yield with the quantum yields of state 1 and 2 (QY_1 and QY_2), respectively). Combining Eqs. (2.1) and (2.3) with experimental τ_1 and τ_2, it is possible to test whether radiative or non-radiative rates can be assumed to be constant in both states or not. Moreover, Eq. (2.10) can be used to test the two-state hypotheses itself.

The intensity maximum of a single QD time trace can be directly related to QY_1, which is often assumed to be 1 (Grabolle et al., 2008). Hence it is plausible to rewrite QY by I/I_{max}.

In a three-state system with the convention $\sum_{i=1}^{3} p_i = 1$ and a dark state <3> defined by setting $QY_3 = 0$ we get

$$QY = p_1 QY_1 + p_2 \cdot QY_2. \tag{2.11}$$

The population of the bright state as a function of the quantum yield thus depends only on QY_1, QY_2 and the population of the intermediate state

$$p_1 = \frac{QY - p_2 QY_2}{QY_1}. \tag{2.12}$$

By combining the approximation of (2.11) with (2.7a), it is possible to calculate p_1, p_2 and p_3 as function of the QY as follows:

$$\frac{I_1}{I_1 + I_2}(QY) = \frac{1}{1 + \left(\frac{QY_1}{QY - p_2 QY_2} - 1 - \frac{QY_1 p_3}{QY - p_2 QY_2}\right) \cdot QY_2 / QY_1}$$
$$= \frac{1}{1 + \left(\frac{QY_1 - QY_1 p_3}{QY - p_2 QY_2} - 1\right) \cdot QY_2 / QY_1} = \frac{1}{1 + \left(\frac{(1 - p_3) QY_2}{QY - p_2 QY_2} - \frac{QY_2}{QY_1}\right)}. \tag{2.13}$$

By setting $p_3 = 0$, Eq. (2.13) becomes equivalent to Eq. (2.10) for the two-level system. In another limiting case with no intermediate state, we get $p_2 = 0$, the second lifetime component disappears and the respective ratio $I_1/(I_1 + I_2)$ becomes 1, demonstrated in the following:

$$\frac{I_1}{I_{l1} + I_2}(QY) = \frac{1}{1 + \left(\frac{QY_1}{QY} - 1 - \frac{QY_1 p_3}{QY}\right) \cdot QY_2 / QY_1} = \frac{1}{1 + \left(\frac{QY_1}{QY}(1 - p_3) - 1\right)}$$
$$= \frac{1}{1 + \frac{QY_2}{QY_1}\left(\frac{QY_1}{QY} QY / QY_1 - 1\right)} = 1.$$

Our model implies that p_i is a function of QY. However, to date there is no unified physical model to describe $p_i(QY)$ nor is it evident how many states of <*i*> physically exist. To give a reason,

QDs undergo no simple excitation relaxation process since they are prone for electron and hole trapping at the surface, the core-shell interface or the local environment. To name a few physical mechanisms affecting k_r and/or k_{nr} we consider multi-exciton generation (Kambhampati, 2011), Auger recombination (Klimov et al., 2007), electron-phonon coupling (Martin et al., 2008) and charge transfer (Kuçur et al., 2006). How charge trapping and each of these physical mechanisms are interrelated remains under debate as well as the strategy to disentangle all these mechanisms from each other. In conclusion, we choose an empirical approach and approximate $p_i(\mathrm{QY})$ using a polynomial of this type according to

$$p_i = (a_i + b_i\mathrm{QY} + c_i\mathrm{QY}^2). \tag{2.14}$$

Setting boundary conditions $p_2(0) = 0$ gives $a_2 = 0$ and $p_2(1) = 0$ gives $c_2 = -b_2$. Hence, we obtain

$$p_2 = (b_2\mathrm{QY} - b_2\mathrm{QY}^2). \tag{2.15}$$

Setting $p_3(0) = 1$, we get $a_3 = 1$ and for $p_3(1) = 0$ we get $c_3 = -1 - b_3$ resulting in

$$p_3 = 1 + (b_3\mathrm{QY} + (-1 - b_3)\mathrm{QY}^2) \tag{2.16}$$

$$p_1 = 1 - p_2 - p_3 = -(b_2 + b_3)\mathrm{QY} + (b_2 + b_3 + 1)\mathrm{QY}^2. \tag{2.17}$$

The relative fractions as a function of the quantum yield can be obtained by non-linear regression of (2.15) and (2.16) in (2.10) with the restrictions $\sum_i^{i_{max}} p_i = 1$, $0 \geq \pi \geq 1$ as exemplified in Fig. 2.6.

Alternatively the following equation holds if $\mathrm{QY}_1 > \mathrm{QY}_2$ and $\mathrm{QY}_3 \approx 0$:

$$p_1 \approx \left[\frac{I_1}{I_1 + I_2}(\mathrm{QY})\right] \cdot \frac{I}{I_{max}} = \left[\frac{I_1}{I_1 + I_2}(\mathrm{QY})\right] \cdot \frac{\mathrm{QY}}{\mathrm{QY}_1}. \tag{2.18}$$

From (2.11), we get

$$p_2 \approx \frac{\mathrm{QY} - p_1 \cdot \mathrm{QY}_1}{\mathrm{QY}_2} = \left(\frac{\mathrm{QY}}{\mathrm{QY}_1} - p_1\right)\frac{\mathrm{QY}_1}{\mathrm{QY}_2} = \left(\frac{\mathrm{QY}}{\mathrm{QY}_1} - \left[\frac{I_1}{I_1 + I_2}(\mathrm{QY})\right] \cdot \frac{\mathrm{QY}}{\mathrm{QY}_1}\right)\frac{\mathrm{QY}_1}{\mathrm{QY}_2}$$

$$= \left(1 - \left[\frac{I_1}{I_1 + I_2}(\mathrm{QY})\right]\right) \cdot \frac{\mathrm{QY}}{\mathrm{QY}_2} \tag{2.19}$$

$$p_3 \approx 1 - p_1 - p_2. \tag{2.20}$$

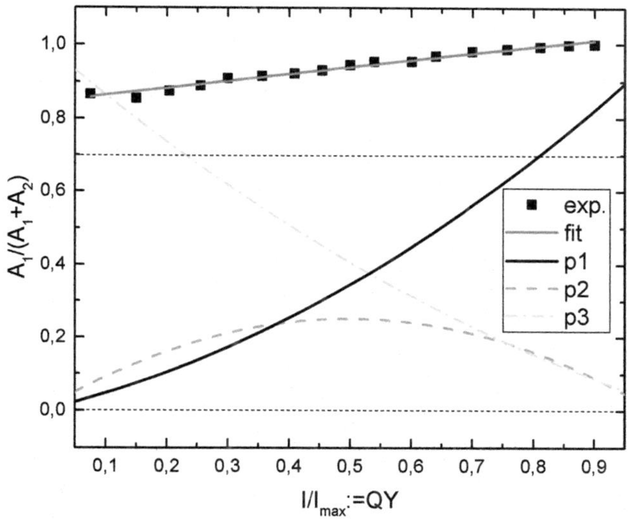

Figure 2.6 Extracting the relative population of the dark state p_3 in a three state system from averaged PLIDs: Experimental data of Fig. 2.5 were supplemented by interpolated data. The empirical functions $p_1(\mathrm{QY})$ and $p_2(\mathrm{QY})$ follow Eq. (2.14) and were derived by non-linear regression of Eq. (2.14) in (2.10). $p_3(\mathrm{QY})$ was subsequently calculated using Eq. (2.20).

2.4 Surface-Chemistry of Quantum Dots Studied at the Ensemble and Single-Molecule Level

CdSe/ZnS quantum dot surface chemistry and its photophysical properties are intimately interwoven. Typically the photoluminescence (PL) is affected by

(i) ligand depletion/exchange
(ii) nanoenvironment (solvent, polymer, water, air, oxygen)
(iii) surfactant (ligand) molecules
(iv) attached dye molecules.

All these properties influence the surface and thus surface related electronic states which are directly related to its spectroscopic properties. Here, we will study ligand depletion after dilution, titration of potential FRET acceptor dye molecules in solution and photo-oxidative effects by careful investigation of PL quenching and enhancement without and after exposure to light. Results from ensemble spectroscopy (cw and pulsed optical excitation) on time scales of minutes to hours are presented. They illustrate the temporal evolution of PL spectra and lifetimes as a function of observation time and/or dye concentration. Moreover, heterogeneities observed in ensemble experiments are analysed in more detail by time-resolved multi-parameter single-particle spectroscopy. Photo-oxidation of the surface and assembly formation with PBI molecules influence the interrelation between PL intensity, lifetime and spectrum which are caused by a redistribution of the population of bright, intermediate and dark states. Furthermore, spectral diffusion reveals information on crystal structure modifications and the quantum-confined Stark effect (QCSE).

2.4.1 Ensemble Spectroscopy of CdSe/ZnS Quantum Dots

2.4.1.1 Ligand depletion and spectral heterogeneity revealed by spectroscopy of diluted CdSe/ZnS QDs

CdSe/ZnS quantum dots (QDs) dissolved in toluene are diluted from a 50 μM stock solution down to 100 nM concentration. The PL intensity decreases continuously for about 40 min after sample preparation as shown in Fig. 2.7.

Quenching effects have been observed in comparable QD-amine ligand systems by other groups (Ji et al., 2008; Koole et al., 2008) and were ascribed to desorption of amine ligands upon dilution generating surface states resulting in PL quenching. The timescale of PL quenching demonstrated in Fig. 2.8b is comparable to the amine ligand depletion observed by Ji et al. for CdSe QDs (Ji et al., 2008).

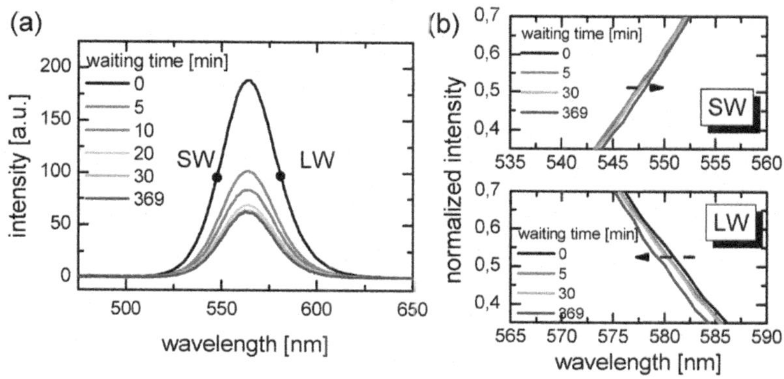

Figure 2.7 (a) Photoluminescence spectra (λ_{exc} = 465 nm) of CdSe/ZnS QDs at different waiting times (after solution in toluene at a concentration of 100 nmol/l). (b) Normalized intensities at half maximum for short (SW) and long (LW) wavelength detection as a function of waiting time. Dashed arrows mark spectral shifts. Between each measurement the sample is kept in the dark. The data on the right show the increase of PL during exposure to daylightr.

Beside PL quenching, we observe an effective blue shift of a few nanometers of the red side of the spectrum but on a slower timescale than the observed intensity decay. Spectral blue shifts are ascribed to the oxidation of the CdSe core that results in shrinkage of the crystal core size (of up to 1 nm) as observed for CdSe/ZnS QDs in air at the single-particle level where blue shift of several tens of nanometres occur (van Sark et al., 2001). Ensemble spectral blue shifts occur on a longer time scale or to a lower extent per time unit compared to the single-particle level, most probably due to the fact that the oxygen is available only in a dissolved form of reduced concentration (compared to air) and first has to diffuse to the surface of the QD.

Ensemble spectra represent a specific size distribution of QDs (Talapin et al., 2001). Hence, an asymmetric spectrum like the one shown in Fig. 2.7a might be caused by an asymmetric size distribution. It turns out that a single Gaussian fit analysis does not match the asymmetric spectral shape shown in Fig. 2.7. However, the QD ensemble spectrum is well described by fitting with two Gaussian line shapes. The temporal evolution of the respective fit results is presented in Fig. 2.8b,c.

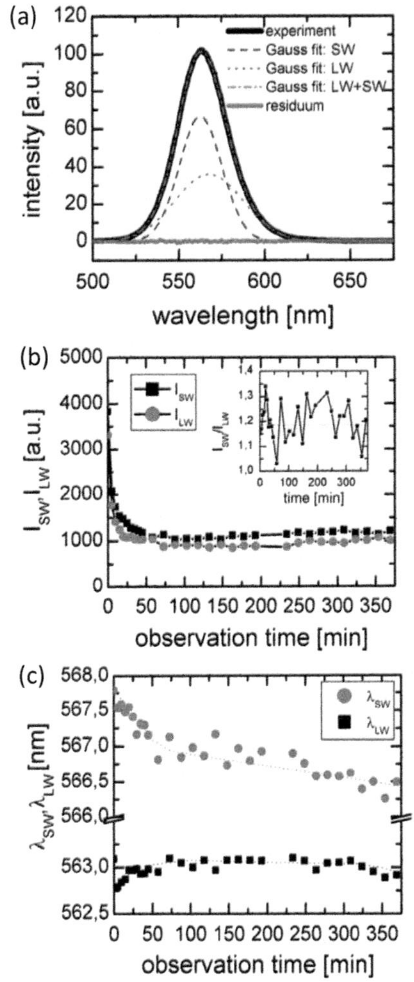

Figure 2.8 (a) Typical CdSe/ZnS QD spectrum in toluene (adopted from Fig. 2.7 at t = 5 min, black line) fitted by two Gaussian line shapes (dashed blue and dotted red). (b) Temporal evolution of the QD PL described by the two-Gaussian fit analysis: integral intensities I_{SW} and I_{LW} of Gaussians shown in (a). The inset shows the amplitude ratio I_{SW}/I_{LW}. (c) peak positions λ_{SW} and λ_{LW}. SW and LW refer to short and long wavelengths to account for the initial spectral position of the maximum of the respective Gaussian line.

Systematic shifts of the peak positions occur, whereas the respective intensities decrease without significant changes in the amplitude ratio (see inset of Fig. 2.8b). Especially λ_{LW} shifts to shorter wavelengths (blue shift), while λ_{SW} remains substantially constant within experimental error, resembling the observations in Fig. 2.7b.

The QD surface coverage with amine ligands is known to decrease upon dilution (Ji et al., 2008). The created vacant surface sites are accessible for a variety of adsorbers, most likely water molecules, dissolved oxygen (Carrillo-Carrión et al., 2009) or solvent impurities. A reduction of unwanted adsorbates is achieved by a more complete passivation of the QD surface with ligands. A decrease of PL quantum efficiency is inhibited upon addition of HDA ligands to the QD solution before dilution as seen in Fig. 2.9.

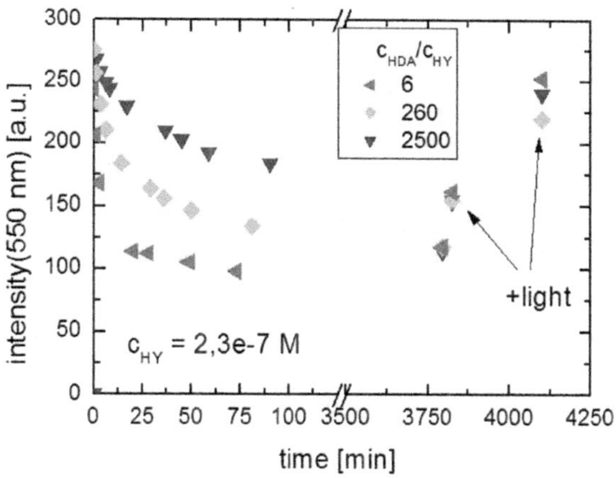

Figure 2.9 PL intensity decrease of CdSe/ZnS QDs upon dilution from 6.5×10^{-5} M to 2.3×10^{-7}M at different concentration ratios, whereat c_{HDA} and c_{QD} (c_{HY}) denote the concentration of the hexadecylamine ligands and QDs, respectively. This effect further supports the finding that PL quenching upon dilution originates in ligand depletion. The data on the right show the increase of PL during exposure to daylight.

2.4.1.2 Non-FRET type PL quenching and spectral blue shifts of CdSe/ZnS QD-dye assemblies

Next, we study the influence of surface-attached functionalized PBI molecules on the photo-physical properties of CdSe/ZnS QDs.

Both entities are exemplarily depicted in Fig. 2.1. This type of dye molecules anchors to the QD surface by self-assembly and subsequently quenches the PL predominantly by a NON-FRET quenching mechanism as recently described in detail by Kowerko et al. (Kowerko et al., 2010). In contrast to ligand depletion, even one single functionalized PBI dye molecule modifies the QD surface (states) in such a way, that the observed PL quenching is distinctly higher than due to the number of displaced ligands (Kowerko et al., 2010). Accordingly, ensemble fluorescence experiments were carried out similar to those of the previous section to investigate how self-assembly with functionalized PBI molecules affects the spectral properties of QDs.

Figure 2.10a demonstrates the PL quenching upon increasing the concentration c_{DPP} of DPP dye molecules. Since the QD PL intensity decreases continuously during time after sample preparation as described before, DPP was only added after a waiting time of about 40 min where ligand association/dissociation shows to be in equilibrium (no further PL intensity changes). The addition of DPP is followed again by enhanced PL quenching increasing with the molar ratio $x = c_{\mathrm{DPP}}/c_{\mathrm{QD}}$. The absolute quenching is not a linear function of x (see Fig. 2.10b). This could be due to either a non-linear concentration dependence of the labelling efficiency of the QD with dye molecules (caused by a limited number of easily accessible or vacant surface sites) getting into saturation already at small x (Kowerko et al., 2010) or the physical nature of the observed non-FRET type quenching resembles the one occurring in FRET type quenching with multiple acceptors (Clapp et al., 2004). A linear behaviour of $I(t_x) - I(t_y)$ would in the latter only linearly depend on x for small FRET efficiencies at small x values.

A partial recovery of PL intensity is achieved by exposing the samples to daylight for a couple of minutes (last data points at $t \approx 240$ min in Fig. 2.10a,c,d). The physico-chemical nature of the related photoinduced enhancement is complex and summarized in detail by Carrillo et al. (Carrillo-Carrión et al., 2009). Mostly such effects result from complex interactions of water molecules or other adsorbates on the surface. Hence, a detailed investigation of the observed PL enhancement is not within the scope of this study, rather we make use of this effect as PL recovery tool. Obviously the remaining absolute quenching after daylight exposure ($I(t_{initial})-I(t_{light})$) is larger than the absolute quenching observed at the minimum of the QD PL after 40 min ($I(t_1)-I(t_2)$) (Fig. 2.10b, below), while the relative quenching $I(t_{light})/I(t_{initial})$ and $I(t_2)/I(t_1)$ is vice versa (Fig. 2.10b, top).

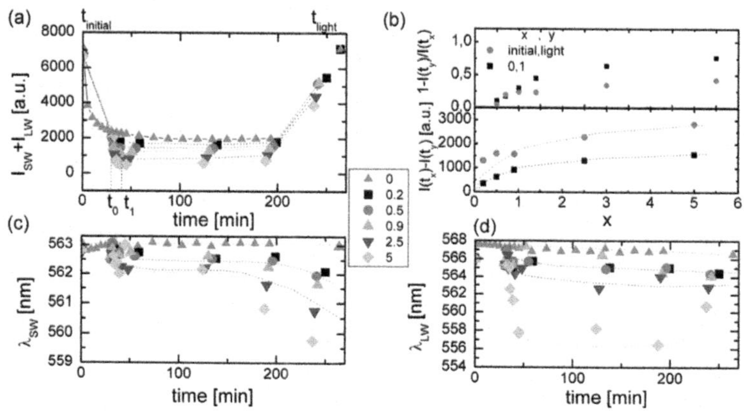

Figure 2.10 Spectral analysis of CdSe/ZnS QDs in toluene by two Gaussian line shapes (as applied in Fig. 2.8a) in absence of DPP dye molecules ($x = 0$) and with increasing molar ratio x (concentration ratio c_{DPP}/c_{QD} with c_{QD} = 100 nM). (a) Decay of the integrated PL intensity upon adding DPP at different x. (b) Relative and absolute QD PL quenching at different reference times caused by DPP depending on the molar ratio x. For detailed explanation refer to text. (c) Spectral shifts of λ_{SW} upon titration with DPP. (d) Spectral shifts of the LW peak upon titration with DPP. λ_{SW} and λ_{LW} refer to short and long wavelengths to account for the initial spectral position of the maximum of the respective Gaussian line shapes. The last data point in all graphs depicts the PL after illumination with daylight for about 30 min. Dotted lines serve in all cases as guide for the eye.

However, in terms of Stern–Volmer type quenching analysis, Fig. 2.10b (top) reveals a comparable but slightly lower quenching rate considering the PL intensities before dilution and after light-induced enhancement. This effect serves as a strong indication that the DPP induced QD PL quenching is not only due to the replacement of several HDA molecules on the QD surface. Otherwise the remaining quenching after photo-enhancement would be expected to be negligible since DPP replaces not more than ten out of more than thousand HDA surface sites. In that case the HDA related quenching is assumed to be a linear function of the HDA surface coverage as discussed already in the previous section (Ji et al., 2008).

In conclusion, even though the anchoring of DPP to the QD surface is in competition with the one of HDA, ensemble spectroscopy clearly indicates a permanently docked QD-DPP fraction. From single-particle experiments Kowerko et al. estimated the non-FRET quenching efficiency of a single PP molecule to be on average about 50% and a labelling efficiency of PP to be 20% at $x = 1$ (Kowerko et al., 2010). Since DPP has two instead of one pyridyl anchoring group the labelling efficiency is assumed to be larger than for PP. Hence the non-FRET-type quenching induced by DPP seems to be comparable to the one of PP and is thus smaller than the expected value of FRET-type PL quenching. However, FRET from the QD to PP and DPP dyes is very small under these experimental conditions contrary to similar QD-PBI-type assemblies studied by Gerlach et al. who demonstrated a turnover of non-FRET to FRET-induced QD PL quenching by adding the non-polar TEHOS (tetrakis(2-ethylhexoxy)silane) to the toluene solution (Gerlach et al., 2013).

For λ_{LW} blue shifts of up to 10 nm for a molar ratio of $x = 5$ are observed that are partly reversible upon illumination of the sample with daylight (see Fig. 2.10d). A blue shift is also observed for λ_{SW} that increases upon time and concentration of DPP, but to smaller extent and non-reversible upon illumination with daylight. In conclusion, λ_{LW} shifts on different timescales, at first a fast blueing occurs for the period of a few minutes during which assembly formation with DPP proceeds accompanied by PL quenching (see Fig. 2.10a) and is followed by continuous red shifting on a much slower timescale of several tens to hundreds of minutes. This effect is enhanced in quantity but not in the related timescales at increased concentrations of DPP molecules. Despite the assembly formation an unknown process increases slightly the PL intensity. Upon further increasing the molar ratio the SW component of the PL intensity is quenched to a slightly larger extent (not shown here).

The inhomogeneity of the QD spectrum indicates different physical origins of the emission. The increased reactivity of the LW-emission upon dye assembly formation suggests that these different types of emission exhibit different efficiencies to PL quenching processes and the net blue shift must not necessarily be only the result of photo-oxidation only. In fact, to consider the emission of QDs of purely excitonic origin may be incorrect. Emissive surface states are considered to contribute to the overall PL emission (Wang et al., 2003). Inerbaev et al. calculated structure and spectra of CdSe QDs based on DFT calculations and reported spectral blue and red shifts that depend on crystal structure and the type of passivation with ligands and oxygen (Inerbaev et al., 2009). Kilina et al. also found strong interrelation between ligands and morphology, electronic structure and optical response in their DFT calculations of $Cd_{33}Se_{33}$ clusters (Kilina et al., 2009). As QD crystal structures and facets are assumed to be heterogeneous (Burda et al., 2005) surface passivation with ligands, ligand depletion as well as oxidation could selectively shift spectra or quench the PL of sub-populations with similar crystal morphology and thus yield spectral asymmetries as observed here for ligand depletion and dye attachment.

2.4.1.3 PL lifetime analysis of diluted CdSe/ZnS QDs

In the beginning of the previous section the ligand depletion induced PL quenching was spectrally analysed. Such quenching may either be caused via reduced radiative or increased non-radiative decay processes. To identify the physical or chemical nature of these processes it is necessary to relate the individual contributions to the observed changes in PL intensity and lifetime. The time-tagged time-correlated single photon counting (tt-TCSPC) method allows monitoring the PL decay time or lifetime τ as a function of the experimental observation time. Ensemble experiments on CdSe/ZnS QD in toluene identical to the ones described in the previous section (see Figs. 2.7–2.10) were repeated in a setup with pulsed excitation and single photon counting detection as described in the experimental section in Chapter 4 in von Borczyskowski and Zenkevich (2016). Figure 2.11 shows the results of decay time analysis of PL quenching induced by ligand depletion subsequent to a dilution process as shown in Fig. 2.8.

The "relative" quantum yield (QY) of a QD solution (Fig. 2.11a) was determined by comparing it to a solution of a standard solution with a known quantum yield (Rhodamine 6G with $QY_{Rh6G} = 95\%$). Methodical details are described elsewhere (Lakowicz, 2006).

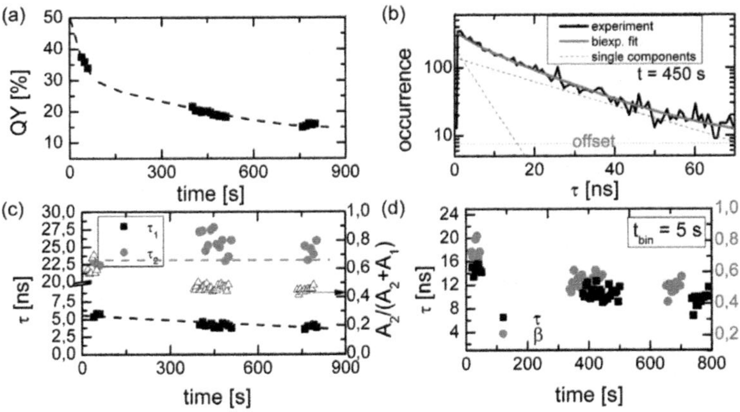

Figure 2.11 (a) Temporal evolution of the PL quantum yield (QY) after sample preparation of a 10^{-7} M concentrated QD sample. (b) PL decay taken at t = 450 s with a binning time of 5 s, fitted exemplarily by a bi-exponential function (components indicated by dotted lines). (c) The respective fit parameters τ_1, τ_2 (squares, circles) and the corresponding amplitude ratio $A_1/(A_2 + A_1)$ (open triangles). (d) Fit results derived from stretched exponential fits. Dashed lines serve as guide for the eye.

A typical PL lifetime histogram of CdSe/ZnS QDs in toluene solution is shown in Fig. 2.11b where the horizontal line indicates an experimental offset determined using Eq. (4.3) of Chapter 4 in von Borczyskowski and Zenkevich (2016) and the dotted lines represent the two exponential functions of the beforehand applied bi-exponential fit Eq. (2.2) with $n_{\max}$ = 2, (Chapter 4 in von Borczyskowski and Zenkevich (2016)). Obviously the PL decay time histogram can be approximated well by applying either a bi- or stretched-exponential fit according to Eqs. (4.3) and (4.5) of Chapter 4 in von Borczyskowski and Zenkevich (2016) resulting in two distinct lifetimes $\tau_2 \approx 25$ (±2) ns and $\tau_1 \approx 5.7$ (±0.2) ns to 3.9 (±0.2) ns. In case of a bi-exponential fit the component τ_1 is shortened by a factor of about 1.5 but τ_2 is not distinctively changing with increasing PL quenching during observation time. The amplitude ratio changes slightly whereas fitting the same data by a stretched-exponential function, τ and β decrease from 15 (±1) ns to 9 (±1) ns and 0.75 (±0.05) to 0.5 (±0.05) while the PL quantum yield is reduced from 40% to about 15%. Quantitatively both observations may not explain the reduction of PL intensity by a factor 3. For clarity the direct PL lifetime–QY relations are illustrated in Fig. 2.12. The goodness of fit for both fit models is comparable but errors are smaller using a stretched exponential fit with only two instead of four fit parameters.

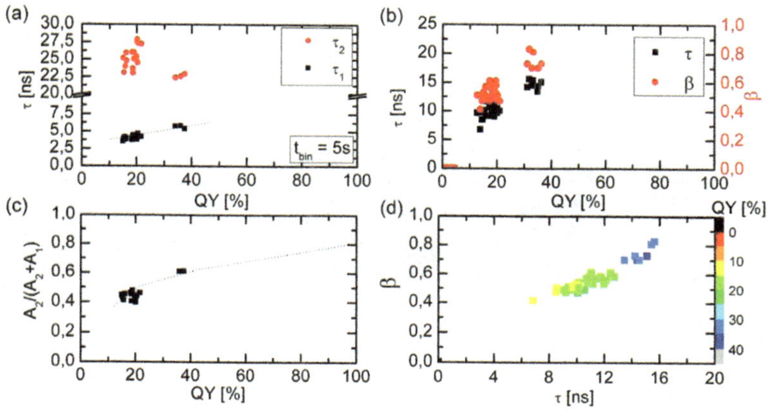

Figure 2.12 PL lifetime–quantum yield relation for QDs described in Fig. 2.11. The PL decay is fitted by a bi-exponential function in (a, c) and by a stretched exponential function in (b, d). We will refer to the physical interpretation of this in the discussion and compare it with single QD PL intensity-lifetime relations showing similar behaviour. An inherent phenomenon of different effects causing a PL intensity reduction in QDs is the appearance of a β–τ relation that resembles the one illustrated in Fig. 2.12d.

2.4.2 Single-Molecule and Single-Particle Spectroscopy

In this subsection time-resolved experiments of single CdSe/ZnS nanocrystals are presented and will be (i) related to the findings from the ensemble data, compared to (ii) photo-degrading and (iii) PP assembled QDs.

Typical spectra of single QDs or QD-PP assemblies and their temporal evolution are presented in Fig. 2.13, where (a) shows the PL spectrum of a single QD and (d) the related spectral diffusion. (b) shows the PL spectrum of another QD while (e) illustrates the spectral shifting to short PL wavelengths (blueing) as a function of observation time. (c) shows a PP assembled QD finding the respective PL spectra in the same confocal spot and the related temporal evolution in (f). In the following we will (i) correlate PL decay time, spectral position, PL intensity and observation time and (ii) compare changes of the correlations upon oxidation and assembly formation with a PP dye molecule.

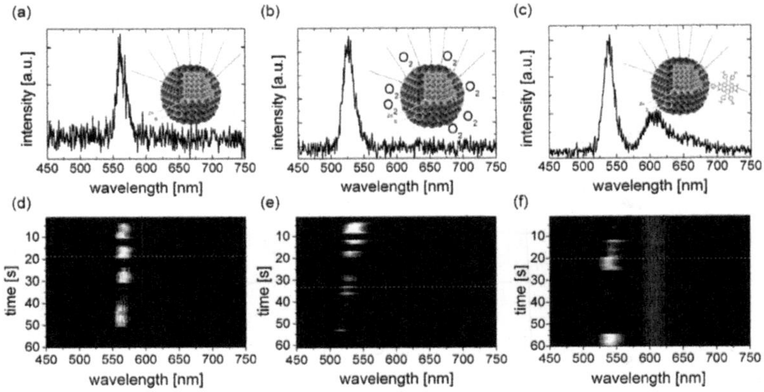

Figure 2.13 (a–c) PL spectra at the beginning of the observation time (t_{bin} = 1 s) and (d–f) related temporal evolution of a (d) single weakly blueing QD, (e) strongly blueing QD and (f) QD-PP assembly. Binning time for each spectrum is 1 s.

Moreover, we will determine the intensity lifetime distribution of single QDs from tt-TCSPC data elucidating to which extent intermediate and dark states contribute to single QD blinking. This enables us to compare whether the relative population of, e.g. the dark state differs significantly for different surface-(photo)chemical effects like ligand depletion, oxidation and dye incorporation. For further experimental details, we refer Chapter 4 in von Borczyskowski and Zenkevich (2016).

2.4.2.1 PL lifetime–intensity relations of single quantum dots

It is well known that single QDs reveal PL intensity fluctuations also denoted as blinking, exemplarily shown in the time trace of Fig. 2.14a. This blinking is accompanied by PL lifetime fluctuations as illustrated in Fig. 2.15 with an applied binning time of 400 ms. The number of photons per time bin essentially influences the accuracy of the determined PL lifetime (Köllner and Wolfrum, 1992). Therefore accumulation of photons by increased bin intervals are necessary, especially if the PL intensity is low. In a first step we derived the PL decay from different PL intensity intervals (see Fig. 2.14). Similar to observations of other groups (Fisher et al., 2004; Schlegel et al., 2002), high intensities are related to long PL lifetimes with a mono-exponential character.

The lower the PL intensity the more the PL decay turns to a more non-exponential character demonstrated in Fig. 2.14b. Additionally it is shown that the PL decay histogram is invariant to observation time at least for high intensities (Fig. 2.14d,e), whereas for low-intensity intervals deviations are small and most probably caused by the slightly different intensity distribution towards lower intensities within the intensity interval (Fig. 2.14c). About 100 single QDs have been investigated and the described observations are similar for all. Individual differences will be described later in this section.

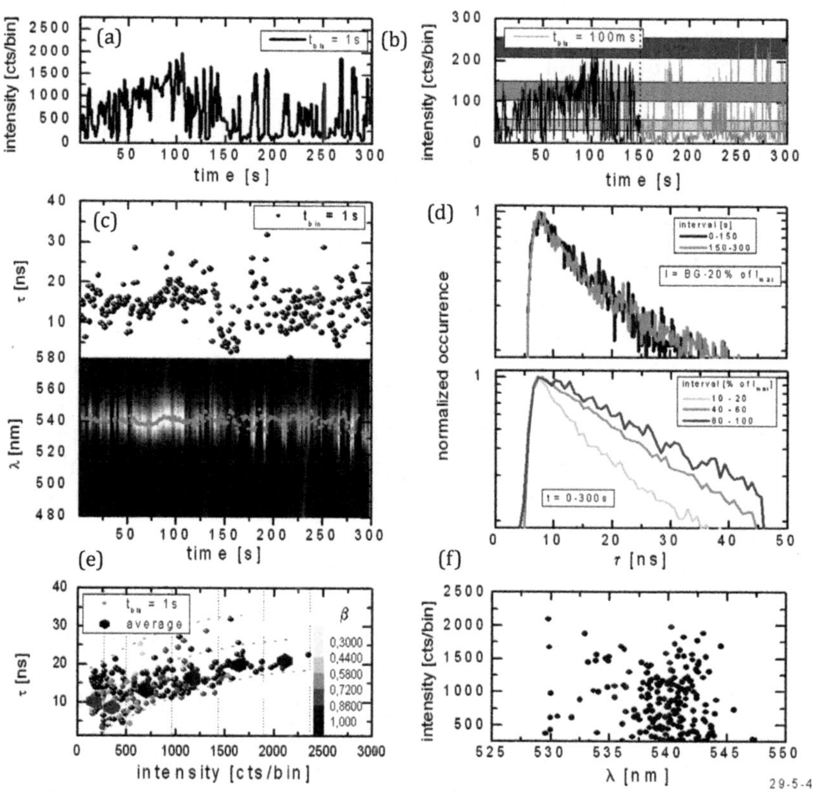

Figure 2.14 (a) Intensity time trace with 0.1 s binning time for different intensity intervals are marked in different grey scales, black (left) and red (right) part separates the time trace into two parts. (b) PL decay obtained by photons from the different grey marked intensity intervals in (a). (c,e) PL decay obtained from the intervals 0–150 s (black), 150–300 s (red) and 0–300 s (grey) of the respective intensity ranges shown in (b).

In Fig. 2.14 the photon arrival time histograms $h(t)$ were obtained from summing over intensity intervals whereas the next subsection is dedicated to the determination of $h(t)$ over short time bins (see Fig. 2.15a). A stretched or bi-exponential function is commonly used in the case $h(t)$ is not described sufficiently by a mono-exponential decay function. Hence, the PL lifetime (τ)-intensity (I) distribution (PLID) requires to be complemented by relating also the stretched exponential parameter β and the amplitude ratio together along with both PL lifetime components in the case of a bi-exponential fit, as shown in Fig. 2.15b–f. However, a non-linear τ–I relation and only slight influences of the PL lifetime at the high intensity range are observed for the majority of molecules. A typical example is illustrated in Fig. 2.15b. Since the PL lifetime and intensity of a single QD fluctuates on time scales below the binning time (t_{bin}) a non-exponential fit has to be applied for the respective decay time histograms with short binning times of below 1 s. The related fit results for each data point of the intensity time trace (shown only for t = 110 s to t = 150 s in Fig. 2.15a) over an observation time of 300 s are presented in Fig. 2.15b.

In general, high PL intensities correlate with longer PL lifetimes and β values close to 1. The fluctuation range of the PL lifetime close to I_{max} is about ±3 ns and almost mono-exponential. The PL intensity dependent τ and β values obtained from the PL decay shown in Fig. 2.14b are added as open symbols in Fig. 2.15b,e. At first the β–τ relation resembles that of the ensemble quenching data shown in Fig. 2.12d since with decreasing PL intensity the PL lifetime shortens and β values decrease continuously. It is clearly seen that on the sub-second time scale τ and β fluctuate substantially but generally follow the trend of the averaged values.

Stretched-exponential functions of the PL decay, where β is in the range of 0.5 to 1 may also be expressed by bi-exponential functions which are physically more intuitive for the interpretation of QD PL fluctuation since bi-exponential PL decay implies two states, a bright and a quenched state with distinct PL lifetimes. The exact physical interpretation of the following results will be given in the discussion. However, the respective PL lifetime distributions are plotted in Fig. 2.15c and the related amplitude ratio ($A_1/(A_2 + A_1)$) is depicted below in Fig. 2.15f.

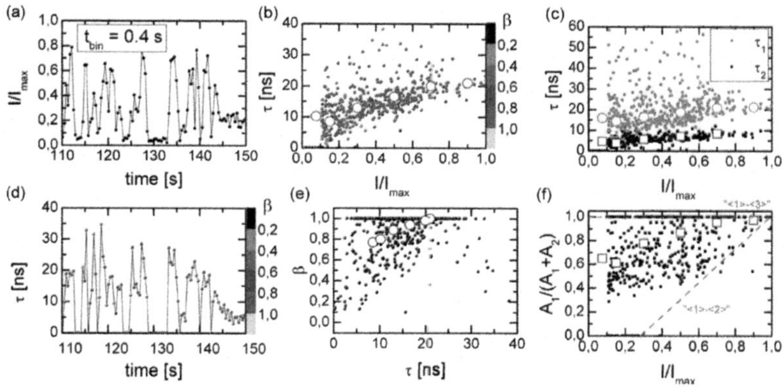

Figure 2.15 (a) Part of the intensity time trace from the single QD shown Fig. 2.14a binned with 400 ms time-resolution and normalized to the maximum PL intensity value occurring during the total observation time of 150 s. The respective PL lifetime fluctuations are shown in (d), where the PL decay of each data point was fitted by a stretched exponential function. The temporal evolution of τ and β is shown for the same period of the time trace as in (a). β is indicated by the colour code on the right axis. (e) shows the corresponding β–τ relation. The relation between fit parameters of the PL decay and intensity with respect to I_{max} for the total observation time of 150 s is shown in (b) for stretched exponential (τ and β) and (c, f) for bi-exponential (τ_1, τ_2, A_1, A_2) fits, respectively. Dotted red lines in (f) refer to limiting cases of a three-state model describing single QD PL intensity fluctuations, elucidated in detail in the discussion section. The big hollow symbols in (b), (c), (e) and (f) refer to fit results obtained by a stretched and bi-exponential fit of the PL intensity interval of the related PL decay shown in Fig. 2.14.

Two distinct lifetime regimes emerge for all QDs in the range of τ_1 = (20 ± 5) ns and τ_2 = (6 ± 3) ns. It is evident that on the sub-second time scale τ_1 and τ_2 also fluctuate but generally follow the trend of the averaged values (open circles). The latter are again obtained from the PL decay of different PL intensity intervals over the complete observation time like the example shown in Fig. 2.14b. Obviously, the amplitude of the short lifetime component increases on average with decreasing PL intensity while on a sub-second time scale the distribution is broad. The upper boundary (horizontal dashed line in Fig. 2.15f) demonstrates that a distinct part of intensity fluctuations exhibits a constant PL lifetime τ_1 whereas the diagonal dashed line represents a lower boundary of the amplitude ratio. The origin of this lower boundary may be related to the increased population of a quenched state with the shortened PL lifetime, here referred to as τ_2, the lifetime of intermediate states. Figure 2.3e visualizes the effect for a two-level system. An analytical description is given thereafter.

Besides variations of the lifetime in the typical range of 1–30 ns, occasionally even longer lifetimes occur prevalently below about 70% of I_{max} (Fig. 2.15b). These long PL lifetimes sometimes cannot be associated with a corresponding spectrum which is either due to a spectrally broad emission or above the detection range of the spectrometer that ends above 750 nm or in the case of spectral detectable emission the peak position determined with a time resolution of 1 s is on average red-shifted against the previous spectrum. Such long-living states have been reported to be spectrally red-shifted in emission (Heinlein et al., 2005; Petrov et al., 2005). We will refer to the nature of such states in the discussion section.

All QDs show PL intermittency correlated by PL lifetime fluctuations as demonstrated exemplarily in Fig. 2.14a. The shape of the PL lifetime–intensity distribution (PLID) shown for an individual QD in Fig. 2.15 resembles that of all investigated QDs but the local data point density may deviate from QD to QD. Photo-oxidized and dye-assembled QDs are quenched in their PL intensity and therefore the density of data points in the respective PLIDs is raised at lower intensity values.

Now we like to investigate spectral diffusion accompanying the already mentioned PL intensity fluctuations of individual QDs with regard to experimental evidence of the QCSE.

Recently, Schmidt et al. presented a sophisticated CPA method to obtain PLIDs from time tagged single photon counting data of single CdSe/ZnS QDs using a combination of change-point analysis (CPA) and clustering (Schmidt et al., 2012). This method is free of any time binning and allows determining the number of states with a maximum likelihood approach. However this method requires large number of photons to allow statistically valid conclusions.

2.4.2.2 Correlation of intensity and spectral fluctuations indicates the quantum confined Stark effect

To obtain single QD spectra of sufficient signal to noise ratio even when the PL intensity is low during a time trace, we increased the binning time to 1 s. Figure 2.16a shows the PL intensity time trace of the QD already mentioned in the previous section. Two typical spectra are shown in Fig. 2.16b and the respective temporal evolution of spectra during the observation time of 300 s is seen in Fig. 2.16c. The red dots mark the centre position of Gaussian line shape function used to fit the spectra. Relating PL peak

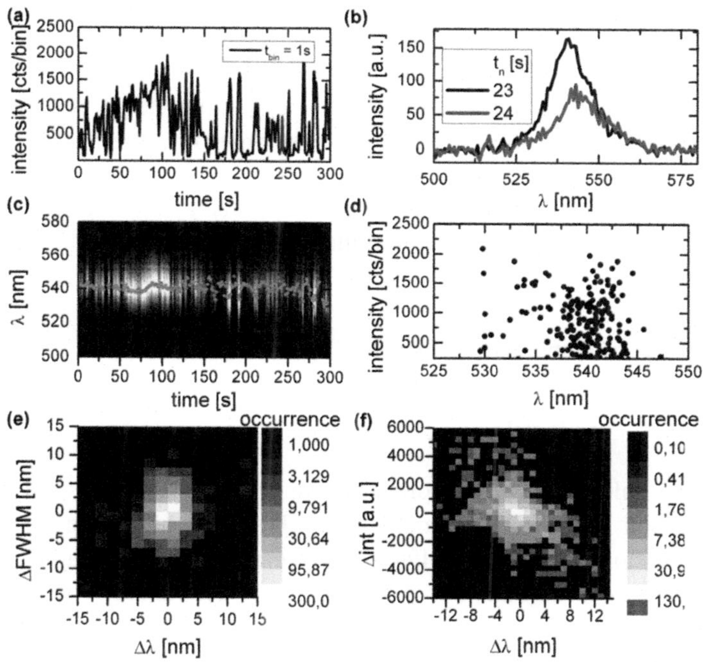

Figure 2.16 Characterization of a single QD by time resolved multi-parameter spectroscopy: (a) intensity time trace with 1 s binning time and (b) related PL spectra taken at 23 and 24 s. (c) Temporal evolution of the QD PL spectra and related peak position fitted by Gaussian lines (red full circles). (d) PL intensity plotted over peak position (1 s binning time). (e, f) Density plots showing the spectral jumps on short time scales (1–5 s) and their related changes in intensity and full width at half maximum for a single QD. For details see text.

position and PL intensity (see Fig. 2.16d) no clear dependence can be found. Intensity fluctuations occur over a broad spectral range of up to 15 nm, but spectral shifts maybe related only on a short time scale with changes of the PL intensity and full width at half maximum. If a QCSE has to be taken into account in consequence of charge trapping at the QD surface or environment, PL intensity decreases accompanied by red shifts and line broadening of the PL spectrum would be expected according to (Empedocles, 1997; Müller et al., 2004). Therefore subsequent spectra like those shown in Fig. 2.16b are analysed for a change in peak position ($\Delta\lambda$), peak full width at half maximum (ΔFWHM) and integrated PL intensity (Δint). All three values are determined for Δt = 1, 2, 3, 4 and 5 s, respectively. The resulting relations between $\Delta\lambda$, ΔFWHM and Δint are shown as density plots in Figs. 2.16e,f. In both images a dependence of these parameters is observable as expected according to QCSE. At the same time the distribution appears quite uncorrelated for spectral shifts below 4 nm. Obviously QCSE is not the only mechanism that modifies PL properties. Possible explanations will be elucidated in the discussion.

2.4.2.3 Photo-oxidation of single CdSe/ZnS QDs investigated by time-resolved single-molecule spectroscopy

In the following, two representative examples of QDs will be discussed showing distinct spectral blue shifts during the observation time. This effect is observed for the majority of the investigated QDs and is nearly absent under vacuum conditions.

Analysing several tens of QDs, it turns out that upon increasing observation time the PL intensity decreases, the lifetime shortens and the spectrum continuously shifts to shorter wavelengths (denoted as blue shift) and is finally "bleached". Such blue shifts of QD under ambient conditions have been assigned to photo-oxidation by several groups before (Ito et al., 2008; Katari et al., 1994; van Sark et al., 2001). The process takes place on different time scales, in most cases starting at the very beginning of the measurement as depicted in Fig. 2.17. The general spectral shift towards shorter wavelengths is superimposed by spectral diffusion in the order of a few nanometres to the red during the first few seconds. This has been attributed to QD ionization and redistributed local electric fields (Neuhauser et al., 2000). We will refer to this point in more detail in the discussion. The average intensity decrease upon blueing differs from particle to particle but tends to approach zero after blue shifts of about 30–50 nm on the time scale of minutes.

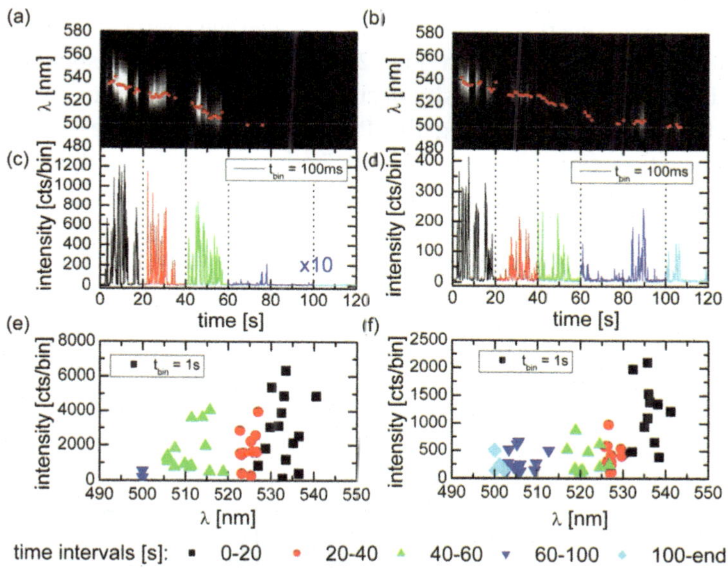

Figure 2.17 Comparison of two typical photo-blueing QDs with (a, b) temporal evolution of the spectra and related spectral peak position fitted by Gaussian lines (red full circles). (c, d) Intensity time traces recorded with 100 ms binning time. (e, f) Correlation of intensity and peak position (t_{bin} = 1 s).

The PL decrease upon increasing observation time can be quantified by the changed distribution of intensities as demonstrated in Fig. 2.18a, where each histogram is normalized to the number of events. Averaging over the colour-marked intervals exemplarily shown in Fig. 2.17c and d, the corresponding PL decay shows an increasing amplitude of the short lifetime component or in terms of a stretched-exponential fit a shortened PL lifetime accompanied by a reduced β value. As already mentioned in the previous sections, both approaches are well suited to quantify PL quenching.

The PL intensity distribution of 100 QDs is shown in Fig. 2.18a separated by the observation time in the colour code of Fig. 2.17. The respective average spectral shift is illustrated in Fig. 2.20a. Even though the average spectral blue shift during the first 60 s is about 5 nm no significant reduction in the PL intensity distribution is observed. A distinctive shift towards lower PL intensities occurs after about 2 min observation time, where the spectral blue shift is on average between 15 and 20 nm. The examples in Fig. 2.17 illustrate that a significant reduction of the PL intensity occurs not until a minimum spectral shift of 10–15 nm. Fitting the PL decay related to the different observation time intervals of all QDs by a stretched-exponential function, the β–τ relation is shown in Fig. 2.18b. Even in the beginning of the measurement during 0–20 s the brightness of individual QDs and the respective τ/β values are broadly distributed. It is obvious that during 0–20 s and 40–60 s the distributions resemble each other, whereas for long observation times of 100–300 s the lifetimes τ appear shortened and β is also reduced indicated by the arrow in Fig. 2.18b.

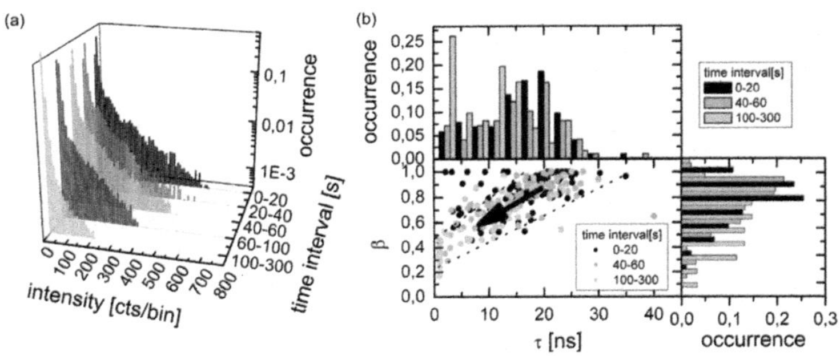

Figure 2.18 (a) Intensity histograms obtained from 100 isolated QDs during different time regimes of observation time, exemplarily shown for those two individual QDs in Fig. 2.17. (b) The respective β-τ relation determined from PL decay at different observation time intervals. The respective probability distribution of τ and β are also shown to emphasize the average shortening of the lifetime accompanied by decrease of β upon observation time under ambient conditions.

Again we like to emphasize the similarity with those observations shown in Fig. 2.12d which are related with ligand depletion and those in Fig. 2.15e related with blinking of single

QDs. Here the PL intensity decrease accompanied by shortened τ and decreased β is ascribed to photo-oxidation identified by spectral blueing. Performing the same experiments under vacuum conditions no significant blue shifts have been observed and the photo-stability was significantly enhanced. Without access to light no significant spectral shifts of the QDs are observed. From that we conclude that this process is in fact photo-induced. Moreover, similar observations have been reported for the same kind of QDs under ambient conditions by van Sark et al. (2001, 2002) and other groups (Ito et al., 2008; Nazzal et al., 2004). This chemical process of photo-oxidation and its consequences for photo-physics will be elucidated in the discussion.

2.4.2.4 Non-FRET PL quenching in single QD-dye assemblies

Another way to selectively modify the QD surface or surface states is the attachment of functionalized dye molecules like PP via self-assembly. In a recent paper we demonstrated self-assembly of pyridyl-substituted PBI molecules and CdSe/ZnS QDs. Even though the spectral overlap and the distance of QD donor and PBI acceptor allow for fluorescence resonance energy transfer (FRET) according to Förster theory, in such assemblies most of the QD quenching is caused by a NON-FRET process (Kowerko et al., 2010). Single-particle spectroscopy of single isolated and PP-assembled QDs facilitate the direct comparison of the lifetime–intensity distribution of such single emitters with and without surface attached dye molecules.

Observations on QD-dye assemblies may be interpreted analogue to the findings for photo-oxidized QDs. Essentially different is the initial presence of PL quenching instead of the gradual process of photo-oxidation on the time scale of seconds to minutes described in the section before. The quantity of quenching varies from particle to particle. In some cases the intensity distribution and the PLID resemble considerably that of free QDs after photo-degradation process like elucidated in the section before.

This effect is demonstrated in the intensity distribution of 90 isolated and 20 PP-assembled QDs obtained from the first 20 seconds of observation time (see Fig. 2.19a). The τ-β relation of assembled QDs follows that of isolated QD (see Fig. 2.19b) but the density of data points is significantly shifted towards shorter lifetimes and smaller β values. The PL quenching resulting from assembly formation obviously does not affect the relative contribution of the strong and weak quenching mechanism behind the PL fluctuations. The τ-β relation of assembled QDs as depicted in Fig. 2.19 is in general indistinguishable from a distribution of photo-oxidized QDs with the basic difference of the occurring blue-shift.

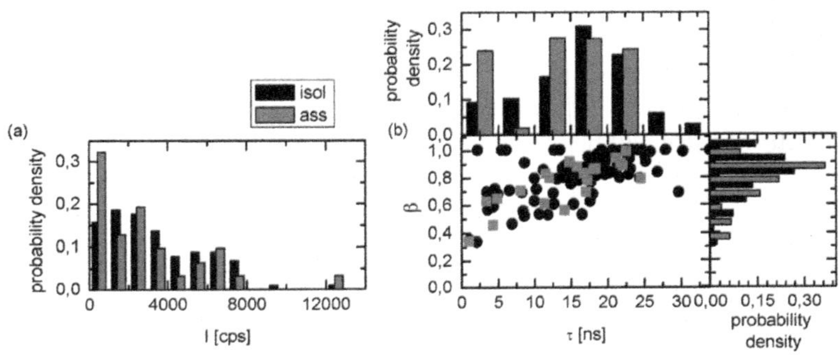

Figure 2.19 (a) Intensity histograms of 90 isolated and 20 PP-assembled QDs. (b) β-τ relation determined from PL decay of the first 20 s of observation time. The respective probability distribution of τ and β are shown to emphasize the average shortening of the lifetime accompanied by a decrease of β of PP-assembled QDs.

In (Kowerko et al., 2010) we described in detail a method allowing for quantification of the fraction of unassembled (isolated) and PP-assembled QDs. In consequence, about 1/3 of our (spectrally) detected assemblies (see Fig. 2.13c,f) are only statistic co-localizations and not real assemblies. Due to this fact, we observe several assembled QD where PL intensity, lifetime and β resemble that of free QDs. On the other hand, there are assemblies where the dye bleached already before the measurement and thus appears in the statistics as isolated QD. This is one reason why the τ–β relation of isolated QDs is distributed relatively broad down to $\tau \approx 5$ ns and $\beta \approx 0.6$. Another reason for the broad distribution of τ and β of isolated QD is the heterogeneity of ligand coverage of individual QDs.

In Fig. 2.20 the temporal evolution of the PL peak position and width averaged for about 100 individual QD spectra from free and PP-assembled QDs is exemplarily shown.

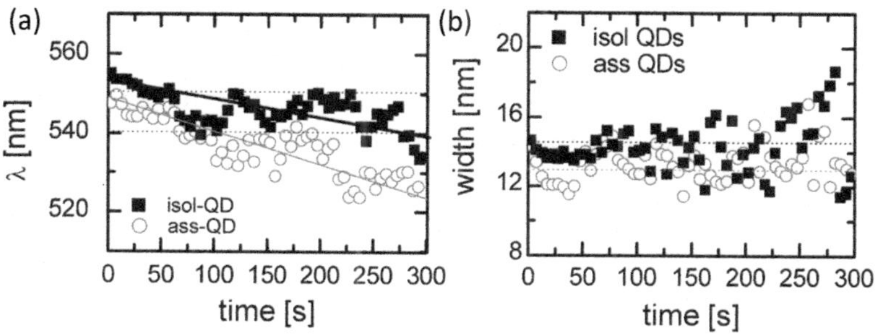

Figure 2.20 Averaged temporal evolution of the QD spectrum separated into Gaussian center (a) and full width at half maximum (FWHM) (b) from 80 isolated and 20 PP assembled QDs. Binning time is 5 s to reduce the influence of spectral diffusion observed on shorter timescales.

In both cases blue shifts during the observation times are significant. Assembled QD are already at shorter emission wavelength at the beginning of the experiment. The time-dependent blue shift is photo-induced, since isolated QD were measured one after another and no systematic blue shift of the initial wavelength was observed for QD recorded at later times after sample preparation. After the strong blueing in the first 1 or 2 min of observation time the process slows down. Dye-assembled QD experience a stronger photo-degradation which might be due to the fact that QDs with a reduced ligand surface coverage enhance access for functional dye molecules like PP or other adsorbates like oxygen at the same time. This would also allow for an increased diffusion of oxygen to the CdSe core of PP-assembled QDs and subsequently facilitates oxidation of the core and causes the observed stronger blue shift. On the other hand, the width of the spectra of assembled QDs is reduced on average which is not necessarily the result of the oxidation. We will refer to that in the discussion section in the context of QCSE.

2.4.3 Comparison of Surface-Related Photophysical Phenomena of Ensemble and Single-Particle Experiments

Finally we like to emphasize that different chemical modifications of the QD surface, like ligand depletion or exchange, photo-oxidation and assembly formation with perylene bisimide dyes all result in PL quenching accompanied by an increased non-exponential character of the photon arrival time histogram. One possible approach to quantify this effect is applying stretched exponential functions to that histogram as has already been shown by other groups (Fisher et al., 2004; Schlegel et al., 2002). Here, the PL lifetime τ is shortened and β decreases below values of 1 in case of PL intensity quenching processes.

As shown in Figs. 2.21a,b,d,e, g the slope of the τ-β relation is comparable for all PL quenching effects induced by processes of different physical or chemical origins. It is evident that all three possible types of quenching are related to the QD surface. As shown by Frenzel et al., dangling bonds at unsaturated surface atoms of CdSe type nanocrystals introduce intra-band states in comparison to those QDs with a completely H-saturated surface (Frenzel et al., 2007). A common observation with respect to PL quenching is the fact that the τ-β relation is very similar in all three scenarios. Small differences within the scatter plots are typical for the QD to QD heterogeneity. One might argue that HDA depletion in the ensemble experiment (overlaid as red circles in Fig. 2.21g) is the only one that exhibits a slightly different shape compared to those of surface immobilized QDs. However, the ligand coverage in solution as well as the dielectric environment are surely different from the situation following spin-coating QDs on a quartz or SiO_2 substrate. For instance, the effect of the local dielectric environment on single QD blinking and PL lifetime statistics has been demonstrated recently (Issac et al., 2005; Schmidt et al., 2012).

The increased non-exponential character of photon arrival time histograms upon augmented PL quenching resembles that of single blinking QD registering only photons of lower intensity intervals in time-binned PL intensity time traces as illustrated in Fig. 2.14b. For details note the similarly increasing slope of Fig. 2.21b with that in Figs. 2.21a,d,e. In Section 2.3.2.2 we presented a formalism describing PLIDs of a three state system, where state <3> is a strongly quenched state of $QY_3 \approx 0$, schematized in Fig. 2.4a. Verberk et al. ascribe "on"-states where I/I_{max} is close to 1 to QDs in a neutral state, "off"-states to a hole in the QD core. Intermediate intensities are related to electron trapping with the hole localized close to the surface (Verberk et al., 2002). Possibly PL quenching mechanism—independent of the type of physico-chemical modification—shift or create traps that act either as electron or hole traps according to the model of Verberk et al.

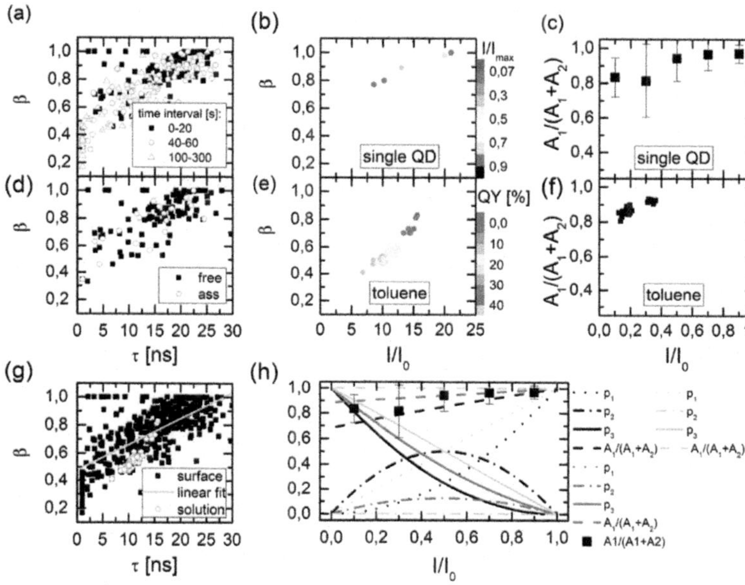

Figure 2.21 Compilation of β-τ relations of (a) photo-bleaching QDs, (b) a single QD at different PL intensity intervals, (d) assembly formation with different PBI dyes; (e) HDA ligand depletion, adapted from Figs. 2.18, 2.15, 2.19 and 2.12, respectively. (c, f) Show the amplitude ratio obtained from fitting the photon arrival time histograms of (b) and (e) with a bi-exponential distribution. (g) Is an overlay of (a), (d) and (e) highlighting data points from the ligand depletion experiment in (e) with red open circles. (h) The black squares show the experimental data from (c), lines show potential population scenarios for the bright <1>, intermediate <2> and dark state <3> using Eqs. (2.7) and the formalism described in Section 2.2.3.2.

However, the slope of the τ-β relation contains information about the relative contribution of the underlying PL quenching mechanisms. Figure 2.21g implies that these do not differ dramatically for each of the three surface-related processes leading to PL quenching, namely ligand depletion, dye adsorption and photo-oxidation. Finally, it is questionable, that fast Auger relaxation subsequent to charge trapping is the only physical origin of PL quenching. Electron–phonon coupling for example has also been mentioned as an additional physical process known to influence the PL quantum yield (Martin et al., 2008). It will be up to future investigations to identify in which way chemical modifications affect the physical origins for PL quenching.

From the formalism describing QD blinking in a simplified three-state system derived in Section 2.3.2.2, we can deduce a quantitative estimate how much the dark state <3> contributes at each level of intensity or quantum yield. Following p_3 in Fig. 2.21h, it is possible to explain single QD blinking processes only via such three-state system with a noticeable large population of the dark state. Galland et al. could show that relative population of $p_2(I/I_0)$ and $p_3(I/I_0)$ can be triggered using spectro-electrochemistry and refer to the underlying blinking processes as type A and type B (Galland et al., 2011). The latter would be described in our model by $A_1/(A_1 + A_2) \approx 1$ as a function of intensity represented by the light grey dashed line in Fig. 2.21h, expressing a dominance of the long lifetime component τ_1 with a nearly mono-exponential photon arrival time histogram at different PL intensities. Blinking here is describable by the scheme in Fig. 2.3a (top). For type A, shortening of PL lifetimes is observed if the photon arrival time histogram is fitted mono-exponentially yielding a bended PLID (Galland et al., 2011; Jones et al., 2009; Zhang et al., 2006). In our three-state approach, we would obtain at least a second lifetime component τ_2 and the amplitude ratio $A_1/(A_1 + A_2)$ would be shifted below 1 as a function of the relative intensity or effective quantum yield, represented by the red and black dashed lines. This case is schematized in Fig. 2.4a. In conclusion, we could show that neither of the three QD surface modifications (ligand depletion, dye adsorption, photo-oxidation) changes the PLID to such an extent like demonstrated by Galland et al. upon applying a voltage, e.g. +0.6 V in electrochemical experiments (Galland et al., 2011). While they were able to eliminate the so-called type A blinking, we find that hexadecylamine ligand depletion, PBI dye adsorption and photo-oxidation alone do not distinctly change the ratio of dark and intermediate states.

2.5 Conclusion

We have demonstrated similarities of photoluminescence intensity lifetime distributions (PLIDs) of three typical surface related chemical modification processes of CdSe/ZnS nanocrystals, namely hexadecylamine ligand depletion, photo-oxidation and assembly formation with functionalized PBI dye molecules. We derived a qualitative formalism to quantitatively describe PLIDs of three-state systems, describing each state with an individual quantum yield QY_i, PL lifetime τ_i and relative population p_i (i = 1,2,3). State <3> was defined as dark state. This is based on the simplified assumption that single QD PL intensity fluctuations are caused by interconversion dynamics between a bright state, weakly and a strongly quenched states. Thus, we were able to demonstrate that dark states contribute considerably while single QDs are in a state of low emission intensity.

Making use of this approach, we do not aim at a microscopic understanding of the obviously involved different (trap) states (as we laid out in Chapters 4 and 6 in von Borczyskowski and Zenkevich (2016) and in recent publications Schmidt et al. (2014) and Zenkevich et al. (2015)) but propose a simplified approach to compare different PL quenching mechanism and, if possible, to discriminate them from each other.

From the similarity of the PLIDs in stretched and bi-exponential data evaluation we conclude that surface modifications investigated here do not severely change the relative forward and backward rates between the bright, the dark and the intermediate states. We assumed in this approach that such modifications do not distinctly shift the energetic distribution or decay rates of surface or intra-band gap states acting as trap sites for charges that are known to increase non-radiative recombination rates by Auger relaxation and/or electron-phonon coupling. The observation of the QCSE related spectral shifts implies that PL quenching is not only caused by opening non-radiative recombination channels. Also in this aspect the presented model is a necessary simplification since not in all complex cases very distinctive spectroscopic approaches are in such a detail feasible as those reported recently (Zenkevich et al., 2015).

Additionally, the fact that radiative rates might also be affected makes the analysis of blinking and quenching even more complex and hard to interpret using only standard optical spectroscopy on QD ensembles. In Chapters 4 and 6 of von Borczyskowski and Zenkevich (2016), we discussed alternative approaches to capture fast interconversion processes of systems with an unknown number of involved states by more elaborated approaches like change point and cluster analysis. First, it remains a daunting challenge to identify rates and states using numerical and statistical approaches. Next, multi-parameter experiments are required to assign these states to their physico-chemical nature. To disentangle electron and/or hole trapping at the core/shell interface or the surface, the generation of multi-excitons, and their quantitative contribution to PL quenching by electron-phonon coupling, Auger recombination or charge transfer, QDs need to be investigated varying excitation power and energy, temperature, electric fields and the local dielectric environments. This will remain among the challenging tasks of the community even after more than two decades of research in this field.

References

Blaudeck, T., Zenkevich, E. I., Cichos, F., and von Borczyskowski, C. (2008). Probing wave functions at semiconductor quantum-dot surfaces by non-FRET photoluminescence quenching. *The Journal of Physical Chemistry C*, 112, pp. 20251–20257.

Blaudeck, T., Zenkevich, E. I., Abdel-Mottaleb, M., Szwaykowska, K., Kowerko, D., Cichos, F., and von Borczyskowski, C. (2012). Formation Principles and Ligand Dynamics of Nanoassemblies of CdSe Quantum Dots and Functionalised Dye Molecules. *ChemPhysChem*, 13, pp. 959–972.

Bullen, C., Mulvaney, P., Uni, V., Park, V., Caesar, S., and Allee, L. (2006). The effects of chemisorption on the luminescence of CdSe quantum dots. *Langmuir*, 22, pp. 3007–3013.

Burda, C., Chen, X., Narayanan, R., and El-Sayed, M. A. (2005). Chemistry and properties of nanocrystals of different shapes. *Chemical Reviews*, 105, pp. 1025–1102.

Carrillo-Carrión, C., Cárdenas, S., Simonet, B. M., and Valcárcel, M. (2009). Quantum dots luminescence enhancement due to illumination with

UV/Vis light. *Chemical Communications (Cambridge, England)*, 35, pp. 5214–5226.

Clapp, A. R., Medintz, I. L., Fisher, B. R., Anderson, G. P., and Mattoussi, H. (2005). Can luminescent quantum dots be efficient energy acceptors with organic dye donors? *Journal of the American Chemical Society*, 127, pp. 1242–1250.

Clapp, A. R., Medintz, I. L., Mauro, J. M., Fisher, B. R., Bawendi, M. G., and Mattoussi, H. (2004). Fluorescence resonance energy transfer between quantum dot donors and dye-labeled protein acceptors. *Journal of the American Chemical Society*, 126, pp. 301–310.

Efros, A. L., and Rosen, M. (1997). Random telegraph signal in the photoluminescence intensity of a single quantum dot. *Phys. Rev. Lett.*, 78, p. 1110.

Empedocles, S. A. (1997). Quantum-confined Stark effect in single CdSe nanocrystallite quantum dots. *Science*, 278, pp. 2114–2117.

Fisher, B. R., Stott, N. E., and Bawendi, M. G. (2004). Emission intensity dependence and single-exponential behavior in single colloidal quantum dot fluorescence lifetimes. 11, pp. 143–148.

Frenzel, J., Joswig, J.-O., and Seifert, G. (2007). Optical excitations in cadmium sulfide nanoparticles. *Journal of Physical Chemistry C*, 111, pp. 10761–10770.

Galland, C., Ghosh, Y., Steinbrück, A., Sykora, M., Hollingsworth, J. A., Klimov, V. I., and Htoon, H. (2011). Two types of luminescence blinking revealed by spectroelectrochemistry of single quantum dots. *Nature*, 479, pp. 203–207.

Gerlach, F., Täuber, D., von Borczyskowski, C., and Taeuber, D. (2013). Correlated blinking via time dependent energy transfer in single CdSe quantum dot-dye nanoassemblies. *Chemical Physics Letters*, 572, pp. 90–95.

Grabolle, M., Ziegler, J., Merkulov, A., Nann, T., and Resch-Genger, U. (2008). Stability and fluorescence quantum yield of CdSe-ZnS quantum dots--influence of the thickness of the ZnS shell. *Annals of the New York Academy of Sciences*, 1130, pp. 235–241.

Heinlein, T., Biebricher, A., Schlüter, P., Roth, C. M., Herten, D.-P., Wolfrum, J., Sauer, M. (2005). High-resolution colocalization of single molecules within the resolution gap of far-field microscopy. *Chemphyschem: A European Journal of Chemical Physics and Physical Chemistry*, 6, pp. 949–955.

Holman, M. W., Liu, R., and Adams, D. M. (2003). Single-molecule spectroscopy of interfacial electron transfer. *Journal of the American Chemical Society*, 125, pp. 12649–12654.

Inerbaev, T. M., Masunov, A. E., Khondaker, S. I., Dobrinescu, A., Plamadă, A.-V., and Kawazoe, Y. (2009). Quantum chemistry of quantum dots: Effects of ligands and oxidation. *The Journal of Chemical Physics*, 131, p. 044106.

Issac, A., Jin, S., and Lian, T. (2008). Intermittent electron transfer activity from single CdSe/ZnS quantum dots. *Journal of the American Chemical Society*, 130, pp. 11280–11281.

Issac, A., von Borczyskowski, C., and Cichos, F. (2005). Correlation between photoluminescence intermittency of CdSe quantum dots and self-trapped states in dielectric media. *Physical Review B*, 71, p. 161302.

Ito, Y., Matsuda, K., and Kanemitsu, Y. (2008). Photoluminescence intermittency in single CdSe nanoparticles: Environment dependence. *Journal of Luminescence*, 128, pp. 868–870.

Ji, X., Copenhaver, D., Sichmeller, C., and Peng, X. (2008). Ligand bonding and dynamics on colloidal nanocrystals at room temperature: The case of alkylamines on CdSe. *Journal of the American Chemical Society* 10, pp. 7012–7021.

Jones, M., Lo, S. S., and Scholes, G. D. (2009). Quantitative modeling of the role of surface traps in CdSe/CdS/ZnS nanocrystal photoluminescence decay dynamics. *Proceedings of the National Academy of Sciences of the United States of America*, 106, pp. 3011–3016.

Kambhampati, P. (2011). Hot excitation relaxation dynamics in semiconductor quantum dots: Radiationless transistions on the nanoscale. *The Journal of Physical Chemistry C*, 115, pp. 22089–22109.

Katari, J. E. B., Colvin, V. L., and Alivisatos, A. P. (1994). X-ray Photoelectron spectroscopy of CdSe nanocrystals with applications to studies of the nanocrystal surface. *The Journal of Chemical Physics*, 98, pp. 4109–4117.

Kilina, S., Ivanov, S., and Tretiak, S. (2009). Effect of surface ligands on optical and electronic spectra of semiconductor nanoclusters. *Journal of the American Chemical Society*, 131, pp. 7717–7726.

Klimov, V. I., Ivanov, S. A., Nanda, J., Achermann, M., Bezel, I., McGuire, J. A., and Piryatinski, A. (2007). Single-exciton optical gain in semiconductor nanocrystals. *Nature*, 447, pp. 441–446.

Köllner, M. (1991). Koellner_Int Phys Conf Ser1991_Best Est Exp Dec Param Des Of SPC exp.pdf.

Köllner, M., and Wolfrum, J. (1992). How many photons are necessary for fluorescence-lifetime measurements? *Chemical Physics Letters*, 200, pp. 199–204.

Koole, R., Schapotschnikow, P., de Mello Donega, C., Vlugt, T. J. H., Meijerink, A., Surface, D., and Donega, C. D. M. (2008). Time-dependent photoluminescence spectroscopy as a tool to measure the ligand exchange kinetics on a quantum dot surface. *ACS Nano*, 2, pp. 1703–1714.

Kowerko, D., Schuster, J., Amecke, N., Abdel-Mottaleb, M., Dobrawa, R., Würthner, F., and von Borczyskowski, C. (2010). FRET and ligand related NON-FRET processes in single quantum dot-perylene bisimide assemblies. *Physical Chemistry Chemical Physics: PCCP*, 12, pp. 4112–4123.

Krengel, U. (2005). Einführung in die Wahrscheinlichkeitstheorie und Statistik: [für Studium, Berufspraxis und Lehramt]. Vieweg-Studium Aufbaukurs Mathematik (8., erw. A). Wiesbaden: Vieweg. Retrieved from http://d-nb.info/976268701/04.

Kuçur, E., Bücking, W., Arenz, S., Giernoth, R., and Nann, T. (2006). Heterogeneous charge transfer of colloidal nanocrystals in ionic liquids. *Chemphyschem: A European Journal of Chemical Physics and Physical Chemistry*, 7, pp. 77–81.

Lakowicz, J. R. (2006). *Principles of Fluorescence Spectroscopy*. Springer.

Martin, J., Cichos, F., Huisken, F., and von Borczyskowski, C. (2008). Electron-phonon coupling and localization of excitons in single silicon nanocrystals. *Nano Letters*, 8, pp. 656–660.

McKinney, S. A., Joo, C., and Ha, T. (2006). Analysis of single-molecule FRET trajectories using hidden Markov modeling. *Biophysical Journal*, 91, pp. 1941–1951.

Messina, T. C., Kim, H., Giurleo, J. T., and Talaga, D. S. (2006). Hidden Markov model analysis of multichromophore photobleaching. *The Journal of Physical Chemistry. B*, 110, pp. 16366–16376.

Müller, J., Lupton, J., Rogach, A., Feldmann, J., Talapin, D., and Weller, H. (2004). Monitoring surface charge movement in single elongated semiconductor nanocrystals. *Physical Review Letters*, 93, p. 167402.

Nazzal, A. Y., Wang, X., Qu, L., Yu, W., Wang, Y., Peng, X., and Xiao, M. (2004). Environmental effects on photoluminescence of highly luminescent CdSe and CdSe/ZnS core/shell nanocrystals in polymer thin films. *The Journal of Physical Chemistry B*, 108, pp. 5507–5515.

Neuhauser, R. G., Shimizu, K. T., Woo, W. K., Empedocles, S. A., and Bawendi, M. G. (2000). Correlation between fluorescence intermittency and spectral diffusion. *Physical Review Letters*, 85, pp. 3301–3304.

Peterson, J. J., and Nesbitt, D. J. (2009). Modified power law behavior in quantum dot blinking: a novel role for biexcitons and auger ionization. *Nano Letters*, 9, pp. 338–345.

Petrov, E. P. P., Cichos, F., Zenkevich, E., Starukhin, D., and von Borczyskowski, C. (2005). Time resolved photoluminescence anisotropy of CdSe/ZnS nanoparticles in toluene at 300 K. *Chemical Physics Letters*, 402, pp. 233–238.

Prezhdo, O. V. (2008). Multiple excitons and the electron–phonon bottleneck in semiconductor quantum dots: An ab initio perspective. *Chemical Physics Letters*, 460, pp. 1–9.

Saba, M., Aresti, M., Quochi, F., Marceddu, M., Loi, M. A., Huang J., and Talapin, D. V., Mura, A., and Bongiovanni, G. (2013). Light-induced charged and trap states in colloidal nanocrystals detected by variable pulse rate photoluminescence spectroscopy. *ACS Nano*, 7, pp. 229–238.

Schlegel, G., Bohnenberger, J., Potapova, I., and Mews, A. (2002). Fluorescence decay time of single semiconductor nanocrystals. *Physical Review Letters*, 88, p. 137401.

Schmidt, R., Krasselt, C., Göhler, C., von Borczyskowski, C. (2014). The fluorescence intermittency for quantum dots is not power-law distributed: A luminescence intensity resolved approach. *ACS Nano*, 8, pp. 3506–3521.

Schmidt, R., Krasselt, C., and von Borczyskowski, C. (2012). Change point analysis of matrix dependent photoluminescence intermittency of single CdSe/ZnS quantum dots with intermediate intensity levels. *Chemical Physics*, 406, pp. 9–14.

Seufert, J., Obert, M., Scheibner, M., Gippius, N. A., Bacher, G., Forchel, A., et al. (2001). Stark effect and polarizability in a single CdSe/ZnSe quantum dot. *Applied Physics Letters*, 79, p. 1033.

Talapin, D. V., Rogach, A. L., Kornowski, A., Haase, M., and Weller, H. (2001). Highly luminescent monodisperse CdSe and CdSe/ZnS nanocrystals synthesized in a hexadecylamine–trioctylphosphine oxide–trioctylphospine mixture. *Nano Letters*, 1, pp. 207–211.

Valeur, B. (2002). *Molecular Fluorescence: Principles and Applications* (Wiley-VCH Verlag GmbH, Weinheim, FRG).

Van Sark, W. G. J. H. M., Frederix, P. L. T. M., Bol, A. A., Gerritsen, H. C., and Meijerink, A. (2002). Blueing, bleaching, and blinking of single CdSe/ZnS quantum dots. *ChemPhysChem*, 3, pp. 871–879.

Van Sark, W. G. J. H. M., Frederix, P. L. T. M., Van den Heuvel, D. J., Gerritsen, H. C., Bol, A. A., van Lingen, J. N. J., et al. (2001). Photooxidation and photobleaching of single CdSe/ZnS quantum dots probed by room-temperature time-resolved spectroscopy. *The Journal of Physical Chemistry B*, 105, pp. 8281–8284.

Verberk, R., van Oijen, A. M., and Orrit, M. (2002). Simple model for the power-law blinking of single semiconductor nanocrystals. *Physical Review B*, 66, p. 233202.

Von Borczyskowski, C. (2016). In: *Tuning Semiconducting and Metallic Nanoparticles* (von Borczyskowski, C., and Zenkevich, E., eds.), Chapter 4, Pan Stanford Publishing.

Vosch, T., Fron, E., Hotta, J., Deres, A., Uji-i, H., Idrissi, A., et al. (2009). Synthesis, ensemble, and single molecule characterization of a diphenyl-acetylene linked perylenediimide trimer. *Journal of Physical Chemistry C*, 113, pp. 11773–11782.

Voznyy, O., Thon, S. M., Ip, A. H., and Sargent, E. H. (2013). Dynamic trap formation and elimination in colloidal quantum dots. *The Journal of Physical Chemistry Letters*, 4, pp. 987–992.

Wang, X., Qu, L., Zhang, J., Peng, X., and Xiao, M. (2003). Surface-related emission in highly luminescent CdSe quantum dots. *Nano Letters*, 3, pp. 1103–1106.

Wang, X., Ren, X., Kahen, K., Hahn, M. A., Rajeswaran, M., Maccagnano-Zacher, S., et al. (2009). Non-blinking semiconductor nanocrystals. *Nature*, 459, pp. 686–689.

Watkins, L. P., and Yang, H. (2005). Detection of intensity change points in time-resolved single-molecule measurements. *The Journal of Physical Chemistry. B*, 109, pp. 617–628.

Würthner, F. (2004). Perylene bisimide dyes as versatile building blocks for functional supramolecular architectures. *Chem. Commun.*, 14, pp. 1564–1579.

Würthner, F., Sautter, A., Schmid, D., and Weber, P. J. A. (2001). Fluorescent and electroactive cyclic assemblies from perylene tetracarboxylic acid bisimide ligands and metal phosphane triflates. *Chemistry*, 7, pp. 894–902.

Yu, W. W., Qu, L., Guo, W., and Peng, X. (2003). Experimental determination of the extinction coefficient of CdTe, CdSe, and CdS nanocrystals, 125, pp. 2854–2860.

Zenkevich, E., Stupak, A., Göhler, C., Krasselt, C., and von Borczyskowski, C. (2015). Tuning electronic states of a CdSe/ZnS quantum dot by only one functional dye molecule. *ACS Nano*, 9, pp. 2886–2903.

Zhang, K., Chang, H., Fu, A., Alivisatos, A P., and Yang, H. (2006). Continuous distribution of emission states from single CdSe/ZnS quantum dots. *Nano Letters*, 6, pp. 843–847.

Chapter 3

Fluorescence Quenching of Semiconductor Quantum Dots by Multiple Dye Molecules

Thomas Blaudeck

Department of Electrical Engineering and Information Technology, Center for Microtechnologies, Technische Universität Chemnitz, Reichenhainer Str. 70, D-09107 Chemnitz, Germany

thomas.blaudeck@zfm.tu-chemnitz.de

3.1 Introduction

During the past decades, colloidal nanocrystals (NCs) that form quantum dots (QDs) from various II–VI semiconductor materials (e.g., CdSe, CdTe, CdS, ZnS) have been introduced to a variety of photophysical applications in the modern technology and in the life sciences. Especially the use of optically active self-assembled heteroaggregates from QDs and dye molecules has opened up routes towards a variety of applications in biomedicine, nanosensorics, and photovoltaics.

In previous publications, we have experimentally discussed the case of multiple acceptors (see, e.g., Blaudeck et al., 2008;

Self-Assembled Organic-Inorganic Nanostructures: Optics and Dynamics
Edited by Eduard Zenkevich and Christian von Borczyskowski

ISBN 978-981-4745-43-7 (Hardcover), 978-981-4745-44-4 (eBook)
www.panstanford.com

Zenkevich et al., 2009; Blaudeck et al., 2012; Zenkevich et al., 2012; von Borczyskowski and Zenkevich, 2014), but not rigorously dealt with a formalism for the quantitative evaluation. In order to determine the basic interaction properties within the QD-dye heteroaggregates, photoluminescence (PL) titration experiments looking at the inhibition (or: quenching) of the QD PL have been a method-of-choice, within the well-established Stern–Volmer formalism as put forward by Stern and Volmer (1919) and taken up by Förster (1948) used as a common analysis framework for evaluation (see, e.g., Valeur, 2002; Lakowicz, 2006). In this chapter, we concentrate on the basic extension of this formalism taking into account the role of multiple, yet a finite number of quenchers in the heteroaggregates. For the rigorous introduction of the basic principles of the Stern–Volmer approach, we refer the reader to Chapter 4.

3.2 Theory of Acceptor Redistribution in Heteroaggregates

3.2.1 Fluorescence Quenching

The phenomenon of PL quenching or inhibition was already reported by G. G. Stokes in the 1850's for quinine dyes and traces of sodium chloride in aqueous sulfuric acid. Theodor Förster (1948) defined it as a non-radiative (reversible) reduction of the PL intensity of a fluorescing chromophore upon its participation in the formation of a weakly-bound aggregate without (i) a change of the chemical constitution and (ii) optical filtering of the excitation light (re-absorption effects). For molecular aggregates, in the early 1980's, works by Tachiya (1982, 1983) as well as Infelta and Grätzel (1983) extended the binary case to lumophores (also called as donor, D) organized in colloidal micelles having n binding sites for quencher molecules (also called acceptors, A). Here, the number of binding sites is limited but no clear saturation of the interaction by a clear upper limit of quenchers per micelle is expected. In general, when multiple quenching processes are applicable regardless whether they are dynamic or static in nature; each one j can be assigned separately to its respective Stern–Volmer constant $K_{\text{SV},j}$ and will depend on the particular concentration of donors and acceptors. Extending the case of micelles, to ensembles of semiconductor QDs having a narrow size distribution, the quenching lumophores are expected to find a certain, yet finite, number of binding sites on the QD surface.

3.2.2 Evaluation of Photoluminescence Quenching Experiments and Modified Stern–Volmer Formalism

Generally, when PL quenching occurs in an ensemble of such heteroaggregates with a multiple stoichiometry of acceptors DA_n and assuming a dynamic equilibrium, the Stern–Volmer constants $K_{sv,j}$ known from the binary case will depend on the molar quencher-to-fluorophore ratio x. Depending on the respective intention of the discussion, this important parameter is synonymously referred to as (i) acceptor-to-donor-ratio, or (ii) molecule-to-NC ratio, or (iii) simply molar ratio. The results from the Stern–Volmer analysis of steady-state optical spectroscopy, e.g., in our earlier publications, were derived from non-monotonous slopes found in the dependence of the interaction parameter K_{sv} on the molar ratio x. Hence, in this modified Stern–Volmer formalism, in contrast to the established approaches, the quencher concentration [A] (square brackets indicate the concentration) is expressed in terms of the molar ratio x. Consequently, in our modified case, the unit of the Stern–Volmer factor K_{sv} is defined as [1] (see, e.g., Blaudeck et al., 2008) instead of [mol^{-1}] (in this case square brackets confine the respective unit (see, e.g., Valeur, 2002). We have to point out that this fact is in complete accordance with the known formalism, as [A] is referred to a given fluorophore concentration [D] which is usually constant during the experiment (*cf.* (Förster, 1951), pages 194 *ff.*). Only this modification allows an extension of the formalism for aggregates with stoichiometry DA_n with $n \in N \geq 2$.

3.2.3 Derivation of the Photoluminescence Quantum Yield of a Donor in Presence of Multiple Acceptors

In our previous experimental work, we probed the luminescence quenching of CdSe or CdSe/ZnS QDs acting as donor (D) by pyridyl-substituted porphyrin molecules as the quencher (A). The transfer rate k_{DA} of the relaxation (not necessarily the radiative one) competes with the intrinsic radiative and non-radiative relaxation rates of the donor, $k_{\mathrm{D}}^{\mathrm{r}}$ and $k_{\mathrm{D}}^{\mathrm{nr}}$, respectively. The transfer rate k_{DA} can be assumed to be a function of the number of acceptor molecules n in the vicinity of the donor denoted as [A]. In order to keep things simple and general, for an ensemble of donors, all quenching processes are assumed to depend identically on the number of acceptors per donor. According to literature, and assuming an instantaneous excitation of the fluorophores at rate k_{exc}, the excited state decay kinetics can be described by

$$\left(\frac{dN_2}{dt}\right)_{\mathrm{D}} = k_{\mathrm{exc}} \cdot N_1 - (k_{\mathrm{D}}^{\mathrm{r}} + k_{\mathrm{D}}^{\mathrm{nr}}) \cdot (N_2)_{\mathrm{D}} = 0 \tag{3.1}$$

for the case of absent quenchers and by

$$\left(\frac{dN_2}{dt}\right)_{\mathrm{DA}} = k_{\mathrm{exc}} \cdot N_1 - (k_{\mathrm{D}}^{\mathrm{r}} + k_{\mathrm{D}}^{\mathrm{nr}} + k_{\mathrm{DA}}(n)) \cdot (N_2)_{\mathrm{DA}} = 0 \tag{3.2}$$

for the case of n present quenchers, respectively. In order to determine the quenching efficiency, the combination of Eqs. (3.1) and (3.2) yields a relation between the PL intensity, the relaxation rates, and the PL decay times of the donor-acceptor system according to

$$\frac{I_{\mathrm{D}}}{I_{\mathrm{DA}}(x)} = \frac{(N_2)_{\mathrm{D}}}{(N_2)_{\mathrm{DA}}} = \frac{(k_{\mathrm{D}}^{\mathrm{r}} + k_{\mathrm{D}}^{\mathrm{nr}} + k_{\mathrm{DA}}(n))}{k_{\mathrm{D}}^{\mathrm{r}} + k_{\mathrm{D}}^{\mathrm{nr}}} = \frac{\tau_{\mathrm{D}}}{\tau_{\mathrm{DA}}(x)} \tag{3.3}$$

Here, in case of a diffusion-limited quenching, τ_{D} and τ_{DA} are the intensity-averaged PL decay times of the donor in absence and presence of quenching molecules, respectively. The message of Eq. (3.3) is that the fluorescence decay time of an ensemble of fluorophores is altered successively with the average number of quenchers per donor (QD), x = [A]/[D], related to the concentration ratio between quenchers and QD in the ensemble. The Stern–Volmer equation describing the fluorescence intensity loss of an ensemble of fluorophores upon increasing concentration of quenchers can be obtained in the case of static quenching (i.e., $k_{\mathrm{DA}}(n) = k_{\mathrm{DA}}(n = 1)$) and an ensemble of fluorophore-quencher assemblies DA_n with a narrow stoichiometric distribution (i.e., $n \approx x$) by

$$\frac{I_D}{I_{DA}(x)} = \frac{\tau_D}{\tau_{DA}(x)} = 1 + K_{SV} \cdot x. \tag{3.4}$$

Equation (3.4) can be related to an empirical analysis of PL quenching experiments assuming a typical (i.e., inversely proportional) dependence of the integrated PL quenching change on the quencher concentration at any given QD concentration.

From their nature, dynamic and static mechanisms of PL quenching can be distinguished (see Chapter 4). In the dynamic case, the relaxation of the donor D from the excited state to the ground state is influenced by its diffusive encounters (i.e., quencher molecules) A. In contrast, static quenching is defined as the transformation of a fluorophore into a non-fluorescing entity by an association not affecting the chemical constitution. A static PL quenching may be described by the formation of a dark complex between the compounds D and A. In this case, no change in the fluorescence decay time will be observed (i.e., $\tau_D = \tau_{DA}$), as only the molecular ground state of the hence non-fluorescing chromophore D is affected.

In Eq. (3.4), for the case of dynamic quenching, the Stern–Volmer factor K_{SV} represents the interaction constant. For the case of purely static quenching, this quantity is referred to as the stability constant. In this case, K_{SV} is a measure for the ratio between the concentrations of the completely dark complexes and the unbound constituents, respectively. Here, the relation $K_{SV} = [DA]/([D] \cdot [A])$ from literature (see, e.g., Valeur, 2002; Lakowicz, 2006) applies. In absence of quencher molecules, the total PL intensity obtained from an ensemble of donor QDs in solution is given by

$$I_D = \frac{k_D^r}{k_D^r + \sum_j k_{D,j}^{nr}} = \frac{k_D^r}{k_D}, \tag{3.5}$$

where k_D^r is the intrinsic radiative rate, and $k_{D,j}^{nr}$ any of a competing set (i.e. of the j-th) intrinsic non-radiative decay rates. Generally, I_D is proportional to the donor PL quantum yield q_D. For an analytical description of quenching upon aggregate formation, we first assume that the donor QDs possess discrete but large number $n \in N$ of binding sites. Thereafter we will discuss possible limitations.

Assuming a non-static interaction process between donor and acceptor (such as Fluorescent Resonant Energy Transfer, FRET), the formation of aggregates DA_n by adsorption of acceptor molecules to the surface of a single donor QD leads to an increasing quenching of donor fluorescence upon increasing the number of acceptor molecules. In a microscopic picture, each single step of adsorption opens one new competitive decay channel with the identical effective rate k_{DA}, which is responsible for the donor quenching. Thus the PL quantum yield of a single donor attached by n acceptors is determined by

$$\Phi_D(n) = \frac{I_{DA}(n)}{I_D} \approx q(n) = \frac{k_D}{k_D + n \cdot k_{DA}}. \quad (3.6)$$

Consequently, for $n \to x$, the average PL intensity of the donor QDs in presence of the acceptor molecules becomes

$$\Phi_D(x) = \frac{I_{DA}(x)}{I_D} \approx q(x) = \frac{k_D}{k_D + x \cdot k_{DA}}. \quad (3.7)$$

3.2.4 Case Study: A Donor with Quasi-Infinite Number of Binding Sites

In solution, the acceptor molecules undergo a diffusion process and encounter the free binding sites on the donor surface. For a semiconductor quantum dot as a donor D, the number of available binding sites is assumed to be countable, yet in principle infinite. Under this assumption, in a dynamic equilibrium, and at a given maximum number of binding sites n^* on the donor quantum dot, the probability of the existence of an aggregate DA_n with particular n molecules A follows a Binomial distribution

$$P_n(n^*) = \binom{n^*}{n} \cdot p^n \cdot (1-p)^{n^*-n}. \quad (3.8)$$

In Eq. (3.8), the introduced quantity p is the probability of a particular donor binding site to become encountered by one acceptor molecule, all of which are assumed statistically equivalent. Assuming, further, that the probability p is equal for all

n acceptors in the aggregates with a stoichiometry DA_n, p can be calculated as

$$p = \frac{[A]}{[D] \cdot n^*} = \frac{x}{n^*}. \tag{3.9}$$

Hence, for $n^* \cdot p \to$ const, the Binomial distribution of Eq. (3.8) merges into the Poissonian distribution (see, e.g., Lakowicz, 2006)

$$P_n(x) = x^n \cdot \frac{e^{-x}}{n!}, \tag{3.10}$$

which is valid if the total number of binding sites available $[D] \cdot n^*$ is comparable to the total number of molecules present [A]. This is true if a large, quasi-infinite number n^* of sites are available per donor.

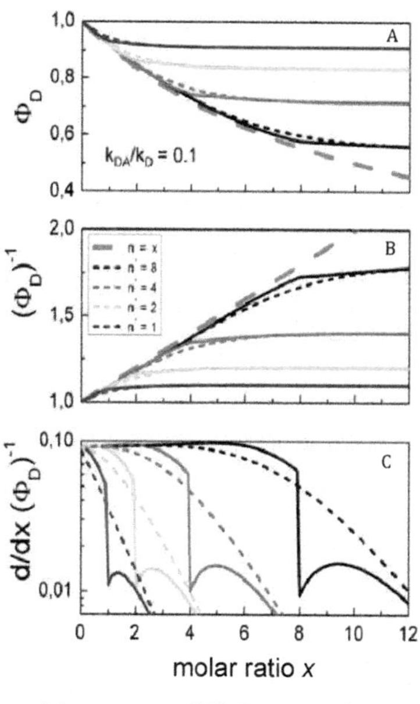

Figure 3.1 (A) Quenching curve, (B) Stern–Volmer plot, and (C) Stern–Volmer function calculated for cases of a limited number of free binding sites (n^* = {1,2,4,8}) available per QD. Legend: (*solid lines*) calculation without redistribution of free acceptor molecules according to Eq. (3.12), (*short-dashed lines*) calculation with a *complete* redistribution according to Eq. (3.13), (*long-dashed lines*). Calculation with $x = n$ not accounting for a Poissonian distribution of aggregates DA_n according to Eq. (3.6).

3.2.5 Case Study: A Donor with a Finite Number of Binding Sites (Redistribution Model)

In any experiment of steady-state optical spectroscopy, the total fluorescence intensity $\Phi_D(x)$ obtained from an ensemble of aggregates DA_n with a static distribution of n at a given molecule-to-NC ratio x is

$$\Phi_D(x) = \sum_{n=0}^{n^*} q_n \cdot P_n(x), \tag{3.11}$$

that is, the weighted sum of the contributions of each individual aggregate. For any non-infinite number of binding sites per QD, in practice, all aggregates with $n > n^*$ are degenerate in the stoichiometry DA_n. Due to this fact, Eq. (3.11) becomes

$$\Phi_D(x) = \sum_{n=0}^{n^*-1} q_n \cdot P_n(x) + q_{n^*} \cdot \left(1 - \sum_{n'=0}^{n^*-1} P_n'(x)\right). \tag{3.12}$$

The free acceptor molecules can then be thought to be subject to redistribution over the free binding sites with respect to their availability, with the probability of filling a site on a particular aggregate stoichiometry DA_n proportional to the probability of its occurrence in the ensemble. The redistribution of acceptors leads to correction term to be added to Eq. (3.12), yielding

$$\Phi_D(x) = \sum_{n=0}^{n^*-1} q_n \cdot P_n(x) + q_{n^*} \cdot \left(1 - \sum_{n^{(1)}=0}^{n^*-1} P_{n^{(1)}}(x)\right) + \sum_{n^{(2)}=0}^{n^*-1} (q_{n^{(2)}+1} - q_{n^{(2)}}) \cdot \mathbf{N} \cdot \frac{P_{n^{(2)}}(x)}{\sum_{n^{(3)}=0}^{n^*-1} P_{n^{(3)}}(x)}. \tag{3.13}$$

Hence,

$$\mathbf{N} = \min\left\{ \sum_{n^{(4)}=n^*}^{\infty} (n^{(4)} - n^*) \cdot P_{n^{(4)}}(x) \middle| \sum_{n^{(4)}=n^*}^{\infty} (n^* - n^{(4)}) \cdot P_{n^{(4)}}(x) \right\} \tag{3.14}$$

is a quantity describing the number of possible steps in a redistribution process, becoming constant either when all acceptors are bound or all donor binding sites are encountered, whatever comes first. Figure 3.3 shows the dependencies of Eqs. (3.7),

(3.12) and (3.13) on x for an overall quenching rate one order of magnitude smaller than the intrinsic decay rate, i.e., $k_{DA} = 0.1 \cdot k_D$. In consequence of Eqs. (3.11) to (3.14), for high molar ratios x, the final level of quenching (cf. Figs. 3.1a,b) is related to the upper limit of binding sites per donor QD n^* via

$$\lim_{x\to\infty} \Phi_D(x) = \frac{1}{1 + n^* \cdot (k_{DA}/k_D)}. \quad (3.15)$$

Figures 3.1a,b contain the calculated PL intensity and the Stern–Volmer plot, respectively. Figure 3.1c shows a representation of the Stern–Volmer function, in which the upper limit of binding sites n^* can be seen as a discontinuity described by a vertical asymptote. It further turns out that the kinetics of the assumed redistribution has a significant influence on the calculated values of $\Phi_D(x)$ at low molecule-to-NC ratios x.

3.3 Application of the Redistribution Model to Experiments

Figure 3.2 contains the experimental Stern–Volmer plots for the quenching of semiconductor QDs by pyridyl-substituted porphyrin molecules (*cf.* (Blaudeck, 2007) and Chapter 4 for experimental details)). Three samples were titrated differently and in three different titration regimes: phase I in the range $x = 0.0\ldots3.2$; phase II in the range $x = 4.0\ldots6.4$ and phase III in the range $x = 7.2\ldots9.6$. For the study reported here, the experimental procedure was altered with respect to the methodology in previous experiments (*cf.* (Zenkevich et al., 2005; Blaudeck et al., 2008; Blaudeck et al., 2012)). Now, three ensembles of CdSe/ZnS QDs (capped by TOPO ligands) were at first titrated by the pyridyl-substituted porphyrin (m-Pyr)$_4$-H_2P, ensemble #1 was continuously titrated by (m-Pyr)$_4$-H_2P, ensemble #2 firstly with (m-Pyr)$_4$-H_2P, then, secondly, further diluted by toluene and thirdly again titrated with (m-Pyr)$_4$-H_2P, and ensemble #3 firstly titrated with (m-Pyr)$_4$-H_2P and then secondly and thirdly by tetraphenyl porphyrin Ph_4-H_2P. The latter molecule does not contain functional pyridyl groups and does hence not give rise to any assembly formation

(Blaudeck, 2007). Dilution effects were carefully taken into account.

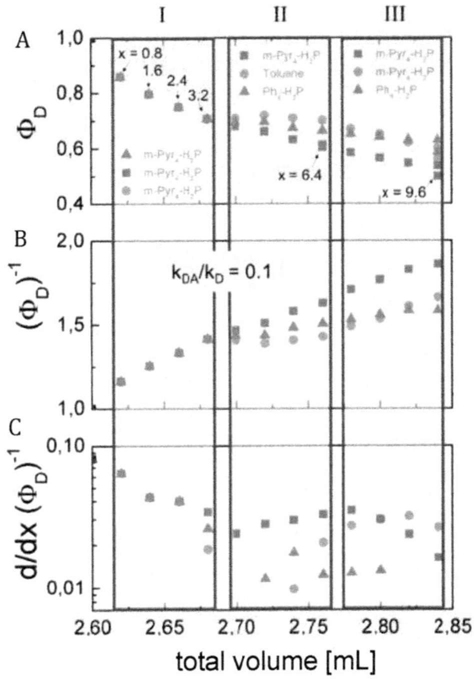

Figure 3.2 (A) Quenching curve, (B) Stern–Volmer plot, and (C) Stern–Volmer function experimentally obtained for titration of CdSe/ZnS colloidal QDs with pyridyl-substituted porphyrin dye molecules (m-Pyr)$_4$-H_2P. Three samples were titrated differently in three titration regimes: Sample #1 (black squares) incurred PL quenching by subsequent addition of (m-Pyr)$_4$-H_2P in all three regimes (I: x = 0.0…3.2; II: x = 4.0…6.4; III: x = 7.2…9.6). Sample #2 incurred quenching by (m-Pyr)$_4$-H_2P in regime I, then dilution by toluene in regime II, and finally further quenching by (m-Pyr)$_4$-H_2P in regime III. Sample #3 incurred quenching by (m-Pyr)$_4$-H_2P in regime I (like sample #2), then dilution by toluene but adding a corresponding amount of chemically inert tetraphenyl-porphyrin Ph_4-H_2P in regime II, and continuing that in regime III. Dilution effects were recalculated from the experimental data.

Sample #1 represents the ideal case, but the increasing importance of interactions with the TOPO ligand shell for higher molar ratio x screens the effect of immediate quenching. Sample #2 (dilution by toluene at x = 3.2) was taken as an example for the perturbation of the dynamic equilibrium between QDs, ligand molecules (TOPO) and porphyrin dye molecules, intended to allow an enforcement of the redistribution. Sample #3 (addition of Ph_4-H_2P which is considered not binding to the QDs) is intended to maintain the dynamic equilibrium constant, which should inhibit the redistribution of dye molecules to form heteroaggregates with other stoichiometry.

The observations can be summarized as follows:

(i) For sample #1, a comparison with the calculations of the redistribution model verifies that the Stern–Volmer factor K_{sv} is dependent of the molar ratio x and hence rather named Stern–Volmer function.

(ii) For sample #2, the slight recovery of the PL intensity in phase II (addition of toluene at x = 3.2) clearly indicates the perturbation of the dynamic equilibrium.

(iii) For sample #3, the constance of the PL intensity in phase II upon addition of non-binding (Ph4-H2P (and only slight drop in phase III obviously due to interactions of the ligand shell, even with the tetrapyrrolic scaffold) can explain the unchanged PL intensity by an inhibition of the redistribution of porphyrin molecules (*cf.* (Blaudeck et al., 2012).

A comparison of Figs. 3.1 and 3.2 prompts the following conclusion: Irrespective of the long-term interactions between quencher molecules and the ligand shell, the assumption can explain the non-monotonous shape of the Stern–Volmer plots while assuming (i) a finite number n^* of binding sites per QD and (ii) a partial redistribution of the quencher molecules in the ensemble of heteroaggregates DA_n. The vertical asymptote of the Stern–Volmer function as predicted by Eq. (3.13) and shown in Fig. 3.1c will never be followed, as this effect is already outnumbered by the distribution of the upper limits of the number of binding sites n^* of the QDs due to size distribution and a heterogeneous surface morphology.

3.4 Conclusion

Our results show that the hypothesis of a finite number of binding sites on semiconductor quantum dots (QDs) indicated by the finite surface area and the competition with the ligand shell (e.g., TOPO; *cf.* Blaudeck et al., 2012, and Chapter 4) is reasonable as it can qualitatively explain the photoluminescence quenching characteristics. When described in terms of the Stern–Volmer formalism, they show an unsteady shape with increasing molar ratio between dye molecules (as quenching acceptor) and quantum dots (as donor). Here, in the respective mathematical description, a transition from the well-known Stern–Volmer constant to a Stern–Volmer function of the molar ratio is necessary.

On the molecular level, the results can be explained in terms of a partial redistribution of the quencher molecules amongst the heteroassemblies present in the ensemble, assuming a Poissonian distribution. We take these results as an encouragement that a careful differential steady-state photoluminescence analysis, at least in a statistic ensemble, can be a tool to trace back to interaction effects in heteroassemblies even on the molecular level.

References

Blaudeck, T. (2007) Self-assembly of functionalized porphyrin molecules on semiconductor nanocrystal surfaces, http://nbn-resolving.de/urn:nbn:de:bsz:ch1-200701451.

Blaudeck, T., Zenkevich, E. I., Abdel-Mottaleb, M., Szwaykowska, K., Kowerko, D., Cichos, F., and von Borczyskowski, C. (2012). Formation principles and ligand dynamics of nanoassemblies of CdSe quantum dots and functionalised dye molecules, *ChemPhysChem,* **13**(4), pp. 959–972.

Blaudeck, T., Zenkevich, E. I., Cichos, F., and von Borczyskowski, C. (2008) Probing wave functions at semiconductor quantum-dot surfaces by non-FRET photoluminescence quenching, *J. Phys. Chem. C,* **112**, pp. 20251–20257.

Förster, T. (1948). Zwischenmolekulare Energiewanderung und Fluoreszenz, *Ann. Phys.,* **437**, pp. 55–75 (in German).

Förster, T. (1951). *Fluoreszenz Organischer Verbindungen* (Vandenhoeck & Rupprecht, Göttingen) (in German).

Infelta, P. P., and Grätzel, M. (1983). Dynamics of fluorescence quenching in micellar systems, *J. Chem. Phys.,* **78**, pp. 5280–5281.

Lakowicz, J. R. (2006). *Principles of Fluorescence Spectroscopy,* 3rd ed. (Springer Science).

Stern, O., and Volmer, M. (1919). Über die Abklingungszeit der Fluoreszenz, *Phys. Z.,* **20**, pp. 183–188 (in German).

Tachiya, M. (1982). Kinetics of quenching of luminescent probes in micellar systems. II, *J. Chem. Phys.,* **76**, pp. 340–348.

Tachiya, M. (1983). On the kinetics of luminescence quenching in micellar systems, *J. Chem. Phys.,* **78**, pp. 5282–5282.

Valeur, B. (2002). *Molecular Fluorescence: Principles and Applications* (Wiley-VCH, Weinheim).

von Borczyskowski, C., and Zenkevich, E. I. (2014). Formation principles and excition relaxation in semiconductor quantum dot-dye nanoassemblies. In *Quantum Dot Molecules, Lecture Notes in Nanoscale Science and Technology* **14** (Wu, J., and Wang, Z. M., eds.), Chapter 4, Springer Science, pp. 77–148.

Zenkevich, E. I., Blaudeck, T., Heidernätsch, M., Cichos, F., and von Borczyskowski, C. (2009). Effects of electron tunneling and nonresonance quenching of photoluminescence in semiconducting CdSe/ZnS and CdSe nanocrystals by porphyrin molecules in joint complexes, *Theor. Exp. Chem.,* **45**, pp. 23–34.

Zenkevich, E. I., Blaudeck, T., Milekhin, A., and von Borczyskowski, C. (2012). Size-dependent non-FRET photoluminescence quenching in nanocomposites based on semiconductor quantum dots CdSe/ZnS and functionalized porphyrin ligands, *Int. J. Spectrosc.,* **2012**, p. 971791.

Zenkevich, E. I., Cichos, F., Shulga, A., Petrov, E. P., Blaudeck, T., and von Borczyskowski, C. (2005) Nanoassemblies designed from semiconductor quantum dots and molecular arrays, *J. Phys. Chem. B,* **109**, pp. 8679–8692.

Chapter 4

Static and Dynamic Quenching of Quantum Dot Photoluminescence by Organic Semiconductors and Dye Molecules

Ines Trenkmann,[a] Thomas Blaudeck,[b] and Christian von Borczyskowski[c]

[a]*Institute of Physics, Experimental Sensor Science, Technische Universität Chemnitz, Reichenhainer Str. 70, D-09107 Chemnitz, Germany*
[b]*Department of Electrical Engineering and Information Technology, Center for Microtechnologies, Technische Universität Chemnitz, Reichenhainer Str. 70, D-09107 Chemnitz, Germany*
[c]*Institute of Physics, Center for Nanostructured Materials and Analytics, Technische Universität Chemnitz, Reichenhainer Str. 70, D-09107 Chemnitz, Germany*

borczyskowski@physik.tu-chemnitz.de, thomas.blaudeck@zfm.tu-chemnitz.de, ines.trenkmann@physikphysik.tu-chemnitz.de

4.1 Introduction

Due to their unique properties, colloidal semiconductor quantum dots (QDs) have attracted great interest in the recent years in many different areas, also for the integration into organic electronic and optoelectronic devices. Several studies on the influence of the organic surrounding on the (optical) properties of nanocrystals are available and often show a decrease (quenching) of the QD

Self-Assembled Organic-Inorganic Nanostructures: Optics and Dynamics
Edited by Eduard Zenkevich and Christian von Borczyskowski

ISBN 978-981-4745-43-7 (Hardcover), 978-981-4745-44-4 (eBook)
www.panstanford.com

photoluminescence (PL) intensity arising from the QD-(organic) matrix interactions. However, still no clear picture of the underlying processes exists (Blaudeck et al., 2012; von Borczyskowski and Zenkevich, 2014).

In order to shed light on these elementary interaction processes, we study the effect of *N,N′*-diphenyl-*N,N′*-bis(3-methylphenyl)-(1,1′-biphenyl)-4,4′-diamine (TPD), which is an often used and well-studied organic hole transporting material, and pyridyl-substituted porphyrin dye molecules on CdSe/ZnS core/shell quantum dots (QDs) by steady-state and time-resolved PL spectroscopy. We use the well-established Stern–Volmer formalism to analyze static and dynamic quenching processes and compare, reported here for the first time and thus extending the limits of ensemble experiments, their relative contributions with respect to PL quenching of QDs.

PL quenching has been investigated in detail at various molar ratios x = [quencher]/[QD] in which the square brackets denote the concentration of the QDs and the organic material or dye molecules, which act as quencher. Additionally, we compared the PL quenching by TPD for CdSe/ZnS QDs passivated by two different organic shells (ligands) to analyze the influence of the surfactant molecules on the QD-(organic) matrix interactions.

4.2 Experimental

We carried out two series of ensemble QD measurements to study the static and dynamic PL quenching of colloidal quantum dots by organic semiconductors and dye molecules.

To study the influence of an organic semiconductor ensemble measurements were performed on two different types of colloidal core/shell CdSe/ZnS QDs, namely (d_{CdSe} = 3.0 nm; 2 ZnS capping monolayers) passivated with trioctylphosphine oxide (TOPO) and the other one (d_{CdSe} = 4.5 nm; 2 ZnS capping monolayers) with a mixture of trioctylphosphine oxide and hexadecylamine (TOPO/HDA) ligands, which were obtained from life technologies™, USA, and Sigma-Aldrich, Germany, respectively.

We have studied the influence of the organic hole transport material *N,N′*-diphenyl-*N,N′*-bis(3-methylphenyl)-(1,1′-biphenyl)-4,4′-diamine (TPD, from Sigma-Aldrich, Germany) on the optical properties of the CdSe/ZnS QD. Steady-state photoluminescence

(PL) spectra were recorded with a UV/Vis Cary 100 Scan (Varian Inc., USA). Time-resolved measurements were performed with a home-built setup (excited at 465 nm) described in Graaf et al. (2011). The TPD concentration was gradually increased by titration of TPD to a solution of QDs in toluene. All measurements were carried out in dried toluene (<0.0005% H_2O, Merck KGaA, Germany) with QD concentrations varied in the range 3×10^{-9}.

The second series of ensemble measurements were performed on CdSe core quantum dots (d_{CdSe} = 3.2 nm) passivated with TOPO from Evident Technology Inc., USA. The influence of the dye tetra-meta-pyridyl-porphyrin ($(m\text{-}Pyr)_4\text{-}H_2P$) on the optical properties of the CdSe QDs was studied by steady-state PL measurements with a Shimadzu-RF-5001-PC spectrofluorometer and by time-resolved measurements with a home-built setup (excited at 532 nm) described in Blaudeck (2007). The $(m\text{-}Pyr)_4\text{-}H_2P$ concentration was gradually increased by titration of $(m\text{-}Pyr)_4\text{-}H_2P$ to a solution of QDs in toluene. The measurements were carried out in toluene (Fluka, Germany) with QD concentrations in the range of 3×10^{-7}.

4.3 Results and Discussion

4.3.1 Decrease of Photoluminescence with Observation Time

After preparing the samples by diluting CdSe/ZnS QDs in toluene PL spectra were recorded as a function of waiting time while storing the samples in darkness between consecutive measurements. The Pl spectra reveal a decrease (quenching) of the PL emission intensity (Blaudeck et al., 2012). The decreasing (integrated) intensity can be approximated by exponential decay functions as shown in Fig. 4.1. The fitted decay times and amplitudes are given in Table 4.1.

The radiative recombination of excitons and consequently the quantum yield of the QDs are affected by the binding of ligands to unsaturated surface atoms. This is accompanied by changes of the energies of surface related exciton states (Fischer et al., 2012; Frenzel et al., 2007; Mews, 2007). The binding of ligands arises from the electrostatic interactions between partial charges of "(functional) binding" groups of ligand molecules and unsaturated surface atoms (Schapotschnikow et al., 2009; Wawrzynczyk et al., 2013). Diluting QDs is followed by a desorption of the formerly bounded ligands until a thermal equilibrium between the bound and free ligand molecules in the solution is reached (Blaudeck et al., 2012; Ji et al., 2008; Kowerko et al., 2010). The related exchange kinetic depends besides other factors on the strength of the electrostatic interactions. The binding energy of phosphine oxides (e.g., TOPO) is with 313.6 kJ/mol about 3.6 higher than the one of amines (86.8 kJ/mol) (Schapotschnikow et al., 2009), which results in a slower desorption of TOPO ligands characterized by a longer bleaching time.

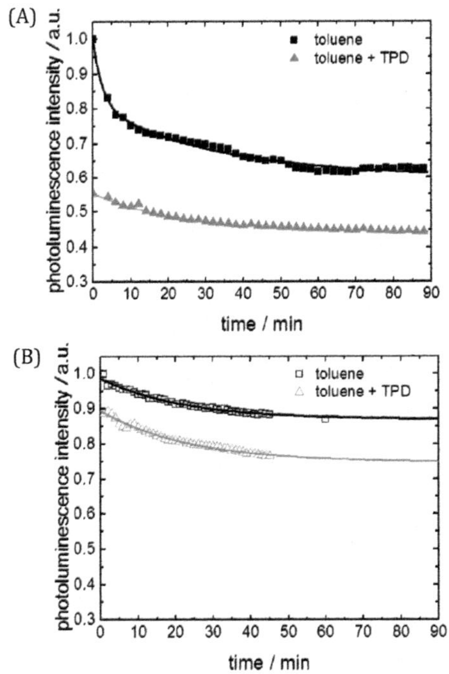

Figure 4.1 Decrease of the (integrated) PL intensity of CdSe/ZnS QDs capped with TOPO (A) and TOPO/HDA (B) molecules in toluene and in toluene with TPD (at a molar ratio (A) x = 655 and (B) x = 275) as a function of the waiting time after sample preparation. The initial PL intensity (of the earliest measurement time of about 1 min) after adding TPD is normalized to the initial PL intensity in toluene. The lines represent exponential decay functions. The fitted decay times and amplitudes are given in Table 4.1.

Table 4.1 Comparison of the PL quenching for CdSe/ZnS quantum dots capped with TOPO and TOPO/HDA ligands in toluene and in toluene with TPD as a function of the waiting time after sample preparation

Ligand	Solvent	A_1	$\tau_{q,1}$ (min)	A_2	$\tau_{q,2}$ (min)	I_{end}
TOPO	Toluene	0.18 ± 0.02	2.4 ± 0.5	0.22 ± 0.01	35 ± 4	0.60 ± 0.01
TOPO	Toluene + TPD	—	—	0.11 ± 0.01	23 ± 1	0.44 ± 0.01
TOPO/ HDA	Toluene	—	—	0.13 ± 0.01	22 ± 2	0.87 ± 0.01
TOPO/ HDA	Toluene + TPD	—	—	0.15 ± 0.01	25 ± 3	0.75 ± 0.01

The measurements as a function of waiting time after sample preparation (Fig. 4.1) show that the change of PL intensity of QDs with a TOPO shell is determined by a slow part with a decay time of $\tau_{q,2}$ = 35 min and a fast part with a decay time of $\tau_{q,1}$ = 2.4 min. In comparison, the measurements of the QDs with a TOPO/HDA shell reveal an "accelerated" PL decrease. While the decay time of the slow component decreases by about one third ($\tau_{q,2}$ = 23 min)—which is in the same order as the change of binding energy—the fast part becomes probably faster than the experimental time resolution ($\tau_{q,1}$ = 1 min). For both QD/ligand systems, the increased amount of the organic semiconductor TPD results in an enlargement of the PL quenching. For both types of ligands the initial PL intensity decreases immediately after adding TPD compared to the measurements in toluene. Obviously the related decay time is so fast that it drops below the time resolution ($\tau_{q,1}$ < 1 min). The further decrease of the PL intensity is determined by the slow component $\tau_{q,2}$. In all cases, we identify a remaining (or extremely slow) decaying PL intensity.

To analyze the influence of TPD onto the PL of the QDs quantitatively steady-state and time-resolved measurements of the PL have been performed as a function of the molar ratio x as will be discussed later on. However, it is already evident at this point that TPD induces additional PL quenching. A first inspection of the data indicates that the PL quenching of TOPO/HDA capped QDs is less effective than for TOPO capped QDs.

4.3.2 Stern–Volmer Formalism for Photoluminescence Quenching

From the variation of PL with observation time and its dependence on the kind of ligand it is obvious that the interaction of QDs with some (partly unknown) quenchers results in a decrease of the PL intensity. Such effects have been repeatedly observed and studied (see, e.g., Blaudeck et al., 2012). To understand such processes quantitatively it is mandatory to control the kind and concentration of quenching entities as has been reported several times by our group during the last decade (von Borczyskowski and Zenkevich, 2014). The Stern–Volmer (SV) model is as has also been shown in Chapter 3 an appropriate formalism for the analysis of the underlying processes (Lakowicz, 2006). It provides access to both the so-called "dynamic" and "static" quenching scenario. Dynamic quenching is a rate changing process with increases the non-radiative recombination rates of the excited luminescent state and results in a decrease of PL lifetime and the related intensity. Static quenching leads to the formation of non-luminescent (or non-absorbing) complexes and consequently to the decrease of the PL intensity. Steady-state measurements can, e.g., reveal that the PL intensity of an ensemble is 50% reduced by the interaction with the quencher. However, it cannot be distinguished whether the PL intensity of all QDs is quenched by 50% (dynamic quenching) or whether the PL intensity of 50% of the QDs is completely quenched (static quenching) or if a combination of dynamic and static quenching occurs (Lakowicz, 2006). Time-resolved measurements can reveal changes of the PL lifetime, which can be used to clarify which kind of quenching mechanism applies. The characteristic SV plots for the intensity I and lifetime τ in the case of dynamic, static and combined quenching are depicted schematically in Fig. 4.2.

If dynamic quenching applies (see Fig. 4.2A) the interaction of the quencher Q with the QDs leads to a new or enhanced non-radiative decay rate k_D (e.g., excitation energy (FRET) or charge transfer) for the emitting state (e.g., exciton) and therefore to a decrease of the lifetime τ and the intensity I, which depend on the concentration of the quencher c_Q. The SV equation for this case is

$$\frac{I_0}{I} = 1 + K_D c_Q = 1 + k_D \tau_0 c_Q = \frac{\tau_0}{\tau}, \tag{4.1}$$

where I_0 and τ_0 represent the intensity and the PL lifetime of the QDs without quencher and K_D is the SV constant for dynamic quenching.

In the case of static quenching (see Fig. 4.2B) the decay time of the QD PL remains constant and the decreasing PL intensity is caused by the decreasing fraction c_F of emitting QDs. Thereby the concentration $c_{F,0}$ of the QDs in the solution is the sum of the concentration of emitting (c_F) and completely quenched QDs (c_{FQ}). The SV equation becomes

$$\frac{I_0}{I} = 1 + K_S c_Q = 1 + \frac{c_{FQ}}{c_F c_Q} c_Q = \frac{c_{F,0}}{c_F} \tag{4.2}$$

with the SV constant K_S for static quenching.

For coexisting dynamic and static quenching (see Fig. 4.2C) the SV equation can be expressed using Eqs. (4.1) and (4.2) by

$$\frac{I_0}{I} = (1 + K_D c_Q)(1 + K_S c_Q) = \frac{\tau_0}{\tau}\frac{c_{F,0}}{c_F} = 1 + K_{app} c_Q \tag{4.3}$$

with the effective SV constant K_{app} which is determined by the dynamic and static part according to

$$K_{app} = (K_D + K_S) + (K_D K_S) c_Q \tag{4.4}$$

and depends on the quencher concentration c_Q.

In a heterogeneous ensemble with i species (e.g., due to different environments of the QDs) or if the QDs have i distinguishable binding sites for the quencher (see Fig. 4.2D) the SV equation can be written as

$$\frac{I_0}{I} = \left[\sum_i \left(\frac{f_i}{1 + K_{app,i} c_Q}\right)\right]^{-1}, \tag{4.5}$$

where f_i is the relative fraction of the binding sites or species i, and $K_{app,i}$ is the (apparent) SV constant of the sites or species i, which is determined by a dynamic and/or static part. If only one binding site or species f_1 is accessible for the quencher and the remaining binding sites or species $(1 - f_1)$ are inaccessible and have no influence on the observed PL quenching, Eq. (4.5) can be simplified to

$$\frac{I_0}{I} = \left[\left(\frac{f_1}{1 + K_{\text{app},1} c_Q}\right) + (1 - f_1)\right]^{-1}. \tag{4.6}$$

The SV constant $K_{\text{app},1}$ depends on the fraction of accessible binding sites or species f_1 and the quencher concentration c_Q according to (Trenkmann, 2015)

$$\int K_{\text{app},1}(c_Q) dc_Q = \frac{f_1}{\frac{I}{I_0} - (1 - f_1)} - 1 \tag{4.7}$$

and decreases with increasing quencher concentration (see Fig. 4.2D).

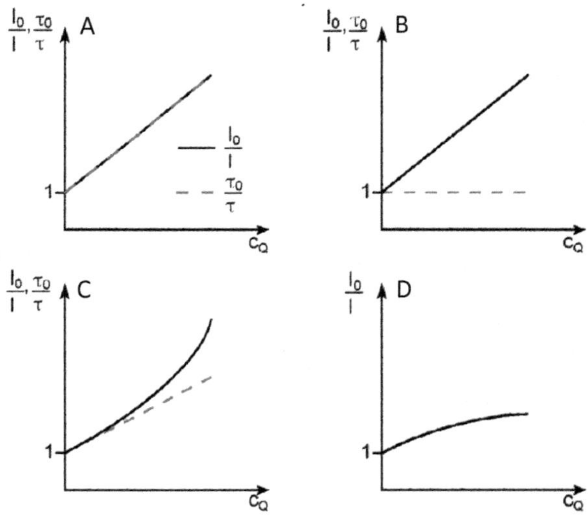

Figure 4.2 Characteristic SV plots of the intensity I and the lifetime τ as a function of the quencher concentration c_Q for dynamic (A) and static quenching (B), for combined dynamic and static quenching (C) and for a heterogeneous ensemble with two different species or bindings sites (D) according to (Lakowicz, 2006).

In the following sections, we examine quantitatively the relative contributions of static and dynamic quenching of QD luminescence which has in case of QDs to the best of our knowledge not yet been reported in literature but which is of importance to understand the elementary interaction processes in molecular aggregates formed from molecules and QDs.

4.3.3 Photoluminescence Intensity as a Function of TPD Concentration

Figure 4.3 shows that titration by TPD does not influence the spectral shape of the QD PL independent of the kind of ligands, but quenches the PL intensity. The same is observed for the absorption spectra (not shown here). Already, Leatherdale and Bawendi (2001) have reported the influence of the permittivity ε of solvents and ligands on the energy of the lowest absorption maximum. Upon a complete exchange of the QD environment from toluene (ε = 2.38) to TPD (ε = 2.95) the permittivity increases by $\Delta\varepsilon$ = 0.57. In the study from Leatherdale and Bawendi (2001) for CdSe QDs (with radii of 0.7 and 1.1 nm) passivated with tributylphosphine oxide the same modification of the solvent permittivity results in a solvatochromatic shift of the lowest absorption band of about 2 and 1 meV, respectively. This shift decrease further with increasing core radius and the deposition of an inorganic shell around the QD core. Therefore, the expected solvatochromatic effect by the modification of the relative permittivity of the QD environment by adding TPD to the solution on the presently used CdSe/ZnS quantum dots can be safely neglected.

Figure 4.4 shows that the SV plots of the PL intensities of both QD/ligand systems resemble the characteristic behavior for PL quenching in a heterogeneous ensemble with two different species or binding sites (as shown in Fig. 4.2D). The red curves represent fits of the experimental data according to Eq. (4.6). For QDs with TOPO and TOPO/HDA ligands, this fits result in 0.58 and 0.44 for the fraction f_1 of the accessible binding sites or species and 0.02 and 0.08 for the (apparent) SV constant $K_{app,1}$, respectively.

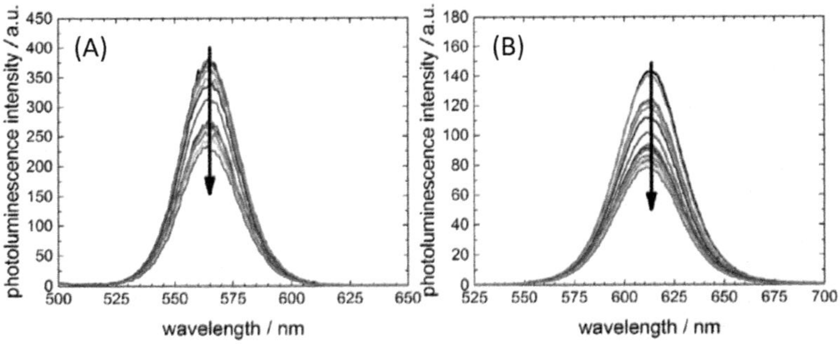

Figure 4.3 PL spectra of CdSe/ZnS quantum dots with TOPO (A) and TOPO/HDA (B) ligands in toluene upon titration with TPD. The arrows indicate the intensity change caused by the stepwise increasing molar ratio x = [TPD]/[QD] from x = 0 up to 5000 (A) and 1400 (B), respectively.

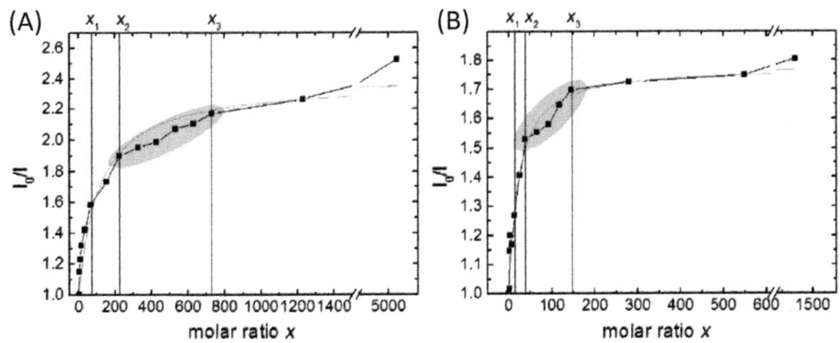

Figure 4.4 SV plot of CdSe/ZnS QDs with TOPO (A) and TOPO/HDA (B) ligands in toluene as a function of molar ratio x. The red curves represent fits of the experimental data according to Eq. (4.6). The gray highlighted regions indicate where the experimental data deviate systematically from the fits. x_1 (71 (A); 12 (B)), x_2 (227 (A); 38 (B)) and x_3 (730 (A); 146 (B)) mark typical molar ratios as will be discussed further on in the text.

We assume that the attachment (binding) of TPD to the QD surface can be qualitatively compared to the binding of ligands to the QD surface and arises from interactions of the quencher molecules to dangling bonds of surface atoms (ions) (e.g., Zn^{2+} or S^{2-} binding sites). TOPO and HDA are supposed to bind to Zn^{2+} ions (Blaudeck et al., 2012; Kowerko et al., 2010). For a (nearly) spherical QD all binding sites are identical for each type of surface atom, but the number of quenchers bound to the QD surface is limited up to a maximum coverage which, due to steric hindrance, depends on the size of quencher molecules and the QD radius (Kowerko et al., 2010; Blaudeck et al., 2012). For this reason a quencher can only bind to a fraction f_1 of the dangling bonds at the surface.

If we compare the experimental data with the fit both QD/ligand systems reveal a deviation from this fit, highlighted by the gray area between the marked molar ratios x_2 and x_3. For both ligands x_3 is three times higher than x_2 and the molar ratios of the QDs with TOPO are six times higher than for the QDs with a TOPO/HDA shell. The maximum PL quenching by TPD seems to be higher for QDs with TOPO ligands as compared to TOPO/HDA ones.

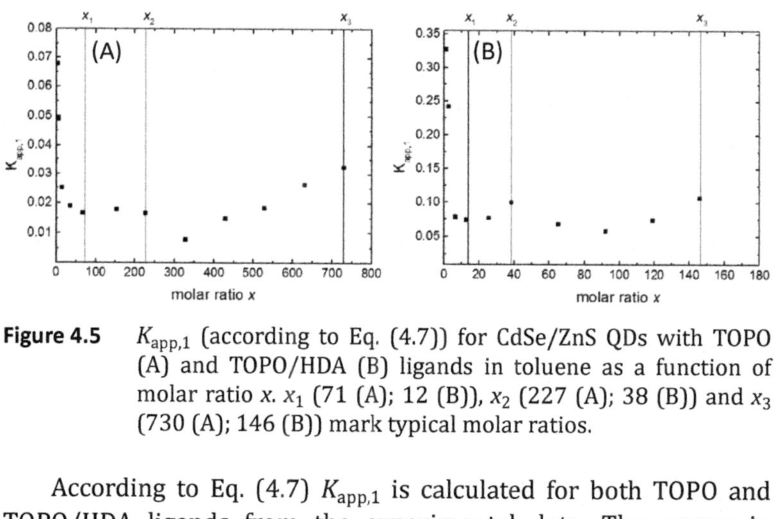

Figure 4.5 $K_{app,1}$ (according to Eq. (4.7)) for CdSe/ZnS QDs with TOPO (A) and TOPO/HDA (B) ligands in toluene as a function of molar ratio x. x_1 (71 (A); 12 (B)), x_2 (227 (A); 38 (B)) and x_3 (730 (A); 146 (B)) mark typical molar ratios.

According to Eq. (4.7) $K_{app,1}$ is calculated for both TOPO and TOPO/HDA ligands from the experimental data. The curves in Fig. 4.5 clearly differ from the expected simple decrease of $K_{app,1}$ with the increase of the molar ratio. Additionally, the magnitude of $K_{app,1}$ is by about one order of magnitude larger in case of TOPO/HDA as compared to TOPO. Figure 4.6 shows the ratio $K_{app,1}$(TOPO/HDA)/$K_{app,1}$(TOPO) and illustrates that the SV constant $K_{app,1}$ of the QDs with TOPO/HDA is about 4.3 larger than the one for TOPO passivation. As mentioned above is the binding strength of phosphine oxides about 3.6 higher than the one of amines. For this reason we conclude that the difference of the SV constants qualitatively reflected the different strengths of the electrostatic interactions of the ligands with the surface ions of the QDs.

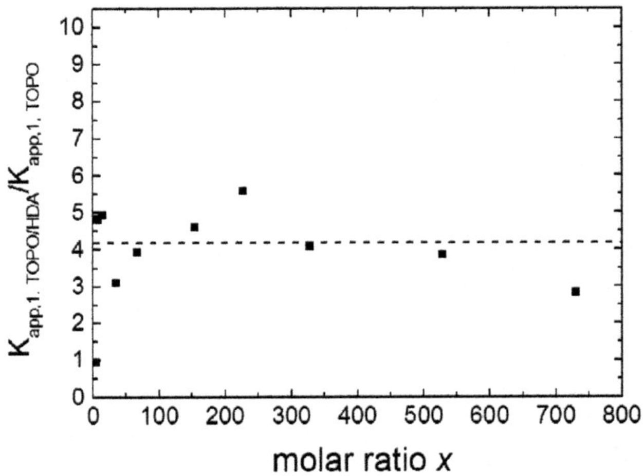

Figure 4.6 Ratio of the SV constants $K_{app,1}$ for CdSe/ZnS QDs with TOPO/HDA and TOPO ligands. The dotted line marks the mean value at (4.2 ± 1.5).

4.3.4 Photoluminescence Decay Time as a Function of TPD Concentration

In the following we discuss the influence of the TPD concentration on the PL decay time which will provide a direct access on the contribution of dynamic quenching. Since QDs containing both types of ligands show qualitatively similar PL quenching behavior upon increasing x, we focus in the following sections on the QDs passivated with TOPO ligands.

The PL decay time was measured for each molar ratio and fitted with a tri-exponential decay function according to

$$I(t) = A_1 \exp\left(-\frac{t}{\tau_1}\right) + A_2 \exp\left(-\frac{t}{\tau_2}\right) + A_3 \exp\left(-\frac{t}{\tau_3}\right). \tag{4.8}$$

We define the average PL lifetime τ_m as (Lakowicz, 2006)

$$\tau_m = \frac{A_1\tau_1^2 + A_2\tau_2^2 + A_3\tau_3^2}{A_1\tau_1 + A_2\tau_2 + A_3\tau_3}. \tag{4.9}$$

The variation of the PL decay time τ_m as a function of the molar ratio is illustrated in Fig. 4.7. The experimental values clearly show contrary to PL intensity data only a moderate decrease of the average PL lifetime τ_m with increasing molar ratio x. In fact, three concentration regions can be identified. In the first region, up to the molar ratio x_1, the average decay time decreases slightly, this is caused by dynamic PL quenching.

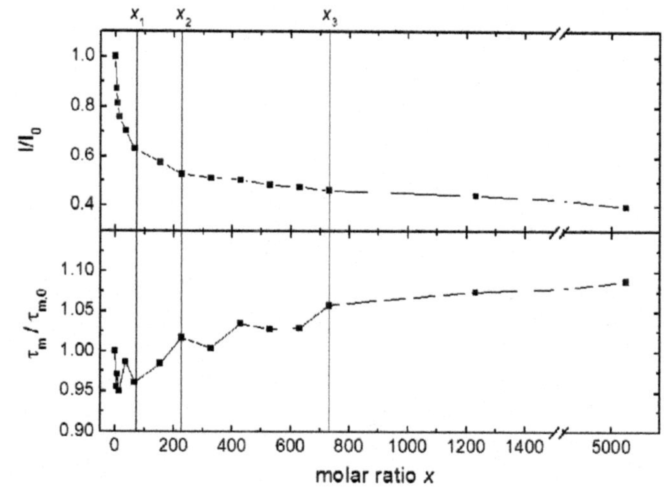

Figure 4.7 Normalized PL intensity I and normalized average PL decay time τ_m of TOPO capped CdSe/ZnS QDs as a function of the molar ratio x. x_1 = 71, x_2 = 227 and x_3 = 730 mark typical molar ratios.

The comparison with the normalized PL intensity in this region reveals that the decrease of PL intensity is higher than the change of the average decay time. According to Eq. (4.1) the increment in both diagrams should be equal if only dynamic quenching applies. Therefore we argue that in the first concentration region the observed PL intensity decrease is caused by dynamic and static quenching. In the second region, until molar ratio x_3, the average PL decay time increase to a value above the PL decay time $\tau_{m,0}$ at x = 0 and remains almost constant upon further increase of the molar ratio.

To explain this unexpected variation of the average decay time τ_m we will have a closer look on the fitted individual decay times τ_i and the related amplitudes A_i, which are depicted in

Fig. 4.8. With increasing molar ratio, all three decay times initially decrease until molar ratio x_1. Upon further increasing the molar ratio x, the decay times remain constant, but simultaneously an increase of the amplitude A_3 of the longest time τ_3 (blue triangles) is observed, while the amplitudes of the two shorter lifetimes decrease. This increase of A_3 effectively results in the observed increase of the average lifetime τ_m in Fig. 4.7 at high molar ratios.

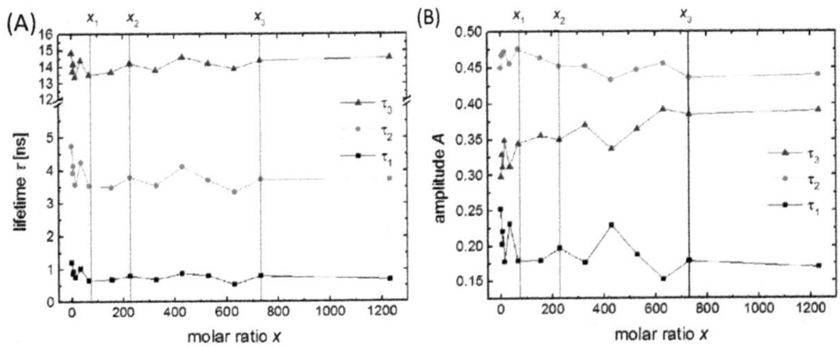

Figure 4.8 Fitted PL lifetimes τ_i (A) and amplitudes A_i (B) for CdSe/ZnS QDs with TOPO ligands in toluene as a function of the molar ratio x. $x_1 = 71$, $x_2 = 227$ and $x_3 = 730$ mark typical molar ratios.

4.3.5 Deconvolution of Static and Dynamic PL Quenching

To understand the interdependence of dynamic and static quench processes on the PL, the related quenching efficiencies Q_{stat} and Q_{dynam} are calculated according to Zhang et al. (2009) by

$$Q_{stat} = \frac{\left(A_{1,0} - A_1 \frac{c_F}{c_{F,0}}\right)\tau_{1,0} + \left(A_{2,0} - A_2 \frac{c_F}{c_{F,0}}\right)\tau_{2,0} + \left(A_{3,0} - A_3 \frac{c_F}{c_{F,0}}\right)\tau_{3,0}}{A_{1,0}\tau_{1,0} + A_{2,0}\tau_{2,0} + A_{3,0}\tau_{3,0}} \quad (4.10)$$

and

$$Q_{dynam} = \frac{A_{1,0}(\tau_{1,0} - \tau_1) + A_{2,0}(\tau_{2,0} - \tau_2) + A_{3,0}(\tau_{3,0} - \tau_3)}{A_{1,0}\tau_{1,0} + A_{2,0}\tau_{2,0} + A_{3,0}\tau_{3,0}}. \quad (4.11)$$

We obtain the fraction $c_F/c_{F,0}$ of emitting QDs from the relation between dynamic and overall quenching as

$$\frac{I_0}{I} = \frac{\tau_0}{\tau} \times \frac{c_{F,0}}{c_F}. \tag{4.12}$$

Figure 4.9 shows the fraction $c_F/c_{F,0}$ of emitting QDs and both quenching contributions Q_{stat} and Q_{dynam} separately.

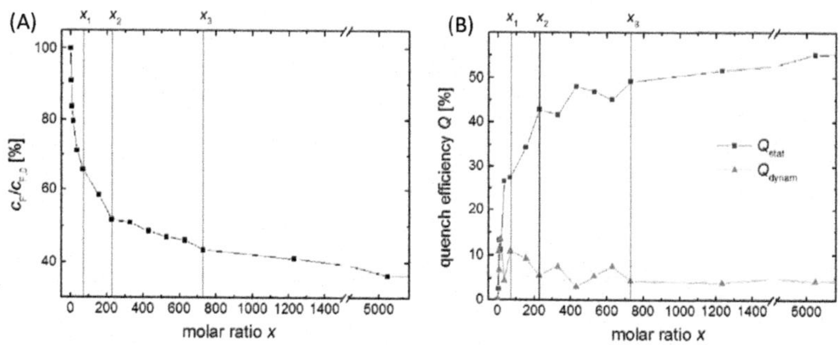

Figure 4.9 Fraction $c_F/c_{F,0}$ of the emitting TOPO capped CdSe/ZnS QDs (according to Eq. (4.12)) (A) and the static Q_{stat} and dynamic quenching efficiency Q_{dynam} (according to Eqs. (4.10) and (4.11)) as a function of the molar ratio x. $x_1 = 71$, $x_2 = 227$ and $x_3 = 730$ mark typical molar ratios.

It is clear that in the first concentration range the static and dynamic quenching efficiencies increase up to the molar ratio x_1 followed by a decrease of the dynamic contribution, while the static one further increases until it becomes nearly constant at molar ratio x_3. The fractions of the static and dynamic contributions of the overall observed PL quenching are compared in Fig. 4.10A. From this comparison, it becomes evident that the PL quenching is mainly caused by static quenching as is quantitatively shown in Fig. 4.10B. However, below $x \approx 14$ the ratio Q_{stat}/Q_{dynam} becomes smaller than 1. This implies that dynamic quenching is larger than static quenching. Since FRET can be excluded since donor energies are lower than acceptor ones, charge transfer remains as a possible mechanism for dynamic quenching. According to reported energy diagrams (Bixby et al., 2012), an optically activated hole transfer from the QD to the hole conducting TPD remains a possible quenching mechanism.

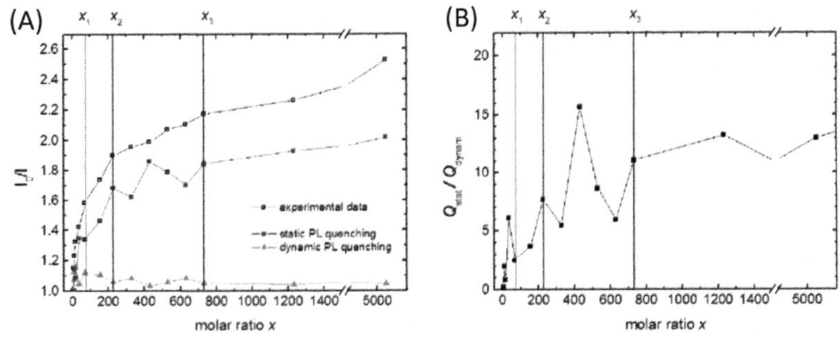

Figure 4.10 PL intensity I/I_0 changes due to static and dynamic quenching (A) and the ratio Q_{stat}/Q_{dynam} of static to dynamic quench efficiency (B) of TOPO capped CdSe/ZnS as a function of the molar ratio x. $x_1 = 71$, $x_2 = 227$ and $x_3 = 730$ mark typical molar ratios.

4.3.6 Ligand-Replacement Model for PL Quenching

Finally we like to discuss qualitatively what the origin of the PL quenching by TPD molecules is. Compared to many recently reported experiments (Blaudeck et al., 2012; von Borczyskowski and Zenkevich, 2014) the molar ratios x of TPD is in our experiments about 100-times larger in order to cause a similar quenching efficiency. Probably the main reason for this observation is the absence of functionalized groups which enables the attachment to the QD surface. TPD has 2 methyl groups which have a small positive local charge density contrary to the negative charge density of TOPO, amine and often used pyridyl-type anchoring groups of dye molecules (see next section). Comparing the molecular volumes (see Fig. 4.11) of ligand and TPD molecules TPD has to replace about 1 TOPO ligand but 2 HDA ligands in order to become attached to surface. In any case TOPO and HDA ligands bind much more strongly to Zn^{2+} or Cd^{2+} ions than TPD might bind to S^{2-} or Se^{2-} ions. Moreover, TPD has a similar dielectric constant and prefers opposite as compared to TOPO and HDA to be solvated in toluene. These reasons qualitatively explain why the probability to replace the ligands by TPD molecules is highly reduced compared to recently reported measurements (Blaudeck et al., 2012). Nevertheless, at sufficiently high TPD concentrations the ligand molecules will be replaced by TPD which prefers to passivate hole instead of electron surface states.

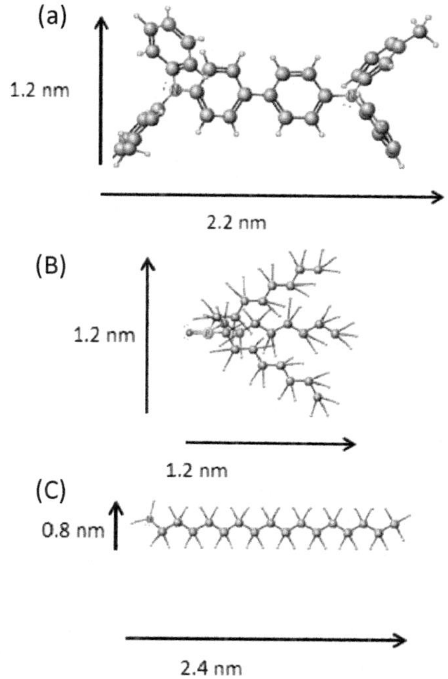

Figure 4.11 Molecular schemes of TPD (A), TOPO (B) and HDA (C).

We have marked a few typical molar ratios x in the previous figures. Up to $x_1 = 71$ both I/I_0 and τ/τ_0 decrease with increasing x and the ratio Q_{stat}/Q_{dynam} increases from about 1 to 3 (see Fig 4.10B). Molar ratios x_2 to x_3 (see gray area in Fig. 4.4) mark a transition range in which $K_{app,1}$ suddenly decreases before rising again (see Fig 4.5A). We take this as an indication that some major "structural" changes take place in the ligand shell, e.g., a complete removal of the TOPO ligand shell by a TPD shell. This might explain why τ_m is from there on rising since the nature of surface states is changing (e.g., from hole to electron type states). Analog effects have been observed at a ligand-type phase transition upon temperature variation (Zenkevich and von Borczyskowski, 2015). Interestingly, such a "structural" transition is also observed for a TOPO/HDA ligand shell (see Figs. 4.4B and 4.5B) but at a much lower x ($x_{2/3}$ (TOPO)/$x_{2/3}$(TOPO/HDA)) $\approx 5 \pm 2$ which is not far from the inverted ratio of related binding energies and thus is in line with a complete ligand exchange in the range $x_2 \div x_3$.

The question remains, what are the microscopic reasons for PL quenching? FRET from CdSe/ZnS to TPD can be ruled out since optical transitions of TPD are much higher compared to those of the QD. Recently Bixby et al. (2012) argued that the enhanced extrinsic hole density in a TPD matrix (which is an organic hole transporting material) is responsible for the observed longer on-times of single CdSe/ZnS QDs compared to those in a dielectric self-trapping matrix such as PMMA (Issac et al., 2005; Issac et al., 2012). If this suggestion is valid we would expect a reduced PL quenching. In any case photoinduced charge transfer might energetically be possible via hole transfer from TPD to CdSe upon excitation of CdSe/ZnS. However, barriers are obviously too high to allow for such a process (Bixby et al., 2012).

Finally, non-FRET quenching has been identified in many cases (Blaudeck et al., 2012; Kowerko et al., 2010; Blaudeck et al., 2008; von Borczyskowski and Zenkevich, 2014). The present experiments on TPD-induced quenching (and also that related to porphyrin, see later on) show that static quenching increases strongly with x across the total molar ratio range and is always more effective than dynamic quenching. How can static quenching be effective in a QD? We have already discussed that absorption and PL spectra are not noticeable changed upon increasing TPD concentration. The obviously reduced quantum yield can be explained by a less effectively populated of the lowest emissive exciton state from the higher excited states upon attachment of quenchers (TPD) to the QD surface. This assumption is reasonable since we have already shown in Chapter 6 of von Borczyskowski and Zenkevich (2016) that non-radiative pathways, especially when exciting elevated QD states, are already present in the absence of any additional quenchers. Such a non-radiative bypass has been suggested by theoretical models via coupling of exciton states to electronic states of ligands (Fischer et al., 2012) or even to electronic states of dye molecules (Kilin et al., 2007). Such electronic couplings will modify electronic states of QDs and will open new non-radiative pathways bypassing the emissive exciton state which naturally appears as PL quenching without necessarily a strong change in PL decay times and can therefore be identified with static quenching. In fact, this has been observed recently (Schmidt et al., 2012; Schmidt et al., 2014) and is also explained in von Borczyskowski and Zenkevich (2016).

Additionally, also the remaining dynamic quenching might have a similar origin since also the lowest emitting states is with respect to PL energy and dynamics by only one attached dye molecule (Zenkevich et al., 2015) in qualitative agreement with model calculations (Kilin et al., 2007). It might also happen that the relative contributions of the three PL decay components change systematically as a function of quencher attachment (Zenkevich and von Borczyskowski, 2015). Non-radiative pathways upon hot electron and exciton generation might be as effective as direct population of the emissive state.

Since we have now explained different microscopic origins of dynamic and static quenching, the question still remains, why is the static one much larger than the dynamic one? This might be qualitatively explained again by inspecting model calculations which show that energetically high electronic states mix much more effectively with ligand or molecular states (Kilina et al., 2009). Whether our explanation holds should be explored by variation of the excitation energy which has not been reported so far in case of quenching processes.

To get more insight into the competition of dynamic and static quenching, we will discuss in the following section the influence of functionalized dye molecules in self-aggregated nanoassemblies for which strong quenching has already been observed at very low molar ratios *x*.

4.3.7 Static and Dynamic PL Quenching by Functionalized Porphyrin Molecules

We have investigated in the previous sections the quantitative relationship between static and dynamic quenching and have given a qualitative explanation of the respective origins. However, the molar ratio of quenchers needed to detect a strong quenching say to about 50% are 2 orders of magnitude higher as compared to recent studies (Zenkevich et al., 2005). In the following, we present a reinvestigation of those recent experiments with respect to a comparison of dynamic and static quenching processes. To proceed we discuss titration experiments of CdSe core QDs capped with TOPO with the dye tetra-meta-pyridyl-porphyrin ($(m\text{-Pyr})_4\text{-}H_2P$).

As discussed in previous studies the dye molecules can be attached to the CdSe surface due to the formation of bonds of the pyridyl lone pair and Cd^{2+} (Blaudeck et al., 2012). In contrast to TPD, discussed above, the spectral overlap of CdSe QD PL spectra with the absorption spectra of (m-Pyr)$_4$-H$_2$P allows for a non-radiative energy transfer (FRET) from an excited QD to a (m-Pyr)$_4$-H$_2$P molecule (Zenkevich et al., 2005; Blaudeck et al., 2008). The PL spectra of a CdSe ensemble shown in Fig. 4.12 exhibit by direct excitation of the CdSe QDs upon titration with (m-Pyr)$_4$-H$_2$P a strong decrease of the QD PL intensity while the intensity of the dye PL increases. We have shown recently that FRET can almost be neglected since non-FRET processes are the origin of about 90% of the total PL quenching (Zenkevich et al., 2005; Blaudeck et al., 2008). Similar observations have been observed for diimide quencher dyes (Kowerko et al., 2010).

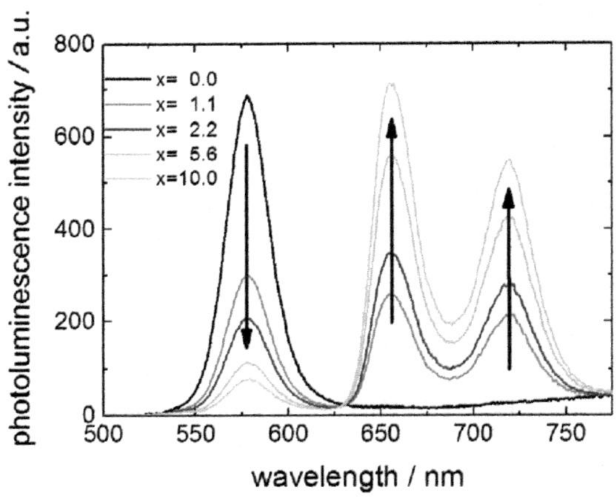

Figure 4.12 PL spectra following titration of (m-Pyr)$_4$-H$_2$P to CdSe core QDs capped with TOPO in toluene. The arrows indicate the intensity change caused by the stepwise increasing molar ratio x = [(m-Pyr)$_4$-H$_2$P]/[QD].

For a further analysis the change of the time resolved PL decay time was measured upon increasing (m-Pyr)$_4$-H$_2$P concentration. The decay in the ns range was fitted with a tri-exponential decay according to Eq. (4.8) (see Fig. 4.13). The results are shown in Fig. 4.14 for three molar ratios x.

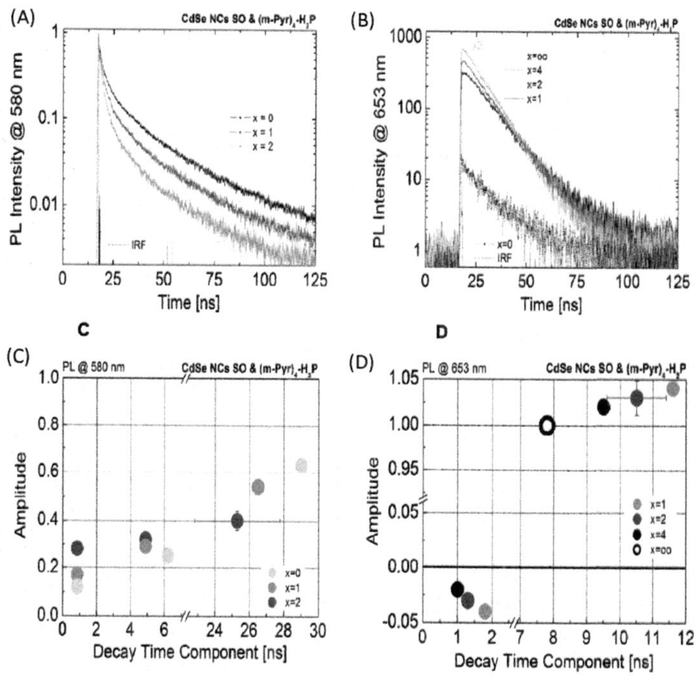

Figure 4.13 Time-resolved PL emission traces (top) and fitted decay times and amplitudes (left column) for CdSe QD (λ_{em} = 580 nm) and (m-Pyr)$_4$-H$_2$P molecules (right column; λ_{em} = 653 nm) at various molar ratios x = [(m-Pyr)$_4$-H$_2$P]/[QD] in toluene at 295 K and λ_{exc} = 532 nm. IRF corresponds to the system response. Top right: x = 0 corresponds to the spurious PL of the QD at the (m-Pyr)$_4$-H$_2$P emission wavelength and $x = \infty$ to a sample with only (m-Pyr)$_4$-H$_2$P. Adapted from Blaudeck (2007).

The average PL decay time τ_m was calculated using Eq. (4.9). The SV plots of the PL intensity and the PL decay time as a function of the molar ratio x are illustrated in Fig. 4.14.

The comparison of both dependencies proofs that the change of the QD PL intensity is larger than the variation of the average PL decay time τ_m. Therefore we conclude that the observed PL quenching arises both from dynamic processes and static processes. (Remaining FRET processes are in principle included in the dynamic quenching process.)

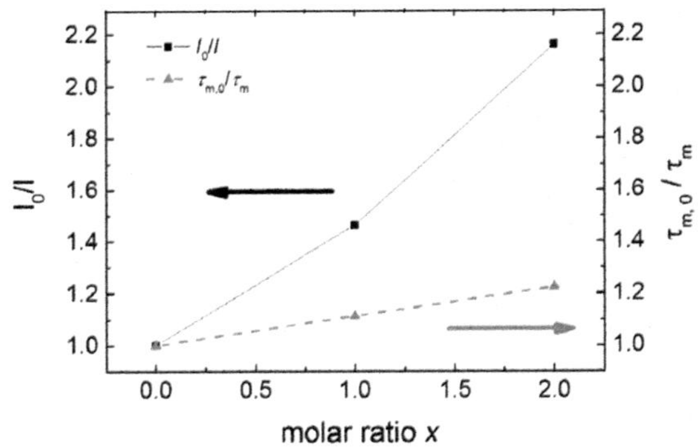

Figure 4.14 SV plot of the PL intensity I (squares) and average lifetime τ_m (triangles) of CdSe QDs capped with TOPO in toluene as a function of the molar ratio x. Data for I_0/I have been selected only for comparable concentrations from Blaudeck et al. (2008).

As already observed for the CdSe/ZnS QD-TPD system, the change of average PL decay time τ_m is determined by a variation of the PL decay times and the respective relative amplitudes. The PL decay times $\tau_{2/3}$ decrease by about 13% and 21%, respectively, and the relative amplitude of the longest decay time A_3 drops about 36% while A_1 and A_2 slightly increase (see Fig. 4.15). In order to elaborate the relationship of dynamic and static processes on the PL quenching in more detail, the related fraction of emitting QDs $c_F/c_{F,0}$ and the quenching efficiencies are calculated according to Eqs. (4.10–4.12). The obtained results are depicted in Fig. 4.16 for each molar ratio.

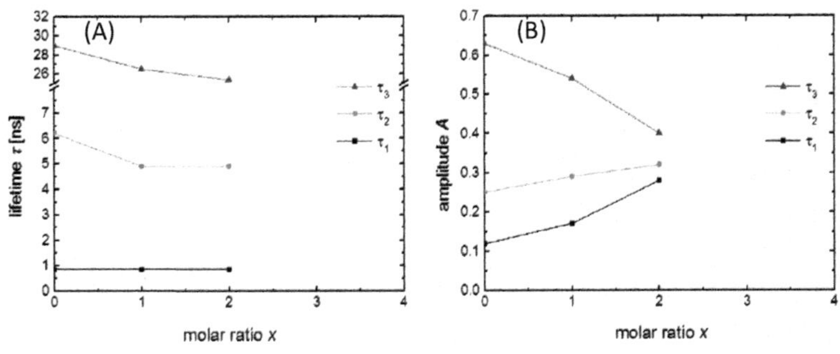

Figure 4.15 Fitted PL decay times τ_i (A) and amplitudes A_i (B) for CdSe QDs capped with TOPO in toluene (from Fig. 4.13) as a function of the molar ratio x.

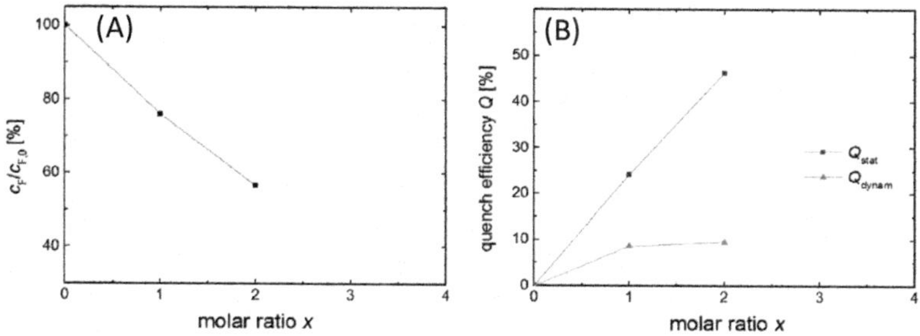

Figure 4.16 Fraction of the emitting TOPO capped CdSe QDs $c_F/c_{F,0}$ (according to Eq. (4.12)) (A) and the static Q_{stat} and dynamic quench efficiency Q_{dynam} (according to Eqs. (4.10) and (4.11)) as a function of the molar ratio x.

At first glance it is surprising that the value of the static quenching efficiency is higher as the dynamic one. Therefore, it is important to remember that according to the definitions of the efficiencies (see Eqs. (4.10) and (4.11)) only changes of the PL decay times cause dynamic quenching, whereas the drop of the fraction $c_F/c_{F,0}$ of emitting QDs and the variation of relative amplitudes of the fitted decay times result in static quenching.

The fractions of the static and dynamic contributions of the overall observed PL quenching and the ratio Q_{stat}/Q_{dynam} of static to dynamic quench efficiency are depicted in Fig. 4.17. This makes clear that the PL decreases is as in the case of TPD quenchers mainly caused by static quenching. We like to emphasize this finding since in literature the efficiency of FRET processes E_{FRET} and the associated energy transfer rate k_{FRET} are often calculated from the observed PL intensity quenching and/or changes of the average PL decay time τ_m. Since both values are in fact often caused by static quenching this approach leads to an overestimation of E_{FRET} and k_{FRET}. Extrapolating the quenching ratio below molar ratios of $x \approx 0.5$ indicates that at these very low molar ratios the dynamic quenching becomes larger than the static one. This implies that PL quenching is in the very low molar ratio range probably predominantly influenced by FRET.

A comparison of the titration experiments with the dye (m-Pyr)$_4$-H$_2$P and the organic hole transporting material TPD for comparable molar ratios x reveals that the ratio Q_{stat}/Q_{dyn} as well as the static quenching efficiency Q_{stat} (see Figs. 4.9B, 4.10B, 4.16B, and 4.17B) are in the case of (m-Pyr)$_4$-H$_2$P considerably larger than for TPD. The analysis of the experiments with TPD shows on the one hand that PL quenching is dominated by static quenching process. However, the influence of dynamic processes at small molar ratios (x < 14) is larger than the one of static processes since $Q_{stat}/Q_{dynam} < 1$ (see Fig. 4.10B). This is in agreement with the fact that a much stronger interaction of the functional pyridyl group of (m-Pyr)$_4$-H$_2$P has to be considered as compared to the much weaker interactions of TPD with the QD surface. According to our recent investigations (Zenkevich and von Borczyskowski, 2015), strong interactions with the QD surface modify considerably excitonic states and relaxation dynamics of the QD. Moreover, previous experiments have shown that FRET processes from QDs to (m-Pyr)$_4$-H$_2$P molecules are most efficient (see also Fig. 4.13 (right) below molar ratios x < 2 (Zenkevich et al., 2005). We have explained this by the assumption that only a few easily accessible sites are available for the attachment of (m-Pyr)$_4$-H$_2$P to the QD surface (Blaudeck et al., 2012).

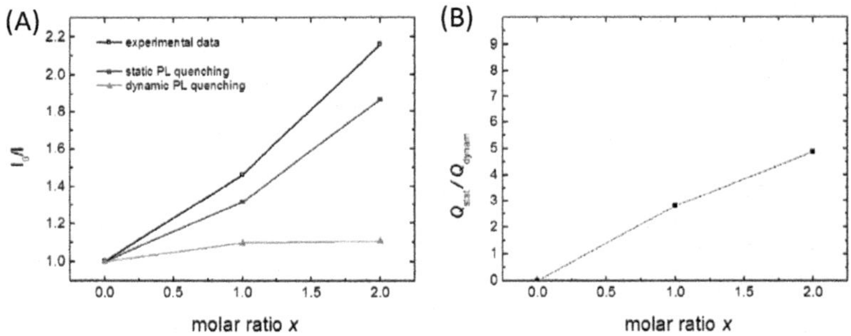

Figure 4.17 PL intensity I_0/I changes due to static and dynamic quenching of TOPO capped CdSe QDs as a function of the molar ratio x (A) and the ratio of Q_{stat}/Q_{dynam} of static to dynamic quench efficiency (B).

4.4 Conclusion

We presented the impact of static and dynamic quenching processes on the PL intensity and the variation of PL decay times of CdSe(/ZnS) QDs upon titration with the organic hole transport material TPD and the dye (m-Pyr)$_4$-H$_2$P. We find that the molar ratios to generate a similar absolute quenching efficiency are by about a factor of 100 smaller for titration by

TPD as compared to $(m\text{-}Pyr)_4\text{-}H_2P$. We argue that this is due to the missing functional groups in case of TPD. We find as already recently reported that FRET processes which have to be considered dynamic quenching processes are very small. This is not surprising since we have already reported that non-FRET quenching is the major quenching mechanism of self-assembled nanoassemblies (Blaudeck et al., 2008; Kowerko et al., 2010; von Borczyskowski and Zenkevich, 2014). However, the presently reported quenching investigation goes a step further since we now show that a major part (also valid for previous investigations) of PL quenching at molar ratios $x > 2$ ($(m\text{-}Pyr)_4\text{-}H_2P$) and $x > 14$ (TPD) is in terms of the Stern Volmer approach due to static instead of dynamic quenching. However, the kind of the dynamic quenching processes is probably completely different for the respective molecular quenchers (FRET for $(m\text{-}Pyr)_4\text{-}H_2P$ and hole transfer for TPD, respectively). The molar range at which such a transition occurs depends strongly on the formation dynamics for assembly formation (Blaudeck et al., 2012). Accordingly, the lower molar ratio for a transition between static and dynamic quenching in case of $(m\text{-}Pyr)_4\text{-}H_2P$ as compared to TPD can be readily explained by the strong "binding" energy of the pyridyl-groups (Blaudeck et al., 2012) as compared to the methyl-groups of TPD. Casting the attachment probability only onto the binding energy, the number of functional molecular attachment groups (4 in case of $(m\text{-}Pyr)_4\text{-}H_2P$ and 2 in case of TPD), and the number of TOPO ligand molecules which have to be replaced by the respective quencher molecule (≈ 3 for $(m\text{-}Pyr)_4\text{-}H_2P$ (Blaudeck et al., 2012) and ≈ 1 for TPD (see Fig. 4.11)), we speculate that the binding energy of $(m\text{-}Pyr)_4\text{-}H_2P$ is about 10 times stronger than the one of TPD. The low molar ratios of a switch-over from dynamic to static quenching indicate that at a given molar ratio not all molecules are included in a nanoassembly with a QD (only 25% to 50% (Blaudeck et al., 2012)). Additionally, for the QDs used in the present experiments FRET to $(m\text{-}Pyr)_4\text{-}H_2P$ will still be enhanced upon attachment of more than 1 dye molecule (Blaudeck, 2007). These arguments suggest that the switch-over is due to a dominating contribution of "static" quenching in case that more than 1 dye becomes attached to a QD. However, even at very low molar ratios "static" quenching has a noticeable influence.

Moreover, we argue that the finding of a "quasi-static" PL quenching can be explained by discrimination between "apparent" quenching due to changes in the population rates from highly excited QD states to the photoluminescent one (quasi "static" quenching, which is in fact also a dynamic one but not of the PL directly) and changes in the PL decay times of the lowest emitting QD state (dynamic quenching). Additionally, subtle changes of the PL energies and the dynamics at the exciton band edge can be identified as a considerable part of the reasons for this dynamic PL quenching, outnumbering FRET or charge transfer. This finding is strongly supported by recent detailed investigations on the tuning of PL properties of QDs (Zenkevich et al., 2015).

Recent density-functional theory calculations further corroborate these arguments revealing that ligand replacements and dye attachments change energies and dynamics of the QD states not only at the band edge but even more effectively in high electronic states due to mixing of molecular and QD states (Kilin et al., 2007; Kilina et al., 2009). These findings should be investigated more deeply via the investigation of quenching processes as a function of excitation energies. Previous experiments have already shown that excitation into energetically high QD states opens new non-radiative pathways diminishing PL intensities as is discussed intensively recently (Schmidt et al., 2014). We have experimentally shown that PL quenching can be strongly enhanced by decreasing the polarity of the matrix which might then result in very effective FRET processes even on a single assembly level (Gerlach et al., 2013).

In spite of all differences between both investigated quencher systems, the observed PL quenching phenomena are mainly influenced by static quenching processes in case that more than one molecule becomes attached to the QD surface. This very general finding should be taken as an motivation to investigate FRET or similar experiments on PL quenching of QDs more carefully which can experimentally achieved by a quantitative comparison of PL (donor) quenching and enhancement of acceptor luminescence. Although a necessary re-interpretation of many respective FRET (and charge transfer) experiments might be disgusting, the presented new insights stimulate and pave a way to investigate interactions at QD surfaces more rigorously.

References

Bixby, T. J., Cordones, A. A., and Leon, S. R. (2012) CdSe/ZnS Quantum Dot Intermittency in N,N′-diphenyl-N,N′-bis(3-methylpheny)-(1,1′-biphenyl)-4,4′-diamine (TPD), *Chem. Phys. Lett.,* 521, pp. 7–11.

Blaudeck, T. (2007) Self-Assembly of Functionalized Porphyrin Molecules on Semiconductor Nanocrystal Surfaces, http://nbn-resolving.de/urn:nbn:de:bsz:ch1-200701451.

Blaudeck, T., Zenkevich, E. I., Abdel-Mottaleb, M., Szwaykowska, K., Kowerk, D., Cichos, F., and von Borczyskowski, C. (2012) Formation Principles and Ligand Dynamics of Nanoassemblies of CdSe Quantum Dots and Functionalised Dye Molecules, *Chem Phys Chem,* **13**, pp. 959–972.

Blaudeck, T., Zenkevich, E. I., Cichos, F., and von Borczyskowski, C. (2008) Probing Wave Functions at Semiconductor Quantum-Dot Surfaces by Non-FRET Photoluminescence Quenching, *J. Phys. Chem. C,* **112**, pp. 20251–20257.

Fischer, S. A., Crotty, A. M., Kilina, S. V., Ivanov, S. A., and Tretiak, S. (2012) Passivating Ligand and Solvent Contribution to the Electronic Properties of Semiconductor Nanocrystals, *Nanoscale,* **4**, pp. 904–914.

Frenzel, J., Joswig, J.-O., and Seifer, G. (2007) Optical Excitations in Cadmium Sulfide Nanoparticles, *J. Phys. Chem. C,* **111**, pp. 10761–10770.

Gerlach, F., Täuber, D., and Von Borczyskowski, C. (2013) Correlated Blinking Via Time Dependent Energy Transfer in Single CdSe Quantum Dot-Dye Nanoassemblies, *Chem. Phys. Lett.,* **572**, 90–95.

Graaf, H., Friedriszik, F., Wagner, C., and Von Borczyskowski, C. (2011) Optical Spectroscopy of Trap States in Amorphous Perylene Derivative Films, *J. Phys. Chem. C,* **115**, pp. 8150–8154.

Issac, A., Von Borczyskowski, C., and Cichos, F. (2005) Correlation between Photoluminescence Intermittency of CdSe Quantum Dots and Self-Trapped States in Dielectric Media, *Phys. Rev. B,* **71**, p. 161302 (R).

Issac, A., Krasselt, C., Cichos, F., and Von Borczyskowski, C. (2012) Influence of the Dielectric Environment on the Photoluminescence Intermittency of CdSe Quantum Dots, *Chem Phys Chem,* **13**, pp. 3223–3230.

Ji, X., Copenhaver, D., Sichmeller, C., and Peng, X. (2008) Ligand Bonding and Dynamics on Colloidal Nanocrystals at room temperature: The case of Alkylamines on CdSe Nanocrystals, *JACS,* **130**, pp. 5726–5736.

Kilin, D. S., Tsemekhman, K., Prezhdo, O. V., Zenkevich, E. I., and Von Borczyskowski, C. (2007) Ab Initio Study of Transfer Dynamics from a Core-Shell Semiconductor Quantum Dot to a Porphyrin Sensitizer, *J. Photochem. Photobiol. A*, **190**, 342–351.

Kilina, S. V., Ivanov, S., and Tretiak, S. (2009) Effect of Surface Ligands on Optical and Electronic Spectra of Semiconductor Nanoclusters, *J. Am. Chem. Soc.*, 131, pp. 7717–7726.

Kowerko, D., Schuster, J., Amecke, N., Abdel-Mottaleb, M., Dobrawa, R., Würthner, F., and Von Borczyskowski, C. (2010) FRET and Ligand Related NON-FRET Processes in Single Quantum Dot-Perylene Bisimide Assemblies, *Phys. Chem. Chem. Phys.*, **12**, pp. 4112–4123.

Lakowicz, J. R. (2006) *Principles of Fluorescence Spectroscopy*, 3nd ed. (Springer Science).

Leatherdale, C. A., and Bawendi, M. G. (2001). Observation of Solvatochromism in CdSe Colloidal Quantum Dots, *Phys. Rev. B*, **63**, p. 165315.

Mews, A. (2007) Surface Chemistry of Semiconductor Nanocrystals, *Z. Phys. Chem.*, **221**, pp. 295–306.

Schapotschnikow, P., Hommersom, B., and Vlugt, T. J. H. (2009) Adsorption and Binding of Ligands to CdSe Nanocrystals, *J. Phys. Chem. C*, **113**, pp. 12690–12698.

Trenkmann, I. (2015) Mikroskopie und Optische Spektroskopie an Heterogenen Nano- und Mikrostrukturen: Halbleiter-Nanokristalle, Molekulare Farbstofffilme und funktionalisiere Hybridstrukturen, Technische Universität Chemnitz.

Trenkmann, I., Göhler, C., Krasselt, C., and von Borczyskowski, C. (2016). Chapter 6. In *Tuning Semiconducting and Metallic Nanoparticles* (Von Borczyskowski, C., and Zenkevich, E., eds.), Pan Stanford Publishing.

Von Borczyskowski, C., and Zenkevich, E. I (2014). Formation Principles and Excition Relaxation in Semiconductor Quantum Dot-Dye Nanoassemblies. In *Quantum Dot Molecules, Lecture Notes in Nanoscale Science and Technology*, **14**, (Wu, J., and Wang, Z. M., eds.), Chapter 4, Springer Science, pp. 77–148.

Wawrzynczyk, D., Bednarkiewicz, A., Nyk, M., Strek, W., and Samoc, M. (2013) Ligand-Dependent Luminescence of Ultra-Small Eu^{3+}-Doped $NaYF_4$ Nanoparticles, *J. Nanopart. Res.*, **15**, p. 1707.

Zenkevich, E., Cichos, F., Shulga, A., Petrov, E. P. Blaudeck, T., and Borczyskowski, C. von (2005) Nanoassemblies Designed from Semiconductor Quantum Dots and Molecular Arrays, *J. Phys. Chem. B*, **109**, pp. 8679–8692.

Zenkevich, E. I., Stupak, A. P., Göhler, C., Krasselt, C., and Von Borczyskowski, C. (2015) Tuning Electronic States of a CdSe/ZnS Quantum Dot by Only One Functional Dye Molecule, *ACS Nano,* **9**, pp. 2886–2903.

Zhang, Y., Jing, P., Sun, Y., Su, H., Wang, Y. A., Kong, X., Zhao, J., and Zhang, H. (2009) Photoluminescence Quenching of CdSe/ZnS Quantum Dots by Hole Transporting Materials, *J. Phys. Chem. C,* **113**, pp. 1886–1890.

Chapter 5

Selected Applications of QDs and QD-Based Nanoassemblies

Eduard Zenkevich[a] and Christian von Borczyskowski[b]

[a]*Department of Information Technologies and Robototechnique, National Technical University of Belarus, Prospect Nezavisimosti 65, 220013 Minsk, Belarus*

[b]*Institute of Physics, Center for Nanostructured Materials and Analytics, Technische Universität Chemnitz, Reichenhainer Str. 70 D-09107 Chemnitz, Germany*

zenkev@tut.by, borczyskowski@physik.tu-chemnitz.de

5.1 Introduction

To date, it is well documented that nanotechnology is playing a pivotal role in advancing nano/bio/info technology by creating new interfaces between multiple disciplines. It provides new opportunities for high-density integration in information technology, engineering as well as for other major medical needs to advance human health (Bhushan, 2012; Lyshevski, 2014; Somasundaran, 2015). Based on evident progress in the synthesis of colloidal semiconductor QDs with relatively narrow size distributions within a few percent, rational shape-engineering, compositional modulation, electronic doping, and tailored surface chemistries, these nanoobjects as well as nanoassemblies based

Self-Assembled Organic-Inorganic Nanostructures: Optics and Dynamics
Edited by Eduard Zenkevich and Christian von Borczyskowski

ISBN 978-981-4745-43-7 (Hardcover), 978-981-4745-44-4 (eBook)
www.panstanford.com

on QDs and functionalized organic compounds (including proteins) have become an important class of nanomaterials with great potential for applications ranging from electronic and optoelectronic devices (Haghi et al., 2015; De Mello Donegá, 2014; Rogers et al., 2014; Vajtai, 2013; Rogach, 2008; Gur et al., 2007; Wang et al., 2010; Coe-Sullivan et al., 2003; Baldenebro-Lopez et al., 2013), including quantum dot solar cells (Wang et al., 2010; Kim et al., 2013; Semonin et al., 2011; Wang et al., 2011), to material science (Kovalenko et al., 2015; Gaponenko, 2010; Klimov, 2000; Schmitt et al., 2011), plasmonics (Lou, et al., 2014), sensorics (Frasko et al., 2010; Jin and Hildebrandt, 2012; Lemon et al., 2013, 2015; Hadar et al., 2015), and nanobioscience/nanomedicine (Nicolini, 2010; He and Ma, 2014; Shatkin, 2012; Mattoussi et al., 2012; Ji et al., 2015a, 2015b; Liang et al., 2010; Mocatta et al., 2011; Moyano et al., 2012; Guo et al., 2015; Liu et al., 2013). Finally, the 2014 Nobel Prize in chemistry devoted to the recent revolution in super-resolution optical microscopy opened a new and extremely wide range for applications of blinking QDs as luminescent label in optical microscopy (Lidke et al., 2005).

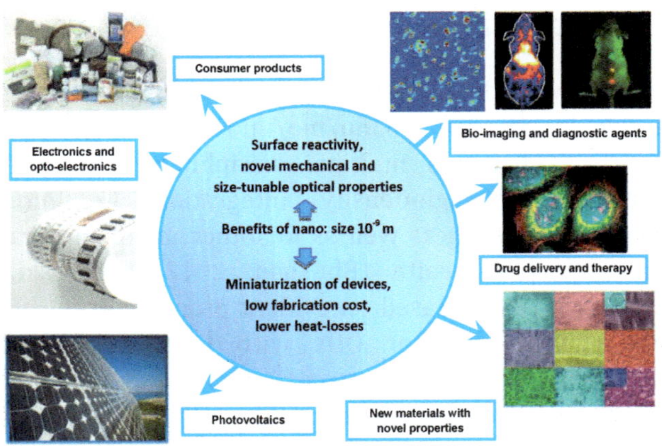

Scheme 5.1 Benefits of QD-based nanomaterials and some of their applications.

On the basis of numerous data presented at least in references cited above as well as of those shown and discussed in this book, we present some principal properties which make QDs and nanoassemblies so special. On one hand, nanometer size is characterized by a large surface-to-volume ratio, which leads to a high surface chemical sensitivity, and reconstructions on the QD surface result in advanced mechanical properties. On the other hand, a strong confinement of excitons possesses advanced optical properties and size-tunable optical characteristics. Taken together, these peculiarities of (colloidal) semiconductor QDs and successes in organo-metallic chemistry determine a wide field of their applications in nanotechnology as seen from Scheme 5.1.

From the basic point of view, the science and technology of QDs and QD-based materials involve on every length scale, from the molecular to the macro, surface and interfacial phenomena that can be tuned by varying the surface and interfacial energy and by varying the specific chemical interactions and chemical groups populating such surfaces and interfaces (Bullen and Mulvaney, 2006; Kalyuzhny and Murray, 2005; Knowles et al., 2012; Munro et al., 2007; Ji et al., 2008; Koole et al., 2008; Blaudeck et al., 2012; Zenkevich et al., 2015). The state of the art in the research on colloidal QDs focusing on the most recent works (published within the last few years) was presented in a comprehensive review of Kovalenko et al. (2015). The progress in QD surface chemistry relies on experimental techniques giving insight into the composition of QD surfaces, a conceptual framework to describe surface termination or coordination, and rational methods to modify QD surfaces. In fact, it is the effect of the surface chemistry on the individual optical properties of QDs and on QD-based optoelectronic devices and as optical labels especially in biology and biomedicine (Petryayeva et al., 2013; Yi and Papaefthymiou, 2013) that will continue to motivate research related to organic/inorganic surface capping and simulation of QD surfaces and the entire QD entity using density functional theory. The same point of view has been pointed out by us upon studies of QD-based nanoassemblies (see Chapters 2 and 3 in von Borczyskowski and Zenkevich (2016) and 1–4 in this book). Correspondingly, QD surface chemistry is considered as gateway to applications of quantum dots (Bucak and Rende, 2013; Somasundaran, 2015). In addition, in the review Kovalenko et al. (2015), new nanomaterials based on QDs and their novel applications have been discussed such as non-metallic plasmonic QDs (Comin and Manna, 2014; Liu and Swihart (2014) for photothermal therapy (Hessel et al., 2011), electrochromic devices (Llordes et al., 2013; Runnerstrom et al., 2014), long-range-ordered nanocrystal assemblies (ordered superstructures of polyhedral QDs) (Arciniegas et al., 2014), QD-derived thermoelectrics (Zebarjadi et al., 2012), nanocrystal lasing (Grim et al., 2014), infrared photodetectors (Deng, et al., 2014) or theranostics applications (QD platform combining medical diagnostics/analysis with subsequent treatment) (Stasiuk et al., 2011; Jing, et al., 2014).

The intention of this semi-review chapter is not a thorough description of all applications of QD and QD–dye nanoassemblies, which is widely presented in a lot of publications and books cited above. Rather the present material should be considered as a more comparative characterization of some applications of QDs and functionalized multicomponent nanoassemblies on their basis which, in some respect, may be connected with the main results and ideas discussed in this book. It is expected to provide a short and multidisciplinary reference to few fields relevant to the general area of nanotechnology and the recently developed super-resolution microscopy to which the Nobel Prize in chemistry was awarded in 2014. This chapter is aimed at graduate students, researchers, and practitioners. For more detailed information on the concrete application topics, we refer the reader to recent reviews and books cited above as well as to numerous other literature sources.

5.2 A Brief Overview of the Early History of QDs Applications

The history of nanomaterials began after the big bang when first nanostructures were formed in the early meteorites. Obviously, first hand-made nanotechnologies and nanomaterials appeared as accidental findings of ancient masters and craftsmen without any deep knowledge about nature and structure of these materials. Ancient and premodern technologies could not control material properties on the basis of knowledge of nanometer-scale properties. Nevertheless, the recipes and approaches for nanomaterial production were successful enough to provide admiration and sometimes fear in scientists of their age. At the same time, even with our modern knowledge, the admiration for these first craftsmen of nanomaterials still holds. In this section, we will give a brief overview of the early history of QDs applications.

At least, in human history the first documentation of using nanotechnology based on QDs for cosmetics and hair dyes is connected with ancient Egyptian times (i.e., over 4000 years ago). In those times, some unusual approaches based on lead compounds were used to dye hair and wool black. Recently, based an ancient Egyptian recipes on the use of lead salts for hair darkening, it has been shown via modern experimental abilities (Walter et al., 2006) that during the dyeing procedure inside a hair, lead–lime mix reacts with sulfur present in hair's keratin proteins to form lead sulfide nanocrystals of nano-size (PbS QDs). Some results of these findings are presented in Fig. 5.1, which shows the progression of darkening for blonde hair being treated with the ancient hair dye recipe. It was found that during this procedure PbS nanocrystals with diameter of ~5 nm are formed being very similar to PbS QDs that are synthesized by modern approaches. Interestingly, in contrast to modern technology, Egyptian dyeing process uses basic chemistry methods and low-cost natural materials. It is remarkable also that the composition and supramolecular organization of keratins can control PbS

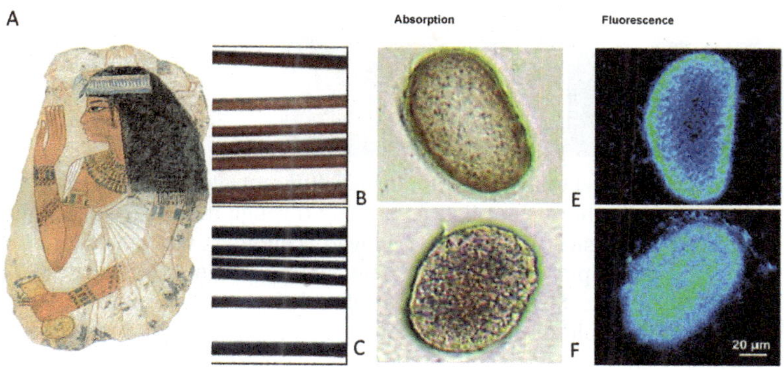

Figure 5.1 Optical macrophotographs of hair (from brown to black color) and microphotographs of the hair cross sections (absorption and fluorescence images) indicating noticeable blackening during treatment with lime and lead oxide in water (after 6 and 72 h). A total adsorbed Pb concentration in bulk samples was measured by X-ray fluorescence. Pb maps of the corresponding treated samples were obtained by SEM-EDX. Spectra show progressive blackening during treatment with increasing portion of PbS nanocrystals (from blue to green). Adapted partly from Walter et al. (2006).

nanocrystal growth inside a hair. Thus, by studying this process of synthetic nanoscale biomineralization in hair, one can better understand how to grow and organize nanoparticles for other applications.

Continuing, there are a number of very interesting examples of ancient objects which have been created using quantum dots (based on metallic or semiconducting materials). One of them is the well-known Lycurgus cup, being a stunning decorative Roman treasure of IV century AD (i.e., over 1600 years ago). This cup is among the most technically sophisticated glass objects produced before the modern era. The main peculiarity of this goblet is the following: when a light source is outside, the cup is green, and when a light source is placed inside, the cup is red (see Fig. 5.2).

Figure 5.2 Left: Lycurgus cup shows different color depending on whether or not light is passing through it. The glass contains nanosize gold–silver alloyed colloidal nanoparticles (an example is shown). Partly adapted from Atwater (2007). Right: Size dependence of color and absorption spectra for gold plasmonic nanoparticles. Adapted from http://www.sigmaaldrich.com/materials-science/nanomaterials/gold-nanoparticles.

Later on it has been shown that the glass of this goblet contains gold–silver alloyed nanoparticles, which are distributed in such a way to make the glass look green in reflected light but, when light passes through the cup, it reveals a brilliant red (Atwater, 2007). In fact, plasmonic excitation (see also Chapter 1 in von Borczyskowski and Zenkevich (2016)) has green and blue absorption spectra of light. When a light source is inside, the glass absorbs short waves (green and blue), while the red component penetrates through the glass, and the cup is red. It should be mentioned also that colloidal gold and silver nanoparticles have been utilized for centuries by artists due to the vibrant colors produced by their interaction with visible light but it has been done on a hunch.

The other example of using metal nanoparticles may be presented by antique stained glasses (colored glass) being applied throughout thousand-year history almost exclusively to the windows of churches, mosques and other significant buildings (one example is shown in Fig. 5.3). At the moment, it is well documented that metal plasmonic nanoparticles are highly efficient in absorbing and scattering light (Ramachandra et al., 2000; Zeng et al., 2013; Comin and Manna, 2014; Low et al., 2014; Runnerstrom et al., 2014). By changing the size, shape, and surface chemistry or aggregation state, the optical properties of the nanoparticles (both semiconductor and metal) can be tuned from the ultraviolet to the visible to the near-infrared regions. By shifting the absorption and scattering, the color of nanoparticle dispersions and films can also be tuned: for example, solutions of spherical gold nanoparticles are ruby red in color due to the strong scattering and absorption in the green region of the spectrum, while solutions of silver nanoparticles are yellow due to the plasmon resonance in the blue region of the spectrum.

Figure 5.3 Example of antique stained glasses containing metal plasmonic nanoparticles: The north transept rose of Chartres Cathedral donated by Blanche of Castile (15th century). Adapted from Eusebius (Guillaume Piolle).

Finally, it should be underlined that the systematic scientific investigations of metal nanoparticles preparation and properties have been carried out for the first time by M. Faraday as early as 1857 (Faraday, 1857). He succeeded to prepare and describe the first pure sample of colloidal gold, which he called "activated gold," as well has shown how to control the color of nanoparticles. More recently, these unique optical and electronics properties of various metal nanoparticles and their compositions have been evaluated and utilized in high technology applications such as organic photovoltaics, sensory probes, nanocatalysis, therapeutic agents, drug delivery in biological and medical applications, and nanoelectronics (Ramachandra et al., 2000; Zeng et al., 2013; Vajtai, 2013; Shao et al., 2015; Guler et al., 2015).

5.3 Semiconductor Nanostructures for Solar Cells and Photovoltaics

Nowadays it is generally agreed that reserves of fossil fuels are limited and their use has serious environmental side effects. Solar energy is one of the renewable, low-carbon resources with both the scalability and the technological maturity to meet ever-growing global demand for electricity (Lewis et al., 2005). Among solar power technologies, solar photovoltaics are the most widely applied, providing 0.87% of the world's electricity in 2013 and sustaining an annual growth rate in cumulative installed capacity of 43% since 2000 (Jean et al., 2015). Encouraged by this realization is the use of cleaner alternative energy, among which dye-sensitized semiconductor solar cells (SCs) are potentially attractive candidates for the lower cost of producing devices which convert the abundant amount of energy from the sun into electricity.

Dye-sensitizer in SCs plays a crucial role as the chlorophyll in plants to harvest solar light and transfer the energy via electron transfer to a suitable semiconductor material (TiO_2 in a lot of cases) to produce electricity. Basically, the functionality of SCs is based on the dye-sensitized photovoltaic effect, where after absorbing a single photon of sufficient energy by a dye molecule an electron may be injected into the semiconductor band gap (Moser, 2010). Sensitization allows for extending the absorption spectrum useable for free carriers generation within a semiconductor. Efficient sensitizers are found within ruthenium polypyridyl dyes. Because TiO_2 does not absorb light at energies smaller than its band gap of 3.2 eV (400 nm), it needs to be sensitized for efficient absorption of the solar light. A dye absorbs visible light and its excited state has an oxidation potential that allows for electron injection into the conduction band of the semiconductor. At the beginning, most sensitization studies have dealt with electrochemical measurements of sensitized wide-band gap semiconductor planar electrodes. However, the development of efficient devices for current collection was impossible due to the low light harvesting capability of single layers of an absorber on the surface and therefore low current densities (Watson and Meyer, 2005).

A breakthrough for these SCs was established by the replacement of the planar electrode by a mesoporous nanocrystalline semiconductor film, possessing a high surface area (O'Regan et al., 1990). In this case, the surface area allowing for a sensitizer to bind on the surface can be three orders of magnitude larger than the geometric area of the sample on a 10 μm-thick sample, which allowed for a drastic increase of power conversion yield up to 12% for SCs (O'Regan and Grätzel, 1991). In fact, Grätzel's group has used for the first time mesoscopic semiconductor architectures for the solar production of electricity and fuels. At the moment, the great majority of devices perspective for SCs employ mesoscopic structures composed of semiconductor nanoparticles (or quantum dots, QDs) as a key substrate element. In nanosized systems, the conduction band is flat within semiconductor particles, and no electric field prevents back electron transfer. Another mechanism should thus prevail, that ensures efficient sustainable charge separation. The general reactions realized in semiconductor dye-sensitized SCs are depicted in Fig. 5.4.

It follows from the Marcus model for the photoinduced electron transfer (Marcus, 1993) that the oxidation potential of the dye excited state should be more negative than the flat band potential of the semiconductor, which is equal to the conduction band potential for quantum dots. If the coupling between the dye and the semiconductor is large enough, direct electron injection into trap states is feasible, even if the conduction band potential lies at a higher energy (Huber et al., 2002). Typically, anchoring the sensitizer to the semiconductor surface is the most practical approach for an efficient sensitization. It should be mentioned in this case, that anchoring of organic sensitizers to a nanocrystalline semiconductor surface requires the existence of specific functional groups in these molecules that can form bonds with the acidic Ti^{4+} surface ions. It means that, like for other QD–dye nanoassemblies (discussed in Chapter 1 of this book), the anchoring conditions and their influence on the energy relaxation processes in the systems under study should be taken into account also.

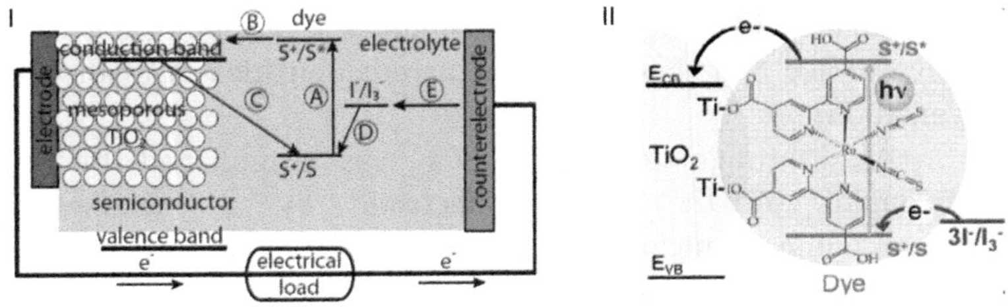

Figure 5.4 Schematic presentation of the composition and processes in (I) and dye sensitization mechanism (II). I: Light is absorbed by a sensitizer which is bound to the surface of a semiconductor (A). Photoinduced electron transfer dye→ TiO_2 (B). The oxidized sensitizer S^+ thus formed can recombine via the injected electron (C) or be regenerated via an external electron donor (D). The competition between these two reactions is crucial to the photon-to-current conversion yield of the SC. The extracted electron can be then used in an external circuit and return to a counter-electrode. The oxidized electron donor transports the positive charge to the counter-electrode where it is reduced (E). II: light absorption (A), electron injection (B), recombination (C), interception of the oxidized dye by a redox mediator (D), redox mediator regeneration at the counter electrode (E).

Basically, photoinduced processes in semiconductor dye-sensitized SCs are realized in competition with a series of unwanted reactions that might occur within the system and lead to a decrease of a power conversion yield. The two major channels for charge annihilation are the recombination of the injected electrons with either the oxidized dye-sensitizer or the oxidized redox mediate. As a consequence, timescales of these reactions and competition between different channels are the key point for efficient current collection. Namely these main reasons should be taken into account upon searching new components for semiconductor dye-sensitized SCs. In a lot of cases, the semiconductor of choice for making SCs is titanium dioxide (TiO_2).

It was shown also that QDs such as CdS or PbS can be efficient sensitizers (Tachibana et al., 2009; Hyun et al., 2008; Anusorn et al., 2008). In fact, low-band gap semiconductor nanostructures can harvest visible light energy if assembled on an electrode surface. Correspondingly, CdS and CdSe QDs being chemically and electrochemically deposited onto TiO_2, SnO_2, or ZnO surfaces are capable to inject excited electrons (via photoinduced electron transfer) and generate photocurrent under visible light irradiation. The systematic study in this direction has shown that the photoresponse of quantum dot-sensitized SCs may be modulated by varying the size and shape of semiconductor nanostructures (between particulate and tubular morphologies, see Fig. 5.5) (Anusorn et al., 2008; Kongkanand et al., 2008). As a consequence, quantum dot-sensitized SCs have emerged as one of the most promising technologies for solution-processed thin-film photovoltaics, leveraging their solution processing and facile band gap driven tailoring (Kim et al., 2013; Semonin et al., 2011; Hillhouse et al., 2009; Guijarro et al., 2010; Hod et al., 2011; Wang et al., 2011; Stolle et al., 2013).

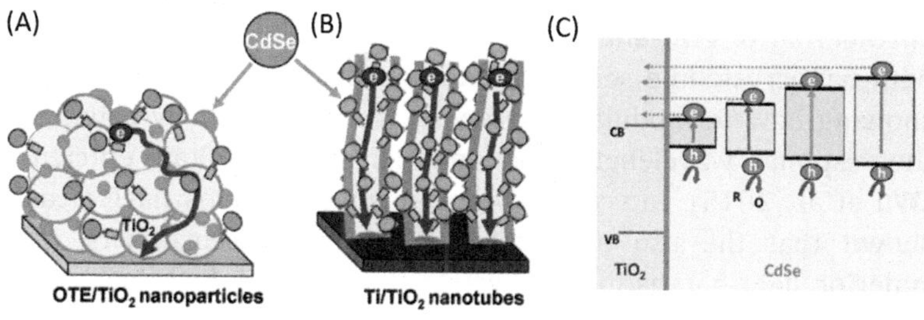

Figure 5.5 Schematic presentation of the directed electron transport through randomly distributed TiO_2 particles (A) and TiO_2 nanotube films modified with CdSe QDs (B) as well as relative energy levels diagram for differently sized CdSe QDs Quantum Dots and TiO_2 supporting architectures (C). The efficiency of the photoinduced electron transfer depends on the energy difference between the two conduction bands shown by arrows (band positions are for reference only and not drawn on a real energy scale). Adapted from Anusorn et al. (2008).

From the practical point of view, dye- and QD-sensitized SCs may be considered as one of the most promising photovoltaic technologies with relatively low cost, low weight, and good flexibility. At the moment, the main aims are devoted to enhance the industrial development of semiconductor sensitized solar cells on a large scale, including portable electronic devices and electricity glass panels. Correspondingly, the optimization of SCs should include at least three main aspects: (i) materials, including not only light-absorbing compounds but also electron and hole conductors and counter electrodes, (ii) control of recombination and band alignment by surface treatments, and (iii) development of absorbing nanocomposites with enhanced light-harvesting and -collecting properties (Mora-Sero et al., 2010; Ruehle et al., 2010).

The practical application of SCs is based on further improvement in power conversion efficiency and long-term stability. Recently, in this direction a significant progress has been established based on the novel concept of the D-A-pi-A motif (e.g., donor–bridge–acceptor structure) for the molecular engineering of organic photosensitizers. Using new organic dyes and tetrapyrrole compounds (porphyrins) based on this motif can not only enhance photovoltaic performance, but also improve durability in SC applications. The basic principles (main of them are discussed in Chapter 1 of this book) are involved in molecular engineering of efficient D-A-pi-A sensitizers and their interaction with nanocrystalline semiconductors, providing a clear road map showing how to modulate the energy bands, rationally extending the response wavelength, and optimizing photovoltaic efficiency (Wu et al., 2015). Interestingly, recent investigations have clearly shown that the use of porphyrin-based photosensitizers with superior light-harvesting properties (like those which have been studied by us (Zenkevich et al., 2005, 2007, 2011, 2012, 2015; Zenkevich and von Borczyskowski, 2009, 2012, 2014, 2015; von Borczyskowski and Zenkevich, 2014)) has enabled the power conversion efficiency of dye-sensitized solar cells (DSCs) to reach 13% being stable for more than 300 h at 60°C under continuous full sun illumination that is comparable to those known for other dyes but with better cell performance and long-term stability (Lau et al., 2015; Mai et al., 2015). In addition, a non-aggregated Zn(II)octa(2,6-diphenylphenoxy) phthalocyanine has been used as a hole-transporting material for perovskite SCs giving a power conversion efficiency of 6.7% for the best cell under optimized configuration (Ramos et al., 2015).

Another approach of SCs preparation (some examples are shown schematically in Fig. 5.6) is based on the implementation of a non-destructive micro-contact printing method which allows for deposition of a thin quantum dot film onto a wide-band gap organic hole transport layer, *N,N′*-bis (3-methylphenyl)-*N,N′*-bis-(phenyl)-9,9-spiro-bifluorene in order to form a unique planar heterojunction photovoltaic device that serves to enhance charge separation. (Arango et al., 2009; Chuang et al., 2015; Jean et al., 2015). This structure is found to produce a much higher photovoltage effect than previously predicted for donor/acceptor heterojunction photovoltaics.

Further development will require reducing exciton and charge transport losses in order to permit efficient charge-generation from thicker QD films with improved absorption. Nanostructured donor/acceptor photovoltaics utilizing small molecule organics or conjugated polymers offer processing advantages that might enable high-throughput and large-area production. Because the device architecture is straightforward and the fabrication techniques are simple, QD tandem cells are easily attained, furthering the prospect for reasonable conversion efficiencies coupled with the potential for scalable manufacturability. However, power conversion efficiencies of these structures have remained low yet, due in large part to low open-circuit voltages.

Finally, it follows from this brief review that the performance of inorganic QD-based photovoltaic and light-emitting devices has become competitive to other state-of-the-art materials. Within future, electricity generated from solar cells will cost less than grid electricity, making it the cleanest, cheapest, and most abundant energy source on the planet. In addition, nanoassemblies based on CdSe, PbSe, and ZnO QDs functionalized by respective dyes (as sensitizers) and capped by necessary ligands (such as NH_2Me, $OPMe_3$) might be considered as prospective systems with pronounced application abilities such as conversion of the absorbed photon to electric current (solar cells, discussed above) or conversion of the photon to photochemical reaction such as H_2 from photoelectrochemical water splitting and reduction of CO_2 to methanol and other products (fuel cells) (Drew et al.,

2005; Sykora et al., 2006). In other words, recent advances as well as predicted prospects of QD-sensitized SCs partly discussed here provide the basis for consideration as "The Next Big Thing" in photovoltaics (Kamat, 2013). In general, the sensitization of wide-band gap nanosized semiconductor nanostructures is relevant for many other applications than SCs. Photochromic and electrochromic devices, photography, xerography, and displays are some of the other affected areas of possible applications (Kamat, 2000; Galoppini, 2004).

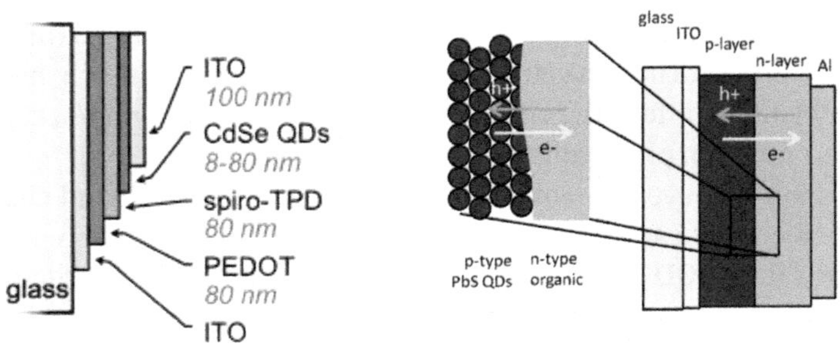

Figure 5.6 Schematic presentation of solar cells with various device architectures formed by a non-destructive micro-contact printing method. Following excitation and exciton diffusion to an interface, dissociation of the exciton produces free carriers that must diffuse to opposite electrodes in order to produce a photocurrent. Adapted from Arango et al. (2009 and Chuang et al. (2015).

5.4 Singlet Oxygen Generation and Biomedical Aspects for Semiconductor Quantum Dots and Their Bioconjugates

One of the important tasks of modern nanotechnology is the development of new nanomaterials and systems, which would significantly enhance the effects of organic drug molecules traditionally applied in biomedicine. In this respect, growing abilities to synthesize and surface-functionalized semiconductor QDs have provided new opportunities in this field as well as in bioanalytical and biological imaging applications. We do not claim to review all published applications where such QDs have been used because this is clearly beyond the scope of this section.

Instead, the intent is to provide a representative description of recent QD-related applications in the field of the photodynamic therapy (PDT) of cancer as far as some of our results (described in Chapter 1 of this book, Section 1.4.1) are directly connected with this direction.

PDT is an effective anticancer procedure that involves the handling of a tumor-localizing photosensitizer (PS) followed by light activation to generate highly cytotoxic reactive oxygen species, particularly singlet oxygen ($^1O_2$ or $^1\Delta_g$), which trigger cell apoptosis and necrosis (Dougherty et al., 1998; Agostinis, et al., 2011; Celli et al., 2010; Lovell et al., 2010; Cheng et al., 2014; Chunbai et al., 2015). The main mechanism for PDT (type II) involves the photoinduced formation of triplet excited molecules of the photosensitizer (via non-radiative intersystem crossing process, $S_1 \gtrsim T_1$) followed by triplet-triplet energy transfer $^3PS \rightarrow (^3O_2)$. In the result, the formation of active singlet oxygen ($^1O_2$) in a long-living $^1\Delta_g$ electronic state occurs (see Fig. 1.36 in Chapter 1 of this book). With respect to PDT effect, namely singlet oxygen $^1O_2$ is obviously considered as the main mediator of photocytotoxicity and causes oxidation and degradation of membranes of malignant cells. It is well known that since a direct excitation from the triplet ground state $^3O_2$ to the singlet excited state $^1O_2$ is forbidden by spin selection rules, oxygen is activated with a sequence of the resonance-exchange energy transfer events involving an excited triplet T_1 state of the photosensitizer. From this point of view, the class of the biologically important tetrapyrrolic pigments (such as porphyrins, chlorins, and phthalocyanines) characterized by high efficiencies of singlet oxygen generation and pronounced accumulation in cancer tissues is commonly used in PDT treatment (Dzhagarov et al., 1987; Dougherty et al., 1998; Zenkevich et al., 1996; Mironov, 2004; Isakau et al., 2008; Parkhats et al., 2009).

In this respect, bioconjugates based on semiconductor QDs with attached functionalized effective organic PSs (especially tetrapyrrolic pigments) may be considered as novel agents for PDT (Bakalova et al., 2004; Samia et al., 2006; Maslov et al., 2012; Rakovich, et al., 2010; Tsay et al., 2007; Juzenas et al., 2008; Kang et al., 2008; Yaghini et al., 2009; Biju et al., 2010; Zenkevich et al., 2011, 2013; Zhdanova et al., 2014; Martynenko et al., 2015; Zenkevich and von Borczyskowski, 2015). Typically, the formation of QD-PS nanoassemblies with participation of tetrapyrrolic pigments is based on at least three approaches: coordination self-assembly (Zenkevich et al., 2011, 2013), covalent bond (Kang et al., 2008) and electrostatic interactions (Martynenko et al., 2015). Some examples of such nanoassemblies are collected in Scheme 5.2, where the directionality of FRET→PS is shown for every nanoassembly as well as the full scheme of photoinduced events is presented also.

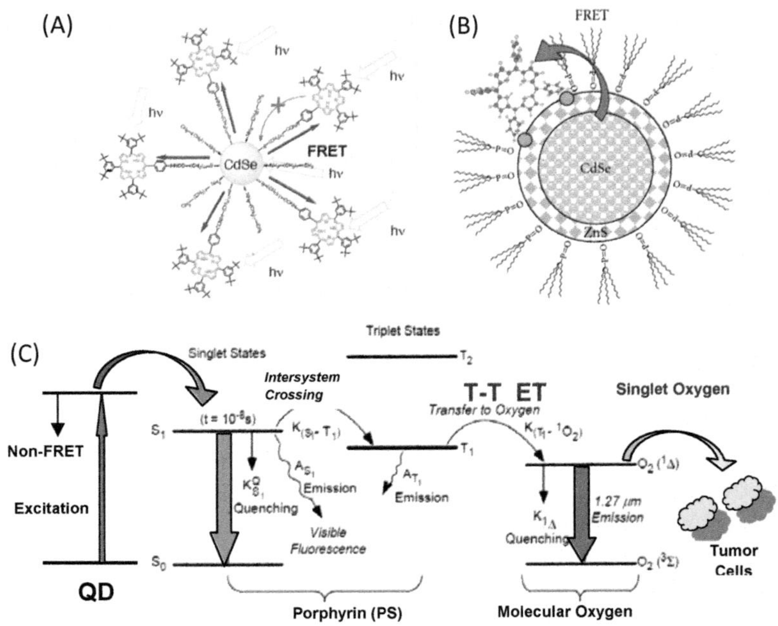

Scheme 5.2 Covalent (a) and coordination (b) interactions of porphyrin molecules with CdSe QD surface and pathways FRET QD→porphyrin (shown by arrows). (C) Full sequence of photoinduced processes in QD–porphyrin nanoassembly leading to singlet oxygen ($^1O_2$ or $^1\Delta_g$) generation.

Although photoinduced generation of singlet oxygen is the mandatory feature for PDT applications, QD–porphyrin nanoassemblies seem to be attractive because of the several reasons (Bakalova et al., 2004; Michalet et al., 2005; Zenkevich et al., 2013). (i) QDs may serve as potential delivery targeting platform for PS molecules (Probst et al., 2013) and poses imaging abilities (Mattoussi et al., 2012; Ji et al., 2015a, 2015b). (ii) QDs are characterized by high efficiencies of the photoluminescence (PL), high photostability, large absorption coefficients, continuous absorption bands for multicolor capability, narrow and symmetric emissions (Rogach, 2008; Gaponenko, 2010; Klimov, 2000; Mattoussi et al., 2012; Haghi et al., 2015; Ji et al., 2015a, 2015b; Guo et al., 2015; Kovalenko et al., 2015). (iii) The photon absorption of QDs can be tuned to the spectral transparency window of human skin. (iv) Taken together, realization of the effective FRET QD→Porphyrin and high yield of singlet oxygen generation by tetrapyrrolic pigments (mentioned above) would give a high positive effect. It means that the $^1O_2$ generation may be enhanced due to Foerster resonance energy transfer (FRET) process QD→PS (Bakalova et al., 2004; Samia et al., 2006; Maslov et al., 2012; Rakovich, et al., 2010; Tsay et al., 2007; Juzenas et al., 2008; Kang et al., 2008; Yaghini et al., 2009; Biju et al., 2010; Zenkevich et al., 2011, 2013; Zhdanova et al., 2014; Martynenko et al., 2015; Zenkevich and von Borczyskowski, 2015).

Nevertheless, some principal arguments should be taken into account in this case. The direct measurements of the efficiencies of $^1O_2$ generation by semiconductor QDs are very rare or contradictory (Bakalova et al., 2004; Samia et al., 2006; Ma et al., 2008; Qi et al., 2011; Samia et al., 2003; Shi et al., 2006; Wang et al., 2004), thus QD full capabilities for PDT have yet to be harnessed. With respect to QD–porphyrin nanoassemblies, the searching and the qualitative (predominantly) detection of FRET has received much attention within the last decade (Samia et al., 2003, 2006; Bakalova et al., 2004; Shi et al., 2006; Dayal et al., 2006; Tokuoka et al., 2006; Rakovich, et al., 2010; Tsay et al., 2007; Juzenas et al., 2008; Kang et al., 2008; Lihitkar et al., 2009; Qi et al., 2011; Maslov et al., 2012; Shibu et al., 2013; Zhdanova et al., 2014; Oluwole et al., 2015; Chambrier et al., 2015; Dadadzhanov et al., 2015). However, as was discussed in Chapter 7 (Sections 7.2.2, 7.2.3, 7.3.3, and 7.4), in a lot of cases FRET QD→Dye (including porphyrin acceptors) is not the only one reason of QD PL quenching in nanoassemblies and namely non-FRET processes are dominant in QD PL quenching (Zenkevich et al., 2005; Blaudeck et al., 2008, 2012; Kowerko et al., 2009, 2010; Kowerko, 2010; von Borczyskowski and Zenkevich, 2014; Zenkevich et al., 2015; Zenkevich and von Borczyskowski, 2015). Correspondingly, in order to select the main prospective QD–porphyrin nanoassemblies for possible PDT applications the direct quantitative analysis of FRET should be done in every case, based on the comparison of the experimental values of FRET efficiencies via the donor (QD) PL quenching and the sensitization of the acceptor (porphyrin) fluorescence (Zenkevich et al., 2005; Blaudeck et al., 2008, 2012; Lemon et al., 2013).

Very recently, we have carried out a comprehensive quantitative experimental analysis of this problem based on steady-state and picosecond time-resolved QD PL and porphyrin fluorescence data combined with the parallel direct measurements of near-IR photoluminescence measurements of singlet oxygen $^1O_2$ emission quantum yields (Zenkevich et al., 2011, 2013; Zenkevich and von Borczyskowski, 2015) (see also Section 1.4.1 in Chapter 1 of this book). The main conclusions are the following. (i) The quantum efficiencies of singlet oxygen generation by individual CdSe/ZnS QDs are very low ($\gamma_\Delta \leq 1.5\%$); thus at ambient temperature QDs are hardly useful for the direct generation of $^1O_2$. (ii) For QD-porphyrin nanoassemblies, the non-linear dependence of the attached porphyrin fluorescence on the energy of laser excitation presents itself as a direct proof of the realization of FRET process QD$\rightarrow$porphyrin competing with both radiative electron-hole recombination and non-radiative Auger processes in the conditions of powerful excitation. (iii) FRET efficiencies Φ_{FRET} obtained from the direct sensitization data for porphyrin fluorescence in QD-porphyrin nanoassemblies are in a good coincidence with FRET efficiencies Φ^{Δ}_{FRET} evaluated via the direct $^1O_2$ emission measurements at low laser excitation. (iv) Efficiencies of $^1O_2$ generation by QD-porphyrin nanoassemblies are higher also than those measured for merely porphyrin molecules at the same molar concentrations. Such quantitative analysis has been done initially, which shows that namely the FRET process QD$\rightarrow$porphyrin is the reason of $^1O_2$ generation by QD-porphyrin nanoassemblies. These conclusions have been recently proven in Lemon et al. (2013) and Martynenko et al. (2015).

It should be noted that water-soluble QD-porphyrin/chlorin nanoassemblies (based on quantum dots and quantum rods) with high efficiencies of $^1O_2$ generation are most attractive for artificial biocompatibility materials used for PDT in vivo

(Wuister et al., 2003; Tsay et al., 2007; Rakovich et al., 2010; Fakhar-e-Alam et al., 2012). Indeed, it has been found very recently (Martynenko et al., 2015), that stable water-soluble nanoassemblies based on ZnSe/ZnS QDs and chlorin e_6 molecules (Ce6) may be used as a successful delivery agent and photosensitizer in PDT against the *Erlich acsite carcinoma* (EAC) cell culture in water solutions (Fig. 5.7). In addition, it was demonstrated that QD–chlorin nanoassemblies show a twofold enhancement of the cancer cell photodynamic destruction as compared to that of alone chlorin e_6 molecules (due to QD sensitization effects) without any noticeable dark or phototoxicity.

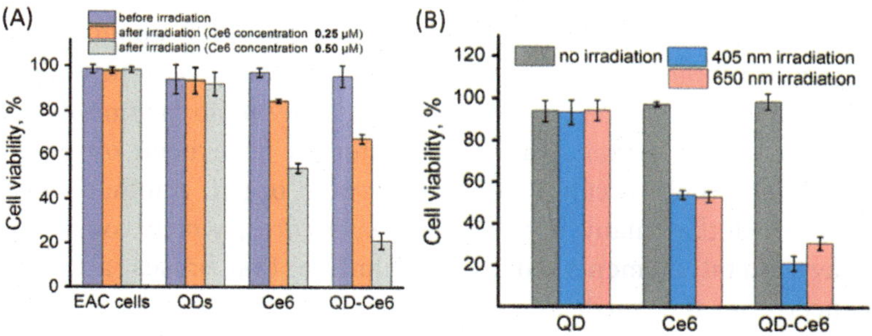

Figure 5.7 Comparative photodynamic cancer cell killing efficiency with QD–chlorin nanoassemblies. A: In vitro photodynamic cancer EAC cell destruction. Histograms of cells treated with QDs, Ce6 and the QD–Ce6 nanoassemblies without and with irradiation by the 405 nm. B: Histograms with cell viability data of EAC cells incubated with QD–Ce6, free Ce6, or QDs, respectively, at Ce6 and QDs concentrations of 0.5 μM in every sample. Adapted from Martynenko et al. (2015).

One should mention also, that semiconductor QDs are one of the best energy donors via two-photon excitation (TPE), which are characterized by a large two-photon absorption (TPA) cross section, and the maximum TPA wavelength can be suitably adjusted via changing the QD size (Schmidt et al., 1996; Pu et al., 2006; Wen et al., 2009; Fowley et al., 2012; Skripka et al., 2013; Chou et al., 2013). In this respect, biocompatible CdSe QD–porphyrin nanoassemblies not only are capable of effective generation of singlet oxygen for PDT through TPA of the QDs followed by FRET to the porphyrin photosensitizer but also are considered as good candidates for "see and treat" PDT in live cell imaging when controlling the molar ratio of dye quenching QDs (Qi et al., 2011) or may be used for oxygen sensing over physiological oxygen ranges (Lemon et al., 2013). For instance, the micelle-encapsulated nanoassemblies based on QDs and Pd-porphyrins (with reasonable FRET efficiencies under both one- and two-photon excitation) have been employed for the lifetime-based quantification of molecular oxygen levels in aqueous media and in vivo (Lemon et al., 2015).

Concluding, though several examples of employing QDs-PS nanoassemblies for PDT have been described, their full capabilities have yet to be harnessed, and QD-photosensitizer nanobioconjugates are still an intriguing option for PDT applications. There are indications in the literature that manufactured nanoscale materials may be distributed in the body in so far unpredictable ways and certain nanoscale materials have been observed to preferentially accumulate in particular organ. Another issue is the fact that many current QDs have cores, which consist of heavy metal elements such as cadmium that raises a major concern in the medical community. Overall, there is a great need for research directed towards increasing the fundamental understanding of QD interactions at the biological and cellular level.

5.5 Quantum Dot-Based Nanoassemblies in Sensing, Imaging, and Diagnostics

In the past two decades it was definitely shown that the ability to rationally design semiconductor QDs of various type and composition with cellular delivery, imaging, sensing and therapeutic functions is one of the most important and challenging tasks in biomedical nanotechnology (Bruchez et al., 1998; Parak et al., 2003; Clapp et al., 2004; Alivisatos, 2004; Ferrari, 2005; Michalet et al., 2005; Kim et al., 2013; Murcia and Naumann, 2005). Correspondingly, the combination of biomedicine and nanotechnology has led to a new generation of nanodevices that make it possible to characterize the chemical, mechanical, and other biological properties, as well as discover novel phenomena and processes occurring at the molecular level. These advances provide science with a wide range of tools for biomedical applications in therapeutic, diagnostic, and preventive medicine (Vo-Dinh, 2007; Moyano and Rotello, 2011). At the moment, nanomaterials based on semiconductor QDs and attached functionalized organic molecules enable the study of chemical and biochemical processes at a level and in dimensions that may not have been envisioned some 20 years ago. In recent years, due to QD inherent small size and the unusual size-dependent optical, magnetic, catalytic, and mechanical properties, a remarkable progress has been achieved in the studies and development of nanosensors and optical (imaging and diagnostics) nanotechnology (Irudayaraj, 2012; Lawrence et al., 2012; Jin and Hildebrandt, 2012; Shatkin, 2012; Mattoussi et al., 2012; Li and Wu, 2013; He and Ma, 2014; Guo et al., 2015; Lemon et al., 2015). It includes many interesting applications, such as the detection of bioterrorist threats, food security, virology, explosive detection (Cusano et al., 2012).

From above discussion it becomes evident that understanding physicochemical and photoinduced processes realized for semiconductor QDs with attached functionalized molecules as well as the interactions of created QD-based nanoassemblies with biomolecules and biosystems is a critical goal in both biomedicine and environmental science. With respect to QD–dye nanoassemblies, these questions have been under the study in our groups and described in Chapters 2 and 3 of Zenkevich and von Borczyskowski (2016) and in Chapters 1–4 of this book. The objective of this section is to highlight some approaches being used for the creation of nanoassemblies based on semiconductor QDs and functionalized organic and/or biological components capable for applications in biomedical nanotechnology. In addition, we are willing to show also some examples of recent developments addressing the bioconjugation of QDs and their surface modification using biocompatible coatings.

Among the different types of fluorescent nanoparticles, semiconductor QDs show the greatest promise as labels in biological imaging applications. To apply fluorescent QDs to biosensing and biomedical imaging applications, it is crucial to develop strategies towards their biofunctionalization (Chan et al., 2002; Mattoussi et al., 2012). Typically, it includes the properly functionalized linkage of biomolecules to QDs (the so-called bioconjugation) and the design of appropriate biocompatible coatings.

Figure 5.8 shows several common strategies for the conjugation of biomolecules to semiconductor QDs, including direct physisorption of biomolecules (A), assisted physisorption using pre-bound molecules (B), chemical linkage of biomolecules to cross-linkers either physisorbed or chemisorbed on the nanoparticle surface (C), direct chemical coupling of biomolecules to nanoparticles (D), and the targeted binding of biotinylated biomolecules to streptavidin coated nanoparticles via biotin–streptavidin coupling (E). It should be mentioned that in spite of the nanosize scale of QDs, their surface area is large enough for the attachment of multiple biomolecules.

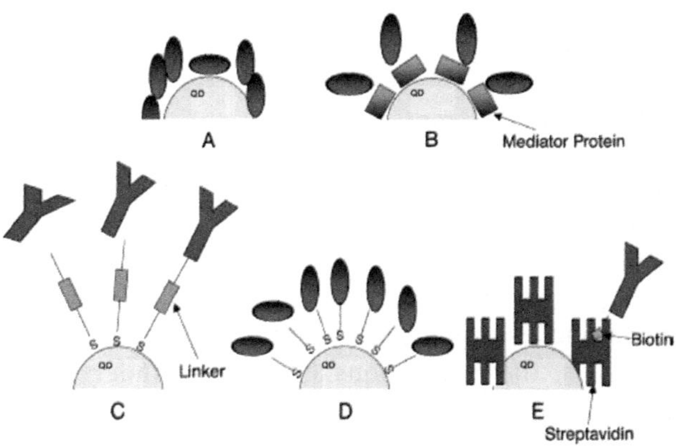

Figure 5.8 Schemes for bioconjugation of biomolecules to semiconductor QDs: (A) direct physisorption; (B) assisted sorption using biocompatible coatings (mediator protein); (C) chemical linkage of biomolecules via cross-linkers on QD surface; (D) direct chemical bond of biomolecules to QD; (E) the targeted binding of biotinylated biomolecules to streptavidin coated QDs via biotin–streptavidin coupling. Adapted from Murcia and Naumann (2005).

In addition, in some cases chemical modification of QDs and polymer-encapsulated CdSe QDs (i.e., attachment of antibodies and other proteins) is realized via electrostatic interactions (Sapsford et al., 2006; Yezhelyev et al., 2008). At the moment, various additional schemes of QDs biofunctionalization are used for the rational design of multifunctional nanoassemblies for a variety of applications, ranging from protein tracking, delivery, imaging and ligand–receptor binding to intracellular sensing (see, e.g., Aldeek et al. (2013), Moyano and Rotello (2011), and Zhan et al. (2013)). It should be mentioned also that namely fluorescent QDs are highly promising probes for bioanalytical applications. In particular, their use in the field of biosensors is attractive because fluorescence-based techniques are extremely sensitive and QD probes show high photostability, thus allowing for long-term observations (Irudayaraj, 2012; He and Ma, 2014; Cusano et al., 2012; Baltazar et al., 2012).

Not willing to give a thorough analysis of applications of QDs and their bioconjugates (presented in references and books cited above as well as recently in Karachevtsev (2015) and Bawa et al. (2016), we would like to present here a few examples of how the main exciton relaxation processes (such as FRET and/or photoinduced charge transfer (CT) discussed in details in previous Chapters of this book and in Zenkevich and von Borczyskowski (2016)) may be exploited in bio-nanotechnology. For the purposes of this review section, we define QD-based biosensing simply as the utilization of specificity of QD-based nanoassemblies to detect and/or quantitatively analyze molecules or biological species.

Numerous studies have shown that QDs (having size tunable narrow emission spectra and a broad absorption spectrum with high extinction coefficients) might significantly improve FRET assay performances in a wide variety of sensing schemes (Sapsford et al., 2006; Medinntz et al., 2003; Somers et al., 2007; Suzuki et al., 2008). In this case, QDs are used as FRET donors being sensitive to detect small analytes by utilizing a common strategy that relies on conjugating QDs to target binding receptors (proteins, antibody fragments or DNA aptamers) (Sapsford et al., 2006; Medintz et al., 2004).

One example of a maltose sensor based on CdSe/ZnS QDs (Medintz et al., 2004; Sapsford et al., 2006) is shown in Fig. 5.9. Here, CdSe/ZnS QDs were surface-functionalized with a maltose-binding protein (MBP) containing β-cyclodextrin-QSY-9 attached to the binding site to quench the QD photoluminescence (PL) via FRET QD$\rightarrow$QSY-9 (Fig. 5.9A). In another QD-based maltose sensor double FRET is used: QD$\rightarrow$Dye Cy3$\rightarrow$QSY-9 (Fig. 5.9B). For the case A, the sensor concept is that maltose displaces the b-cyclodextrin-QSY-9 (due to FRET switching off), thus inducing QD photoluminescence (due to FRET switching off). At the initial stage, the formation of QD-MBP-QSY9 bioconjugate results in QD PL increase. Titration of the bioconjugate with an increasing concentration of maltose leads to the QD PL increase (Fig. 5.9C). Analysis of QD PL titration data (Fig. 5.9D) permitted to estimate a maltose sensing range of ~500 ÷ 100 μM. The same approach has been used for elaboration of nanosensors capable to detect various analytes, including explosive TNT materials as well as DNA or RNA oligomers (Sapsford et al., 2006 and references therein). Very recently, it has been shown also that analyte-dependent conformational changes in a nanosensor based on QD with pH-triggered oligonucleotide may control FRET efficiency (via donor-acceptor distance), which results in a reporter for pH values in individual endosomes of living cells (Kay et al., 2013). Two novel and convenient methods based on distinct changes of fluorescence and in UV-VIS absorption spectra of CdTe QDs and Au nanoparticles have been developed for determination of lincomycin in aqueous solutions (Ge et al., 2015). In addition, semiconductor QDs offer a versatile platform for the development of supramolecular oxygen sensing nanoassemblies based on two-photon excitation of QDs followed by FRET to attached porphyrin molecules (Lemon, 2013, 2015). Finally, single-QD-based nanosensors with FRET and QD-based single-molecule tracking in biological systems (Zhang et al., 2005; Hadar et al., 2015) may be considered as a fascinating emerging research area in nanobiotechnology.

Typically, colloidal QDs have a large fraction of their atoms arrayed on their surface and are usually capped with capping ligands or functionalized organic molecules or bioobjects. As was discussed in Chapter 1 of this book, it makes QD PL properties highly sensitive to attached moieties and surface or interface events. In some cases, changes of QD PL properties may be caused by potential charge transfer (CT) to, or from, these moieties and the surrounding environment. These CT events have been recently studied in covalently assembled CdSe/ZnS QD-dopamine bioconjugates containing the redox active dopamine (Ji et al., 2015a, 2015b) (see Fig. 5.10). Using the combination of steady-state and time-resolved fluorescence along with transient absorption measurements, it was found that QD PL quenching (caused by CT) depends on a few parameters: the QD size (thus the conduction and valence energy levels), the number of dopamine groups per QD-conjugate (valence), and buffer pH (changed from acidic to alkaline). Variation of pH values from acidic to alkaline leads to a decrease in the oxidation potential of the dopamine along with a partial chemical transformation of a few catechols to quinones in the QD-dopamine bioconjugates. More importantly, QD PL quenching efficiency can be strongly influenced by the energy levels of the QDs with substantially higher efficiencies measured for wider band gap dots and vice versa.

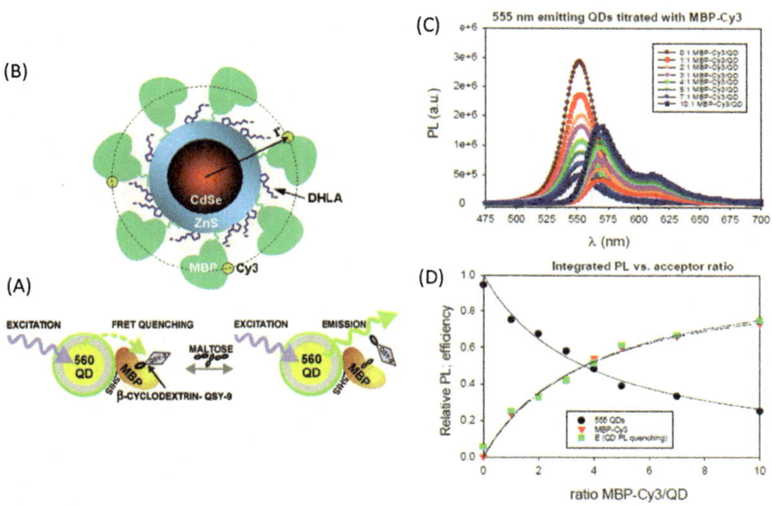

Figure 5.9 Schematic representation of a CdSe/ZnS QDs-Maltose Binding Protein nanosensor based on FRET QD-β-cyclodextrin QSY-9 (A, B), and spectral (C) and kinetic (D) manifestation of its interaction with maltose during titration experiments at ambient temperature. Some details are discussed in the text. Adapted partly from Medintz et al. (2004) and Sapsford et al. (2006).

Thus, covalently assembled CdSe/ZnS QD-dopamine bioconjugates may be considered as perspective nanosensors based on complex CT interactions. In addition, the existence of a polyethylene glycol bridge between QD and dopamine may be used in the realization of flexible redox-active fluorescent platform for a potential integration in biological systems with applications in intracellular sensing and vasculature imaging.

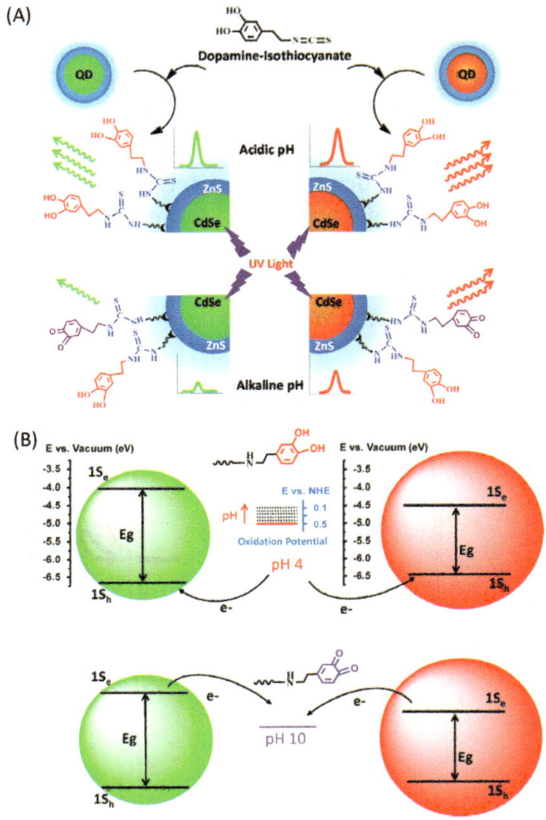

Figure 5.10 Schematic representation of two sets of CdSe/ZnS QD–dopamine bioconjugates (A) and charge transfer events (involving electron and hole carriers) in acidic and alkaline media between QDs and dopamine in the two sets of bioconjugates. (A) Changes of the QD PL intensity are shown in the insets, higher quenching is found for green-emitting QDs (with smaller size and wider band gap) in alkaline buffer. (B): Energy levels for D–A pairs and corresponding CT pathways have been deduced from experimental data and available references. Adapted from Ji et al. (2015a).

Furthermore, it should be mentioned that within last years, semiconductor QD-based theranostic approaches have emerged as an interdisciplinary area, which may be considered as a background for understanding the components, processes, dynamics, and therapies of a disease at a molecular level. In this direction of nanomedicine, semiconductor QDs are exploited as a multifunctional heterogeneous platform combining medical diagnostics, disease analysis and subsequent treatment procedures (Stasiuk et al., 2011; Jing, et al., 2014; Karachevtsev, 2015; Bawa et al., 2016). Finally, it is interesting to note that even photoluminescent and superparamagnetic graphene oxide–iron oxide quantum dots are considered as a novel nanosystem suitable for use in the detection, monitoring, and treatment of diseases (Justin et al., 2016).

5.6 Super-Resolution Microscopy with Quantum Dots

The breakthrough of single-molecule imaging has revolutionized microscopic imaging in biology. However, the great potential of single (organic) molecule measurements is often restricted by photobleaching. In contrast, quantum dots are very stable with respect to photobleaching. Therefore QDs are an alternative for single quantum object imaging. A comprehensive review of the application of QDs in biology has been published recently (Petryayeva et al., 2013). Also, a comparison between the application of molecules and QDs is available (Resch-Genger et al., 2008). However, as we have shown throughout this book, and in von Borczyskowski and Zenkevich (2016) quantum dots are subject to blinking. The blinking of QDs often limits their applications in certain biophysical areas, such as in single particle tracking since the track of the QD is lost during a long dark period (Dahan et al., 2003).

On the other hand, blinking of quantum dots can facilitate achieving optical super-resolution. The Nobel Prize in chemistry was awarded in 2014 to acknowledge developments in super-resolution optical microscopy. One of the approaches to obtain super-resolution is based on stochastic processes in single-molecule detection (Betzig et al., 2006). Blinking QDs fall also into this category and have been used by various approaches to establish super-resolution basically via a method called "super-resolution optical fluctuation imaging" (SOFI) (see Figs. 5.11 and 5.12) and its variants (Lidke et al., 2005; Dertinger et al., 2009; Watanabe et al., 2010; Chien et al., 2010; Wang et al., 2013).

With some of these methods, three-dimensional super-resolution imaging has been achieved up to 8–17 nm in the x–y

plane and 58 to 81 nm in the *z*-direction. Similar resolution is achieved with 3D-STORM ("3D stochastic optical reconstruction microscopy") and related techniques based on single-molecule techniques (Huang et al., 2008; Betzig et al., 2006; Hess et al., 2006; Heilemann et al., 2009). However, molecule based techniques rely on activating a subset of organic dye fluorophores or pairs of fluorophores. In addition, permanent photobleaching is a problem. Additionally, the use of two different lasers or external chemicals to be added is disadvantageous.

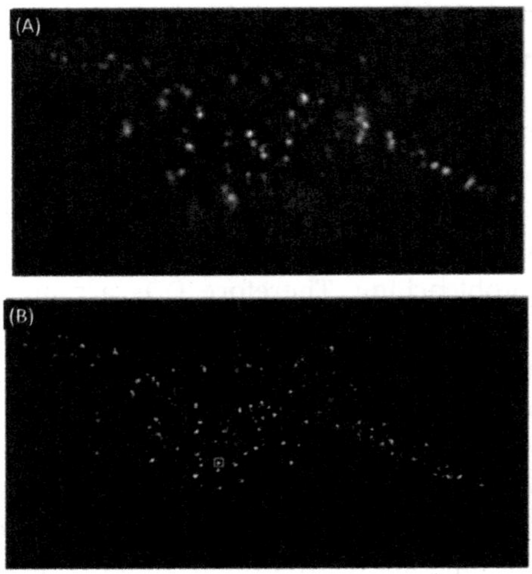

Figure 5.11 Super-resolution using QDs is applied to image epidermal growth factor receptor (EGFR), labeled with quantum dots, at the plasma membrane of resting basal breast cancer cell. (A) Image from normal diffraction limited fluorescence microscopy. (B) Three-dimensional super-resolution image reconstructed with the QDB3 algorithm (Wang et al., 2013).

On the other hand, super-resolution optical fluctuation imaging (SOFI) relies on the intrinsic blinking of the individual fluorophore (QDs). In this method a sequence of blinking events is recorded over time and statistically analyzed. Basically, during luminescence on-time (e.g., of QD1) the position of the resolution limited image of QD1 can be collected with high accuracy. In case nearby QD2 is "off" its eventual photoluminescence does not interfere with the one of the on-intensity of QD1. Since blinking is a statistic process the reverse situation will be eventually found: QD2 will be "on" and QD1 "off." Consequently, super-resolution of a distance d depends on the accuracy how individual positions of luminescence images can be determined. This distance d will at wavelength λ strongly depend on the number m of photons according to (Betzig et al., 2006).

$$d = \frac{\lambda}{2n \sin \alpha \cdot \sqrt{m}}, \tag{5.1}$$

where n is the diffraction constant and α is the aperture angle.

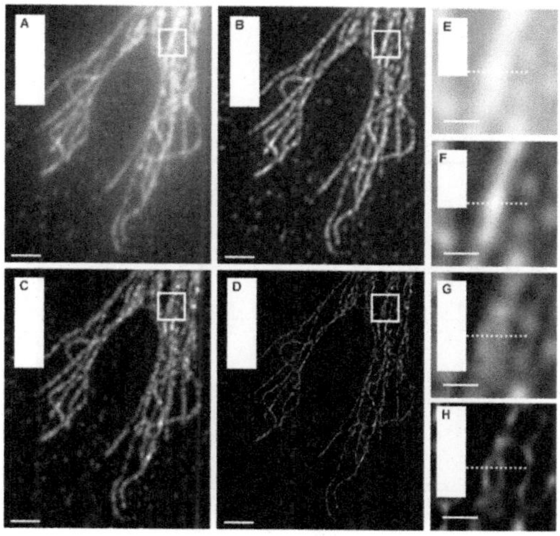

Figure 5.12 SOFI images of cells under various conditions. Wide-field image of QD625 labeled 3T3 cells. (A–D) from top left to bottom right: (A) Original image generated by time averaging all frames of the acquired movie (3,000 frames, 100 ms per frame). (B) The image in A deconvolved. (C) Second order SOFI image. (D) The image in C deconvolved. (E–H) Magnified views of the boxed regions in A–D. (Scale bars: A–D, 2 μm; E–H, 500 nm) (Dertinger et al., 2009).

Super-resolution optical fluctuation imaging (SOFI) needs an impractical number of images for real-time observation. However, real-time nanoscopy can be improved by modifying SOFI via enhancing the blinking of quantum dots. As a consequence, blinking improves the resolution. Watanabe et al. (2010) obtained microscopic images with 90 nm and 80 ms spatial-temporal resolution. This example shows that it is of advantage to enhance blinking (instead of suppressing) in combination with super-resolution.

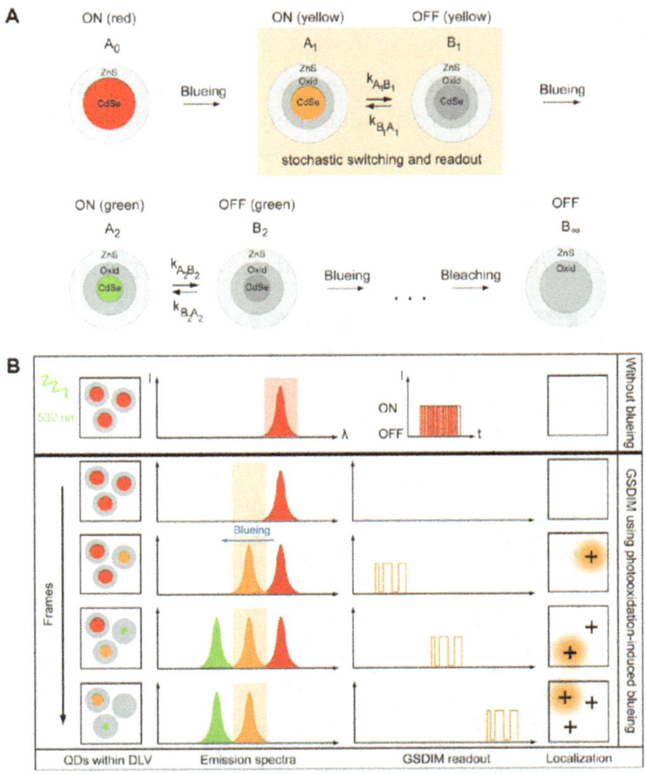

Figure 5.13 (A) Blue shift of emission exemplified for red-emitting quantum dots. Under continuous illumination, the size of the CdSe core decreases by photooxidation, continuously shifting the emission to the blue, ideally across the whole spectrum. Switching between fluorescence "on" and "off" states is apparent during the entire course of blueing until the core is bleached. The "on" and "off" state of a blue-shifted quantum dot marker is used for stochastic switching and readout. (B) Blue shift enables GSDIM ("ground state depletion individual molecule return"). Without blueing, the signal from quantum dots overlaps within a diffraction limited volume (DLV) at a given time point due to rather short "off" periods. Disparate time periods between illumination and onset of blueing separate the fluorescence traces. Successive spectral translation of subsets of quantum dots into the blue-shifted detection window yields a diffraction unlimited image benefitting from the huge number of photons gained from a dot before bleaching or further blueing. (Hoyer et al., 2010).

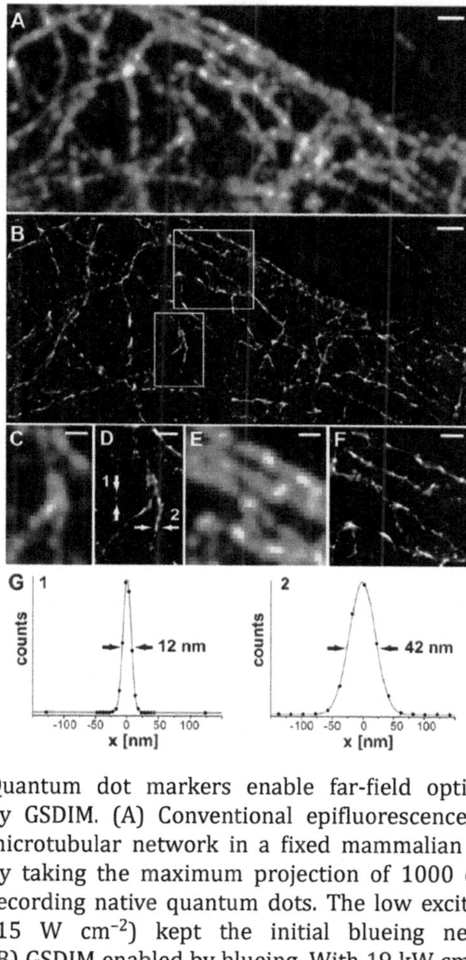

Figure 5.14 Quantum dot markers enable far-field optical nanoscopy by GSDIM. (A) Conventional epifluorescence image of the microtubular network in a fixed mammalian cell generated by taking the maximum projection of 1000 camera frames recording native quantum dots. The low excitation intensity (15 W cm^{-2}) kept the initial blueing negligibly small. (B) GSDIM enabled by blueing. With 19 kW cm^{-2} illumination by a diode laser emitting at 532 nm, a set of 120,000 frames with 15 ms frame time has been recorded. The many frames ensure that >90% of all markers are shifted into the detection channel ranging from 538 to 645 nm and contribute to the image having a pixel size of 20 nm. Scale bars: 1 μm. Regions in boxes are magnified in (D) and (F). The corresponding epifluorescence images are displayed in (C) and (E). The line profiles of a single quantum dot marker and a single microtubule strand are shown in (G1) and (G2), respectively. Black dots represent the data points of the GSDIM image, while the red line is the corresponding Gaussian fit. Scale bars: 500 nm. Adapted from (Hoyer et al., 2010).

A further approach to obtain high resolution is based on the "blueing" of bio-conjugated CdSe/ZnS QDs that is the gradual shift to high PL emission energies based on (controllable) photoinduced oxidation (Hoyer et al., 2010). Figure 5.13 shows such a scheme reaching nearly a sub-diffraction resolution close to the diameter of the QD. The improvement of resolution is demonstrated in Fig. 5.14. These improvements show that surface engineering (with respect to controllable oxidation) of QDs is a further example for tuning QD properties.

5.7 Conclusions

In this chapter, we have described that upon nanoassembly formation, photochemical processes occurring at the interface between inorganic semiconductor QDs and functionalized organic molecules and/or bioobjects are complex and may yield new and unexplored phenomena. The insights gained into understanding what parameters may control the photochemical and optical properties of QDs upon attachment of functionalized moieties and how such attachment onto the QD surface proceeds, are greatly informative, and provide a real basis for using QD-based nanoassemblies in various fields of nanotechnology and biomedicine. The recent progress in nanotechnology gives the possibility to produce a variety of semiconductor QDs and functional nanoassemblies with unique physicochemical, optical, and/or electrical properties, which differ from those of bulk materials of the same composition.

The stimulating development of super-resolution microscopy (Nobel Prize in 2014) placed quantum dot blinking into a new prospective. The stochastic approaches of this new method make extensive use of blinking processes observed for most quantum dots. This is a completely new future for QDs since up to previous attempts avoiding blinking was a mayor goal. At present, it is a desirable to control and engineer blinking, to which the given book hopefully contributes. Nevertheless, interface engineering of QDs is still of mayor importance for at least two reasons, namely to control blinking and to functionalize interfaces in order to incorporate QDs into biological systems which is highly desirable since super-resolution has found an extremely wide application in biology.

We have presented only a limited number of possible applications of QDs and QD-based nanoassemblies. Indeed, nanostructure science and technology is a broad and interdisciplinary area of research and development that has been growing explosively worldwide in the past few years. It has the potential for revolutionizing the ways in which materials, products, and devices are created and the range and nature of functionalities that can be accessed. In this respect, we refer the reader to a very recent publication (Jie and Bawendi, 2015) that describes a colloidal quantum dot spectrometer which, as expected, will be useful in applications where minimizing the size, weight, cost, and complexity of the spectrometer is critical.

To conclude, fundamental research in the field of semiconductor QDs and various nanoassemblies on their basis will certainly continue to lead to novel phenomena, ideas, and concepts as well as for promising devices and applications in future.

References

Agostinis, P., Berg, K., Cengel, K. A., Foster, T. H., Girotti, A. W., Gollnick, S. O., Hahn, S. M., Hamblin, M. R., Juzeniene, A., Kessel, D., Korbelik, M., Moan, J., Mroz, P., Nowis, D., Piette, J., Wilson, B. C., and Golab, J. (2011). Photodynamic therapy of cancer: An update, *J. Clin.*, **61**, pp. 250–281.

Aldeek, F., Safi, M., Zhan, N., Palui, G., and Mattoussi, H. (2013). Understanding the self-assembly of proteins onto gold nanoparticles and quantum dots driven by metal-histidine coordination, *ACS Nano*, **7**, pp. 10197–10210.

Alivisatos, P. (2004). The use of nanocrystals in biological detection, *Nat. Biotechnol.*, **22**, pp. 47–52.

Anusorn, K., Tvrdy, K., Takechi, K., Kuno, M., and Kamat, P. V. (2008). Quantum dot solar cells. Tuning photoresponse through size and shape control of CdSe-TiO_2 architecture, *J. Am. Chem. Soc.*, **130**, pp. 4007–4015.

Arango, A. C., Oertel, D. C., Xu, Y., Bawendi, M. G., and Bulović, V. (2009). Heterojunction photovoltaics using printed colloidal quantum dots as a photosensitive layer, *Nano Lett.*, **9**, pp. 860–863.

Arciniegas, M. P., Kim, M. R., de Graaf, J., Brescia, R., Marras, S., Miszta, K., Dijkstra, M., van Roij, R., and Manna, L. (2014). Self-assembly of octapod-shaped colloidal nanocrystals into a hexagonal ballerina network embedded in a thin polymer film, *Nano Lett.*, **14**, pp. 1056–1063.

Atwater, H. A. (2007). The promise of plasmonics, *Sci. Am.*, **296**, pp. 56–63.

Bakalova, R., Ohba, H., Zhelev, Z., Nagase, T., Jose, R., Ishikawa, M., and Baba, Y. (2004). Quantum dot anti-CD conjugates: Are they potential photosensitizers or potentiators of classical photo-sensitizing agents in photodynamic therapy of cancer?, *Nano Lett.*, **4**, pp. 1567–1573.

Baldenebro-Lopez, J., Flores-Holguin, N., Castorena-Gonzalez, J., Glossman, D., Adegoke, M. O., and Nyokong, T. (2013). Molecular design of copper complexes as sensitizers for efficient dye-sensitized solar cells, *J. Photochem. Photobiol. A Chem.*, **267**, pp. 58–66.

Baltazar, R., Vistas, C. R., and Ferreira, G. N. M. (2012). In: *Nanocomposite Particles for Bio-Applications: Materials and Bio-Interfaces* (Trindade, T., and da Silva, A. L. D., eds.), Taylor & Francis, Boca Raton, FL, pp. 265–282.

Bawa, R., Audette, G. F., and Rubinstein. I. (2016). *Handbook of Clinical Nanomedicine: Nanoparticles, Imaging, Therapy, and Clinical Applications*, Pan Stanford Series on Nanomedicine, ISBN 9789814669207, Pan Stanford Publishing.

Betzig, E., Patterson, G. H., Sougrat, R., Lindwasser, O. W., Olenych, S., Bonifacino, J. S., Davidson, M. W., Lippincott-Schwartz, J., and Hess, H. F. (2006). Imaging intracellular fluorescent proteins at nanometer resolution, *Science*, **313**, pp. 1264–1645.

Bhushan, B. (ed.) (2012). *Encyclopedia of Nanotechnology*. ISBN 978-90-481-9750-7, Springer Science, Business Media B. V., Dordrecht, Heidelberg, New York, London.

Biju, V., Mundayoor, S., Omkumar, R. V., Anas, A., and Ishikawa, M. (2010). Bioconjugated quantum dots for cancer research: Present status, prospects and remaining issues, *Biotechnol. Adv.*, **28**, pp. 199–213.

Blaudeck, T., Zenkevich, E., Abdel-Mottaleb, M., Szwaykowska, K., Kowerko, D., Cichos, F., and von Borczyskowski, C. (2012). Formation principles and ligand dynamics for nanoassemblies of CdSe quantum dots and functionalized dye molecules, *ChemPhysChem*, **13**, pp. 959–972.

Blaudeck, T., Zenkevich, E. I., Cichos, F., and von Borczyskowski, C. (2008). Probing wave functions at semiconductor quantum dot surfaces by non-FRET photoluminescence quenching, *J. Phys. Chem. C*, **112**, pp. 20251–20257.

Bruchez, M. Jr., Moronne, M., Gin, P., Weiss, S., and Alivisatos, A. P. (1998). Semiconductor nanocrystals as fluorescent biological labels, *Science*, **281**, pp. 2013–2016.

Bucak, S., and Rende, D. (2013). *Colloid and Surface Chemistry: A Laboratory Guide for Exploration of the Nano World*, ISBN 9781466553101, CRC Press.

Bullen, C., and Mulvaney, P. (2006). The effects of chemisorption on the luminescence of CdSe quantum dots, *Langmuir*, **22**, pp. 3007–3013.

Celli, J. P., Spring, B. Q., Rizvi, I., Evans, C. L., Samkoe, K. S., Verma, S., Pogue, B. W., and Hasan, T. (2010). Imaging and photodynamic therapy: Mechanisms, monitoring, and optimization, *Chem. Rev.*, **110**, pp. 2795–2838.

Chambrier, I., Banerjee, C., Remiro-Buenamañana, S., Chao, Y., Cammidge, A. N., and Bochmann, M. (2015). Synthesis of porphyrin–CdSe quantum dot assemblies: Controlling ligand binding by substituent effects, *Inorg. Chem.* **54**, pp. 7368–7380.

Chan, W. C. W., Maxwell, D. J., Gao, X., Bailey, R. E., Han, M., and Nie, S. (2002). Luminescent quantum dots for multiplexed biological detection and imaging, *Curr. Opin. Biotechnol.*, **13**, pp. 40–46.

Cheng, Y., Doane, T. L., Chuang, C. H., Ziady, A., and Burda, C. (2014). Near infrared light-triggered drug generation and release from gold nanoparticle cCarriers for photodynamic therapy, *Small*, **10**, pp. 1799–1804.

Chien, F.-C., Kuo, C. W., and Chen, P. (2010). Localization imaging using blinking quantum dots, *Analyst*, **136**, pp. 1608–1613.

Chou, K. L, Won, N., Kwag, J., Kim, S., and Chen, J. Y. (2013). Femtosecond laser beam with a low power density achieved a two-photon photodynamic cancer therapy with quantum dots, *J. Mater. Chem. B*, **1**, pp. 4584–4592.

Chuang, C.-H. M., Maurano, A., Brandt, R. E., Hwang, G. W., Jean, J., Buonassisi, T., Bulović, V., and Bawendi, M. G. (2015). Open-circuit voltage deficit, radiative sub-bandgap states, and prospects in quantum dot solar cells, *Nano Lett.*, **15**, pp. 3286–3294.

Chunbai, H., Demin, L., and Wenbin, L. (2015). Self-assembled core-shell nanoparticles for combined chemotherapy and photodynamic therapy of resistant head and neck cancers, *ACS Nano*, **9**, pp. 991–1003.

Clapp, A. R., Medintz, I. L., Mauro, J. M., Fisher, B. R., Bawendi, M. G., and Mattoussi, H. (2004). Fluorescence resonance energy transfer between quantum dot donors and dye-labeled protein acceptors, *J. Am. Chem. Soc.*, **126**, pp. 301–310.

Coe-Sullivan, S., Woo, W.-K., Steckel, J. S., Bawendi, M., and Bulovic, V. (2003). Tuning the performance of hybrid organic/inorganic quantum dot light-emitting devices. *Org. Electron.*, **4**, pp. 123–130.

Comin, A., and Manna, L. (2014). New materials for tunable plasmonic colloidal nanocrystals, *Chem. Soc. Rev.,* **43**, pp. 3957–3975.

Cusano, A., Arregui, F. J., Giordano, M., and Cutolo, A. (2012). *Optochemical Nanosensors. Series: Series in Sensors*, CRC Press.

Dadadzhanov, D. R., Martynenko, I. V., Orlova, A. O., Maslov, V. G., Fedorov, A. V., and Baranov, A. V. (2015). The formation of molecular aggregates of sulfophthalocyanine in complexes with semiconductor nanocrystals, *Opt. Spectrosc.,* **119**, pp. 738–743.

Dahan, M., Lévi, S., Luccardini, C., Rostaing, P., Riveau, B., and Triller, A. (2003). Diffusion dynamics of glycine receptors revealed by single-quantum dot tracking, *Science,* **302**, pp. 442–445.

Dayal, S., Krolycki, R., Lou, Y., Qiu, X., Berlin, J. C., Kenney, M. E., and Burda, C. (2006). Femtosecond time-resolved energy transfer from CdSe nanoparticles to phthalocyanines, *Appl. Phys. B,* **84**, pp. 309–315.

De Mello Donegá, C. (2014). *Nanoparticles.* (Springer-Verlag, Berlin, Heidelberg).

Deng, Z., Jeong, K. S., and Guyot-Sionnest, P. (2014). Colloidal quantum dots intraband photodetectors, *ACS Nano,* **8**, pp. 11707–11714.

Dertinger, T., Colyer, R., Iyer, G., Weiss, S., and Enderlein, J. (2009). Fast, background-free, 3D super-resolution optical fluctuation imaging (SOFI), *Proc. Natl. Acad. Sci.,* **106**, pp. 22287–22292.

Dougherty, T. J., Gomer, C. J., Henderson, B. W., Jori, G., Kessel, D., Korbelik, M., Moan, J., and Peng, Q. (1998). Photodynamic therapy, *J. Natl. Cancer Inst.,* **90**, pp. 889–905.

Drew, K., Girishkumar, G. K., and Kamat, P. V. (2005). Boosting fuel cell performance with a semiconductor photocatalyst: TiO_2/Pt–Ru hybrid catalyst for methanol oxidation, *J. Phys. Chem. B,* **109**, pp. 11851–11857.

Dzhagarov, B. M., Sagun, E. I., Ganzha, V. A., and Gurinovich, G. P. (1987). Mechanism for triplet state quenching of chlorophyll and related substances with molecular oxygen, *Khimicheskaya Fizika,* **6**, pp. 919–928.

Fakhar-e-Alam, M., Kishwar, S., Siddique, M., Atif, M., Nur, O., and Willander, M. (2012). The photodynamic effect of ZnO nanorods and their ligands with different photosensitizers, *Rev. Nanosci. Nanotechnol.,* **1**, pp. 40–51.

Faraday, M. (1857). Editing the Bakerian lecture: Experimental relations of gold (and other metals) to light, *Phil. Trans. R. Soc.,* **147**, pp. 145–181.

Ferrari, M. (2005). Cancer nanotechnology: Opportunities and challenges, *Nat. Rev. Cancer,* **5**, pp. 161–171.

Fowley, C., Nomikou, N., McHale, A. P., McCarron, P. A., McCaughan, B., and Callan, J. F. (2012). Water soluble quantum dots as hydrophilic carriers and two-photon excited energy donors in photodynamic therapy, *J. Mater. Chem.,* **22**, pp. 6456–6462.

Frasco, M. F., Vamvakaki, V., and Chaniotakis, N. (2010). Porphyrin decorated CdSe quantum dots for direct fluorescent sensing of metal ions, *J. Nanopart. Res.*, **12**, pp. 1449–1458.

Galoppini, E. (2004). Linkers for anchoring sensitizers to semiconductor nanoparticles, *Coordin. Chem. Rev.,* **248**, pp. 1283–1297.

Gaponenko, S. V. (2010). *Introduction to Nanophotonics,* Cambridge University Press, Cambridge.

Ge B., Li Zh., Xi Y., Yang L., Wang R. (2015). CdTe quantum dots and gold nanoparticle based spectral methods for determination of lincomycin. *Optics and Spectroscopy,* **118**, pp. 748–755.

Grim, J. Q., Christodoulou, S., Di Stasio, F., Krahne, R., Cingolani, R., Manna, L., and Moreels, I. (2014). Continuous-wave biexciton lasing at room temperature using solution-processed quantum wells, *Nat. Nanotechnol.*, **9**, pp. 891–895.

Guijarro, N., Shen, Q., Gimenez, S., Mora-Sero., I., Bisquert, J., Lana-Villarreal, T., Toyoda, T., and Gomez, R. (2010). Direct correlation between ultrafast injection and photoanode performance in quantum dot sensitized solar cells, *J. Phys. Chem. C,* **114**, pp. 22352–22360.

Guler, U., Urcan, Shalaev, V. M., and Boltasseva, A. (2015). Nanoparticle plasmonics: Going practical with transition metal nitrides, *Mater. Today,* **18**, pp. 227–237.

Guo, W., Sun, X., Jacobson, O., Yan, X., Min, K., Srivatsan, A., Gang, N., Kiesewetter, D. O., Chang, J., and Chen, X. (2015). Intrinsically radioactive [^{64}Cu]CuInS/ZnS quantum dots for PET and optical imaging: Improved radiochemical stability and controllable Cerenkov luminescence, *ACS Nano,* **9**, pp. 488–495.

Gur, L., Fromer, N. A., Chen, C.-P., Kanaras, A. G., and Alivisatos, A. P. (2007). Hybrid solar cells with prescribed nanoscale morphologies based on hyperbranched semiconductor nanocrystals, *Nano Lett.,* **7**, pp. 409–414.

Hadar, I., Halivni, S., Even-Dar, N., Faust, A., and Banin, U. (2015). Dimensionality effects on fluorescence resonance energy transfer between single semiconductor nanocrystals and multiple dye acceptors, *J. Phys. Chem. C*, **119**, pp. 3849–3856.

Haghi, A. K., Thomas, S., Pourhashemi, A., Hamrang, A., and Klodzinska, E. (2015). *Nanomaterials and Nanotechnology for Composites: Design, Simulation and Applications*, Apple Academic Press.

He, X., and Ma, N. (2014). An overview of recent advance of quantum dots for biomedical applications, *Colloids Surf. B: Biointerfaces*, **124**, pp. 118–131.

Heilemann, M., van de Linde, S., Mukherjee, A., and Sauer, M. (2009). Super-resolution imaging with small organic fluorophores, *Angew. Chem., Int. Ed.*, **48**, pp. 6903–6908.

Hess, S. T., Girirajan, T. P. K., and Mason, M. D. (2006). Ultra-high resolution imaging by fluorescence photoactivation localization microscopy, *Biophys. J.*, **91**, pp. 4258–4272.

Hessel, C. M., Pattani, V., Rasch, M., Panthani, M. G., Koo, B., Tunnell, J. W., and Korgel, B. A. (2011). Copper selenide nanocrystals for photothermal therapy, *Nano Lett.*, **11**, pp. 2560–2566.

Hillhouse, H. W., and Beard, M. C. (2009). Solar cells from colloidal nanocrystals: Fundamentals, materials, devices, and economics, *Curr. Opin. Colloid Interf. Sci.*, **14**, pp. 245–259.

Hod, I., Gonzalez-Pedro, V., Tachan, Z., Fabregat-Santiago, F., Mora-Ser, I., Bisquert, J., and Zaban, A. (2011). Dye versus quantum dots in sensitized solar cells: Participation of quantum dot absorber in the recombination process, *J. Phys. Chem. Lett.*, **2**, pp. 3032–3035.

Hoyer, P., Staudt, T., T. Engelhardt, J., and Hell, S. W. (2010). Quantum dot blueing and blinking enables fluorescence nanoscopy, *Nano Lett.*, **11**, pp. 245–250.

Huang, B., Wang, W., Bates, M., and Zhuang, X. (2008). Three-dimensional super-resolution imaging by stochastic optical reconstruction microscopy, *Science*, **319**, pp. 810–813.

Huber, R., Moser, J. E., Grätzel, M., and Wachtveitl, J. (2002). Real-time observation of photoinduced adiabatic electron transfer in strongly coupled dye/semiconductor colloidal systems with a 6 fs time constant, *J. Phys. Chem. B*, **106**, pp. 6494–6499.

Hyun, B.-R., Zhong, Y.-W., Bartnik, A. C., Sun, L., Abruna, H. D., Wise, F. W., Goodreau, J. D., Matthews, J. R., Leslie, T. M., and Borrelli, N. F. (2008). Electron injection from colloidal PbS quantum dots into titanium dioxide nanoparticles, *ACS Nano*, **2**, pp. 2206–2212.

Irudayaraj, J. M. (2012). *Biomedical Nanosensors*, Pan Stanford Series on Biomedical Nanotechnology, Pan Stanford Publishing.

Isakau, H., Parkhats, M., Knyukshto, V., Dzhagarov, B., Petrov, E., and Petrov, P. (2008). Toward understanding the high PDT efficacy of chlorine e_6-polyvinilpyrrolidone formulations: Photophysical and molecular aspects of photosensitizer-polymer interaction in vitro, *J. Photochem. Photobiol. B. Biol.,* **92**, pp. 165–174.

Jean, J., Brown, P. R., Jaffe, P. R., Buonassisi, R. L., and Bulović, T. V. (2015). Pathways for solar photovoltaics, *Energy Environ. Sci.,* **8**, pp. 1200–1219.

Ji, X., Copenhaver, D., Sichmeller, C., and Peng, X. (2008). Ligand bonding and dynamics on colloidal nanocrystal at room temperature: The case of alkylamines on CdSe nanocrystals, *J. Am. Chem. Soc.,* **130**, pp. 5726–5735.

Ji, X., Makarov, N. S., Wang, W., Palui, G., Robel, I., and Mattoussi, H. (2015a). Tuning the redox coupling between quantum dots and dopamine in hybrid nanoscale assemblies, *J. Phys. Chem. C,* **119**, pp. 3388–3399.

Ji, X., Wang, W., and Mattoussi, H. (2015b). Effects of separation distance on the charge transfer interactions in quantum dot-dopamine assemblies, *Phys. Chem. Chem. Phys.,* **17**, pp. 10108–10117.

Jie, B., and Bawendi, M. G. (2015). A colloidal quantum dot spectrometer, *Nature,* **523**, pp. 67–70.

Jin, Z., and Hildebrandt, N. (2012). Quantum dot nanoparticles for in vitro sensing. In: *Frontiers of Nanoscience,* vol. 4, Nanobiotechnology—Inorganic Nanoparticles vs Organic Nanoparticles, Edited by Jesus M. de la Fuente and V. Grazu, ISBN: 978-0-12-415769-9, pp. 291–306.

Jing, L., Ding, K., Kershaw, S. V., Kempson, I. M., Rogach, A. L., and Gao, M. (2014). Magnetically engineered semiconductor quantum dots as multimodal imaging probes, *Adv. Mater.,* **26**, pp. 6367–6386.

Justin, R., Tao, K., Román, S., Chen, D., Xu, Y., Geng, X., Ross, I. M., Grant, R. T., Pearson, A., Zhou, G., MacNeil, S., Sun, K., and Chen, B. (2016). Photoluminescent and superparamagnetic reduced graphene oxide–iron oxide quantum dots for dual-modality imaging, drug delivery and photothermal therapy, *Carbon,* **97**, pp. 54–70.

Juzenas, P., Chen, W., Sun, Y. P., Coelho, M. A. N., Generalov, R., Generalova, N., and Christensen, I. L. (2008). Quantum dots and nanoparticles for photodynamic and radiation therapies of cancer, *Adv. Drug Deliv. Rev.,* **60**, pp. 1600–1614.

Kalyuzhny, G., and Murray, R. W. (2005). Ligand effects on optical properties of CdSe nanocrystals, *J. Phys. Chem. C,* **109**, pp. 7012–7021.

Kamat, P. V. (2002). Photophysical, photochemical and photocatalytic aspects of metal nanoparticles, *J. Phys. Chem. B,* **106**, pp. 7729–7744.

Kamat, P. V. (2013). Quantum dot solar cells. The next big thing in photovoltaics, *J. Phys. Chem. Lett.*, **4**, pp. 908–918.

Kang, S., Yasuda, M., Miyasaka, H., Umeyama, T., Matano, Y., Yoshida, K., Isoda, S., and Imahori, H. (2008). Multiporphyrin-modified CdSe nanoparticles as efficient light-harvesting materials, *ChemSusChem,* **1**, pp. 254–261.

Karachevtsev, V. A. (2015). *Nanobiophysics: Fundamentals and Applications,* 434 pp. ISBN 9789814613965, Pan Stanford Publishing.

Kay, E. R., Lee, J., Nocera, D. G., and Bawendi, M. G. (2013). Conformational control of energy transfer: A mechanism for biocompatible nanocrystal-based sensor. *Angew. Chem., Int. Ed.,* **52**, pp. 1165–1169.

Kim, J. Y., Voznyy, O., Zhitomirsky, D., and Sargent, E. H. (2013). 25th Anniversary article: Colloidal quantum dot materials and devices: A quarter-century of advances, *Adv. Mater.*, **25**, pp. 4986–5010.

Klimov, V. (2000). In *Handbook of Nanostructured Materials and Nanotechnology* (Nalwa, H. S., ed.), **4**, pp. 451–527, Acadmic Press, USA.

Knowles, K. E., Frederick, M. T., Tice, D. B., Morris-Cohen, A. J., and Weiss, E. A. (2012). Colloidal quantum dots: Think outside the (particle-in-a-) box, *J. Phys. Chem. Lett.*, **3**, pp. 18–26.

Kongkanand, A., Tvrdy, A., Kakechi, K., Kuno, M., and Kamat, P. V. (2008). Quantum dot solar cells. Tuning photoresponse through size and shape control of CdSe-TiO_2 architecture, *J. Am. Chem. Soc.,* **130**, pp. 4007–4015.

Koole, R., Schapotschnikow, P., de Mello Donega, C., Vlugt, T. J. H., and Meijerink, A. (2008). Time-dependent photoluminescence spectroscopy as a tool to measure the ligand exchange kinetics on a quantum dot surface, *ACS Nano*, **2**, pp, 1703–1714.

Kovalenko, M. V., Manna, L., Cabot, A., Hens, Z., Talapin, D. V., Kagan, C. R., Klimov, V. I., Rogach, A. L., Reiss, P., Milliron, D. J., Guyot-Sionnnest, P., Konstantatos, G., Parak, W. J., Hyeon, T., Korgel, B. A., Murray, C. B., and Heiss, W. (2015). Prospects of nanoscience with nanocrystals, *ACS Nano,* **9**, pp. 1012–1057.

Kowerko, D. (2010). Dynamic processes in functionalised perylene bisimide molecules, semiconductor nanocrystals and assemblies. PhD Thesis. Fakultät für Naturwis-senschaften Technische Universität Chemnitz, Germany.

Kowerko, D., Krause, S., Amecke, N., Abdel-Mottaleb, M., Schuster, J., and von Borczyskowski, C. (2009). Identification of different donor-acceptor structures via Förster resonance energy transfer (FRET) in quantum-dot-perylene bisimide assemblies, *Int. J. Mol. Sci.,* **10**, pp. 5239–5256.

Kowerko, D., Schuster, J., Amecke, N., Abdel-Mottaleb, M., Dobrawa, R., Wuerthner, F., and von Borczyskowski, C. (2010). FRET and ligand related non-FRET processes in single quantum dot-perylene bisimide assemblies, *Phys. Chem. Chem. Phys.*, **12**, pp. 4112–4123.

Lau, G., Tsao, P. S., Nok, H., Chenyi, Y., Shaik, Z., Graetzel, M., and Dyson, P. J. (2015). Enhancing the stability of porphyrin dye-sensitized solar cells by manipulation of electrolyte additives, *Chemsuschem,* **8**, pp, 255–259.

Lawrence, W. G., Thacker, S., Palamakumbura, S., Riley, K. J., and Nagarkar, V. V. (2012). Quantum dot-organic polymer composite materials for radiation detection and imaging, *IEEE Trans. Nucl. Sci.,* **59**, pp. 215–221.

Lemon, C. M., Karnas, E., Bawendi, M. G., and Nocera, D. G. (2013). Oxygen Sensing with Quantum Dot-Porphyrin Conjugates *Inorg. Chem.,* **52**, pp. 10394–10406.

Lemon, C., Karnas, E., Han, X., Bruns, O. T., Kempa, T. J., Fukumura, D., Bawendi, M. G., Jain-R. K., Duda, D. G., and Nocera, D. G. (2015). Micelle-encapsulated quantum dot-porphyrin assemblies as in vivo two-photon oxygen sensors, *J. Am. Chem. Soc.*, **137**, pp. 9832–9842.

Lewis, N. S., Crabtree, G. W., Nozik, A. J., Wasielewski, M. R., and Alivisatos, A. P. (2005). Basic energy sciences report on basic research needs for solar energy utilization, *Office of Science, U.S. Department of Energy,* April 2005, pp. 18–21.

Li, J., and Wu, N. (2013). *Biosensors Based on Nanomaterials and Nanodevices*, Series: Nanomaterials and their Applications, ISBN-13: 978-1466551510, CRC Press.

Liang, G.-X., Li, L.-L., Liu, H.-Y., Zhang, J.-R., Burda, C., and Zhu, J.-J. (2010). Fabrication of near-infrared-emitting CdSeTe/ZnS core/shell quantum dots and their electrogenerated chemiluminescence, *Chem. Commun.,* **46**, pp. 2974–2976.

Lidke, K., Rieger, B., Jovin, T., and Heintzmann, R. (2005). Superresolution by localization of quantum dots using blinking statistics, *Opt. Express,* **13**, pp. 7052–7062.

Lihitkar, N. B., Singh, S. B., Singh, J., Srivastava, O. N., Naik, R. H., and Kulkarni, S. K. (2009). Optical investigations of interaction between zinc tetra phenyl porphyrin and CdSe nanoparticles, *Chem. Phys. Lett.,* **483**, pp. 227–232.

Liu, J., Erogbogbo, F., Yong, K.-T., Ye, L., Liu, J., Hu, R., Chen, H., Hu, Y., Yang, Y., Yang, J., Roy, I., Karker, N. A., Swihart, M. T., and Prasad, P. N. (2013). Assessing clinical prospects of silicon quantum dots: Studies in mice and monkeys, *ACS Nano,* **7**, pp. 7303–7310.

Liu, X., Swihart, M. T. (2014). Heavily-Doped colloidal semiconductor and metal oxide nanocrystals: An emerging new class of plasmonic nanomaterials. *Chem. Soc. Rev.*, **43**, pp. 3908–3920.

Llordes, A., Garcia, G., Gazquez, J., and Milliron, D. J. (2013). Tunable near-infrared and visible-light transmittance in nanocrystal-in-glass composites, *Nature,* **500**, pp. 323–326.

Lou, Z., Wang, Z., Huang, B., and Dai, Y. (2014). Synthesis and activity of plasmonic photocatalysts, *ChemCatChem,* **6**, pp. 2456–2476.

Lovell, J. F., Liu, T. W. B., Chen, J., and Zheng, G. (2010). Activatable photosensitizers for imaging and therapy, *Chem. Rev.,* **110**, pp. 2839–2857.

Lyshevski, S. E. (2014). *Dekker Encyclopedia of Nanoscience and Nanotechnology*, 3rd ed, ISBN 9781439891346, CRC Press.

Ma, J., Chen, J.-Y., Idowu, M., Nyokong, T. (2008). Generation of singlet oxygen via the composites of water-soluble CdTe quantum dots-sulfonated aluminium phthalocyanines. *J. Phys. Chem. B.,* **112**, pp. 4465–4469.

Mai, C.-L., Moehl, T., Hsieh, C.-H., Decoppet, J.-D., Zakeeruddin, S. M., Graetzel, M., and Yeh, C.-Y. (2015). Porphyrin sensitizers bearing a pyridine-type anchoring group for dye-sensitized solar cells, *ACS Appl. Mater. Interfaces,* **7**, pp. 14975–14982.

Marcus, R. A. (1993). Electron transfer reactions in chemistry. Theory and experiment, *Rev. Mod. Phys.*, **65**, pp. 599–610.

Martynenko I. V., Kuznetsova, V. A., Orlova, A. O., Kanaev, P. A., Maslov, V. G., Loudon, A., Zaharov, V., Parfenov, P., Gun'ko, Y. K., Baranov, A. V., and Fedorov, A. V. (2015). Chlorin e6–ZnSe/ZnS quantum dots based system as reagent for photodynamic therapy, *Nanotechnology,* **26**, p. 055102.

Maslov, V., Orlova, A., and Baranov, A. (2012). Combination therapy: Complexing of QDs with tetrapyrrols and other dyes. In: *Photo-sensitizers in Medicine, Environment, and Security*, Springer-Verlag.

Mattoussi, H., Goutam, P., and Hyon, B. N. (2012). Luminescent quantum dots as platforms for probing in vitro and in vivo biological processes,

Adv. Drug Deliv. Rev., Biological Interactions of Nanoparticles, **64**, pp. 138–166.

Medintz, I. L., Clapp, A. R., Mattoussi, H., Goldman, E. R., Fisher, B., and Mauro, J. M. (2003). Self-assembled nanoscale biosensors based on quantum dot FRET donors, *Nat. Mater.,* **2**, pp. 630–638.

Medintz, I. L., Konnert, J. H., Clapp, A. R., Stanish, I., Twigg, M. E., Mattoussi, H., Mauro, J. M., and Deschamps, J. R. (2004). A fluorescence resonance energy transfer derived structure of a quantum dot-protein bioconjugate nanoassembly, *P. N. A. S.,* **101**, pp. 9612–9617.

Michalet, X., Pinaud, F., Bentolila, L., Tsay, J., Doose, S., Li, J., Sundaresan, G., Wu, A., Gambhir, S., and Weiss, S. (2005). Quantum dots for live cells, in vivo imaging, and diagnostics, *Science,* **307**, pp. 538–544.

Mironov, A. F. (2004). Modern state of chemistry of photosensitizers on the base of porphyrins and related compounds. In: *Successes of Porphyrin Chemistry* (Golubchikov, O. A., ed.), vol. 4, pp. 271–291, The Scientific Research Institute of Chemistry of St-Peterburg University.

Mocatta, D., Cohen, G., Schattner, J., Millo, O., Rabani, E., Banin, U. (2011). Heavily doped semiconductor nanocrystal quantum dots, *Science,* **332**, pp. 77–81.

Mora-Sero, I., and Bisquert, J. (2010). Breakthroughs in the development of semiconductor-sensitized solar cells, *J. Phys. Chem. Lett.,* **1**, pp. 3046–3052.

Moser, J.-E. (2011). In *Dye Sensitized Solar Cells* (Kalyanasundaran, K., ed.), Chapter 11, EPFL Press.

Moyano, D. F., and Rotello, V. M. (2011). Nano meets biology: Structure and function at the nanoparticle interface, *Langmuir,* **27**, pp. 10376–10385.

Moyano, D. F., Goldsmith, M., Solfiell, D. J., Landesman-Milo, D., Miranda, O. R., Peer, D., and Rotello, V. M. (2012). Nanoparticle hydrophobicity dictates immune response, *J. Am. Chem. Soc.,* **134**, pp. 3965–3967.

Munro, A. M., Plante, I. J.-L., Ng, M. S., and Ginger, D. S. (2007). Quantitative study of the effect of surface ligand concentration on CdS nanocrystal photoluminescence, *J. Phys. Chem. C.,* **111**, pp. 6220–6227.

Murcia, M. J., and Naumann, C. A. (2005). Biofunctionalization of Fluorescent Nanoparticles. In: *Nanotechnologies for the Life Sciences* (Kumar, C. S. S. R., ed.), vol. 1 Biofunctionalization of Nanomaterials. pp. 1–40, Wiley-VCH, Weinheim.

Nicolini, C. (2010). In *Nanotechnology and Nanobiosciences,* Pan Stanford Series on Nanobiotechnology, Chapter 1, vol. 1, Pan Stanford Publishing.

O'Regan, B. C., and Grätzel, M. (1991). A low-cost, high-efficiency solar cell based on dye-sensitized colloidal TiO_2 films, *Nature,* **353**, pp. 737–740.

O'Regan, B. C., Moser, J. E., Anderson, M., and Grätzel, M. (1990). Vectorial electron injection into transparent semiconductor membranes and electric field effects on the dynamics of light-induced charge separation, *J. Phys. Chem.*, **94**, pp. 8720–8726.

Oluwole, D. O., Britton, J., Mashazi, P., and Nyokong, T. (2015). Synthesis and photophysical properties of nanocomposites of aluminum tetrasulfonated phthalocyanine covalently linked to glutathione capped CdTe/CdS/ZnS quantum dots, *Synth. Met.,* **205**, pp. 12–221.

Parak, W. J., Gerion, D., Pellegrino, T., Zanchet, D., Micheel, C., Williams, S. C.,Boudreau, R., Le Gros, M. A., Larabell, C. A., and Alivisatos, A. P. (2003). Biological applications of colloidal nanocrystals, *Nanotechnology*, **14**, pp. R15–R27.

Parkhats, M., Knyukshto, V., Stasheuskii, A., and Dzhagarov, B. M. (2009). Dynamics and efficiency of photosensitized formation of singlet oxygen by chlorin e_6: The influence of the solution pH and polyvinyl-pyrrolidone, *Opt. Spectrosc.,* **107**, pp. 1026–1032.

Petryayeva, E., Algar, W. R., and Medintz, I. L. (2013). Quantum dots in bioanalysis: A review of applications across various platforms for fluorescence spectroscopy and imaging, *Appl. Spectrosc.,* **67**, pp. 215–252.

Probst, C. E., Zrazhevskiy, P., Bagalkot, V., and Gao, X. (2013). Quantum dots as a platform for nanoparticle drug delivery vehicle design, *Adv. Drug Deliv. Rev.,* **65**, pp. 703–718.

Pu, S. C., Yang, M. J., Hsu, C. C., Lai, C. W., Hsieh, C. C., Lin, S. H., Cheng, Y.-M., and Chou, P. T. (2006). The empirical correlation between size and two-photon absorption cross section of CdSe and CdTe quantum dots, *Small,* **2**, pp. 1308–1313.

Qi, Z.-D., Li, D.-W., Jiang, P., Jiang, F.-G., Li, Y.-S., Liu, Y., Wong, W.-K., and Chean, K.-W. (2011). Biocompatible CdSe quantum dot-based photosensitizer under two-photon excitation for photodynamic therapy, *J. Mater. Chem.,* **21**, pp. 2455–2458.

Rakovich, A., Rakovich, T., Kelly, V., Lesnyak, V., Eychmuller, A., Rakovich, Y. P., and Donegan, J. F. (2010). Photosensitizer methylene blue-semiconductor nanocrystals hybrid system for photodynamic therapy, *J. Nanoscience Nanotech.,* **10**, pp. 2656–2662.

Ramachandra, R. C. N., Kulkarni, G. U., Thomasa, P. J., and Edwards, P. P. (2000). Metal nanoparticles and their assemblies, *Chem. Soc. Rev.,* **29**, pp. 27–35.

Ramos, F. J., Ince, M., Urbani, M., Abate, A., Graetzel, M., Ahmad, S., Torres, T., and Nazeeruddin, M. K. (2015). Non-aggregated Zn(II)octa (2,6-diphenylphenoxy) phthalocyanine as a hole transporting material for efficient perovskite solar cells, *Dalton Trans.,* **44**, pp. 10847–10851.

Resch-Genger, U., Grabolle, U. M., Cavaliere-Jaricot, S., Nitschke, R., and Nann, T. (2008). Quantum dots versus organic dyes as fluorescent labels, *Nat. Methods,* **5**, pp. 763–775.

Rogach, A. L. (2008). (ed.), *Semiconductor Nanocrystal Quantum Dots: Synthesis, Assembly, Spectroscopy and Applications*, Springer-Verlag, Wien.

Rogers, B., Adams, J., and Pennathur, S. (2014). *Nanotechnology: Understanding Small Systems*, Series: Mechanical and Aerospace Engineering, CRC Press.

Ruehle, S., Shalom, M., and Zaban, A. (2010). Quantum-dot-sensitized solar cells, *ChemPhysChem,* **11**, pp. 2290–2304.

Runnerstrom, E. L., Llordes, A., Lounis, S. D., and Milliron, D. J. (2014). Nanostructured electrochromic smart windows: Traditional materials and NIR-selective plasmonic nanocrystals, *Chem. Commun.,* **50**, pp. 10555–1057.

Samia, A. C. S., Chen, X., and Burda, C. (2003). Semiconductor quantum dots for photodynamic therapy, *J. Am. Chem. Soc.*, **123**, pp. 15736–15737.

Samia, A. C. S., Dayal, S., and Burda, C. (2006) Quantum dot-based energy transfer: Perspectives and potential for applications in photodynamic therapy, *Photochem. Photobiol.,* **82**, pp. 617–625.

Sapsford, K. E., Pons, T., Medintz, I. L., and Mattoussi, H. (2006). Biosensing with luminescent semiconductor quantum dots, *Sensors,* **6**, pp. 925–953.

Schmidt, M. E., Blanton, S. A., Hines, M. A., and Guyot-Sionnest, P. (1996). Size-dependent two-photon excitation spectroscopy of CdSe nanocrystals, *Phys. Rev. B,* **53**, pp. 12629–12632.

Schmitt, F.-J., Maksimov, E. G., Suedmeyer, H., Jeyasangar, V., Theiss, C., Paschenko, V. Z., Eichler, H. J., and Renger, G. (2011). Time resolved temperature switchable excitation energy transfer processes between CdSe/ZnS nanocrystals and phycobiliprotein antenna from Acaryochloris marina, *Photon. Nanostructures Fundamentals Appl.,* **9**, pp. 190–195.

Semonin, O. E., Luther, J. M., Choi, S., Chen, H. Y., Gao, J. B., Nozik, A. J., and Beard, M. C. (2011). Peak external photocurrent quantum efficiency exceeding 100% via MEG in a quantum dot solar cell, *Science*, **334**, pp. 1530–1533.

Shao, K., Singha, S. S., Clemente-Casares, X., Tsai, S., Yang, Y., and Santamaria, P. (2015) Nanoparticle-based immunotherapy for cancer, *ACS Nano*, **9**, pp. 16–30.

Shatkin, J. A. (2012) *Nanotechnology: Health and Environmental Risks*, 2nd ed., Series: Perspectives in Nanotechnology, CRC Press.

Shi, L., Hernandez, B., and Selke, M. (2006). Singlet oxygen generation from water-soluble quantum dot-organic dye nanocomposites, *J. Am. Chem. Soc.*, **128**, pp. 6278–6279.

Shibu, E. S., Hamada, M., Murase, N., and Biju, V. (2013). Nanomaterials formulations for photothermal and photodynamic therapy of cancer, *J. Photochem. Photobiol. C Photochem. Rev.*, **15**, pp. 53–72.

Skripka, A., Valanciunaite, J., Dauderis, G., Poderys, V., Kubiliute, R., and Rotomskis, R. (2013). Two-photon excited quantum dots as energy donors for photosensitizer chlorine, *J. Biomed. Opt.*, **18**, p. 078002.

Somasundaran, P. (2015). *Encyclopedia of Surface and Colloid Science*, 3rd ed., ISBN 9781466590458, CRC Press.

Somers, R. C., Bawendi, M. G., and Nocera, D. G. (2007). Nanocrystals as sensors, *Green Chem.*, **9**, pp. 403–410.

Stasiuk, G. J., Tamang, S., Imbert, D., Poillot, C., Giardiello, M., Tisseyre, C., Barbier, E. L., Fries, P. H., De Waard, M., and Reiss, P. (2011). Cell-permeable Ln(III) chelate-functionalized InP quantum dots as multimodal imaging agents, *ACS Nano*, **5**, pp. 8193–8201.

Stolle, C. J., Harvey, T. B., and Korgel, B. A. (2013). Nanocrystal photovoltaics: A review of recent progress, *Curr. Opin. Chem.*, **2**, pp. 160–167.

Suzuki, M., Husimi, Y., Komatsu, H., Suzuki, K., and Douglas, K. T. J. (2008). Quantum dot FRET biosensors that respond to pH, to proteolytic or nucleolytic cleavage, to DNA synthesis, or to a multiplexing combination, *J. Am. Chem. Soc.*, **130**, pp. 5720–5725.

Sykora, M., Petruska, M. A., Alstrum-Acevedo, J., Bezel, I., Meyer, T. J., and Klimov, V. I. (2006). Photoinduced charge transfer between CdSe nanocrystal quantum dots and Ru–polypyridine complexes, *J. Am. Chem. Soc.*, **128**, pp. 9984–9985.

Tachibana, Y., Umekita, K., Otsuka, Y., and Kuwabata, S. (2009). Charge recombination kinetics at an in situ chemical bath-deposited CdS/nanocrystalline TiO_2 interface, *J. Phys. Chem. C*, **113**, pp, 6852–6858.

Tokuoka, Y., Yamada, M., Kawashima, N., and Miyasaka, T. (2006). Anticancer effect of dye-sensitized TiO_2 nanocrystals by polychromatic visible light irradiation, *Chem. Lett.*, **35**, pp. 496–497.

Tsay, J. M., Trzoss, M., Shi, L., Kong, X., Selke, M., Jung, M. E., and Weiss, S. (2007). Singlet oxygen production by peptide-coated quantum dot-photosensitizer conjugates, *J. Am. Chem. Soc.*, **129**, pp. 6865–6871.

Vajtai, R., ed. (2013). *Springer Handbook of Nanomaterials*, Springer-Verlag Berlin Heidelberg.

Vo-Dinh, T. (2007). *Nanotechnology in Biology and Medicine: Methods, Devices, and Applications*, CRC Press.

Von Borczyskowski, C., and Zenkevich, E. (2014). Formation principles and exciton relaxation in semiconductor quantum dot–dye nanoassemblies. In: *Quantum Dot Molecules, Lecture Notes in Nanoscale Science and Technology 14. Springer Series in Materials Science* (Wu, J., and Wang, Z. M., eds.), Springer Science+Business Media, New York, vol. **14**, *Quantum Dot Molecules*, pp. 77–148.

Von Borczyskowki, C., and Zenkevich, E., eds. (2016). *Tuning Semi-conducting and Metallic Quantum Dots: Spectroscopy and Dynamics*, **303** p., Pan Stanford Publishing, Singapore.

Walter, P., Welcomme, E., Hallégot, P., Zaluzec, N. J., Deeb, C., Castaing, J., Veyssière, P., Bréniaux, R., Lévêque, J.-L., and Tsoucaris, G. (2006). Early use of PbS nanotechnology for an ancient hair dyeing formula, *Nano Lett.*, **6**, pp. 2215–2219.

Wang, Y., Fruhwirth, G., Cai, E., Ng, T., and Selvin, P. R. (2013). 3D super-resolution imaging with blinking quantum dots, *Nano Lett.*, **13**, pp. 5233–5241.

Wang, S., Gao, R., Zhou, F., Selke, M. (2004). Nanomaterials and singlet oxygen photosensitizers: potential applications in photodynamic therapy, *J. Mater. Chem.*, **14**, pp. 487–493.

Wang, X., Koleilat, G. I., Tang, J., Liu, H., Kramer, I. J., Debnath, R., Brzozowski, L., Barkhouse, D. A. R., Levina, L., Hoogland, S., and Sargent, E. H. (2011). Tandem colloidal quantum dot solar cells employing a graded recombination layer, *Nat. Photon.*, **5**, pp. 480–484.

Wang, M., Moon, S.-J., Xu, M., Chittibabu, K., Wang, P., Cevey-Ha, N.-L., Humphry-Baker, R., Zakeeruddin, S. M., and Graetzel, M. (2010). Efficient and stable solid-state dye-sensitized solar cells based on a high-molar-extinction-coefficient sensitizer, *Small*, **6**, pp. 319–324.

Watanabe, T. M., Fukui, S., Jin, T., Fujii, F., and Yanagida, T. (2010). Real-time nanoscopy by using blinking enhanced quantum dots, *Biophys. J.*, **99**, pp. L50–L52.

Watson, D., and Meyer, G. (2005). Electron injection at dye-sensitized semiconductor electrodes, *Ann. Rev. Phys. Chem.*, **56**, pp. 119–156.

Wen, Y.-N., Song, W.-S., An, L.-M., Liu, Y.-Q., Wang, Y.-H., and Yang, Y.-Q. (2009). Activation of porphyrin photosensitizers by semiconductor quantum dots via two-photon excitation, *Appl. Phys. Lett.*, **95**, p. 143702.

Wu, Y., Zhu, W.-H., Zakeeruddin, S. M., and Graetzel, M. (2015). Insight into D-A-pi-A structured sensitizers: A promising route to highly efficient and stable dye-densitized solar cells, *ACS Appl. Mater. Interfaces,* **7**, pp. 9307–9318.

Wuister, S. F., Swart, I., Driel, F. V., Hickey, S. G., and Donega, C. D. M. (2003). Highly luminescent water-soluble CdTe quantum dots, *Nano Lett.*, **3**, pp. 503–507.

Yaghini, E., Seifalian, A. M., and MacRobert, A. J. (2009). Quantum dots and their potential biomedical applications in photosensitization for photodynamic therapy, *Nanomedicine*, **4**, pp. 353–363.

Yezhelyev, V., Qi, L., O'Regan, R. M., Nie, S., and Gao, X. (2008). Proton-sponge coated quantum dots for siRNA delivery and intracellular imaging, *J. Am. Chem. Soc.*, **130**, pp. 9006–9012.

Yi, D. K., and Papaefthymiou, G. C. (2013). *Nanobiomaterials: Development and Applications. Series: Advances in Materials Science and Engineering*, ISBN 9781439876411, CRC Press.

Zebarjadi, M., Esfarjani, K., Dresselhaus, M. S., Ren, Z. F., and Chen, G. (2012). Perspectives on thermoelectrics: From fundamentals to device applications, *Energy Environ. Sci.,* **5**, pp. 5147–5162.

Zeng, S., Yu, X., Law, W.-C., Zhang, Y., Hu, R., Dinh, X.-Q., Ho, H.-P., and Yong, K.-T. (2013). Size dependence of Au NP-enhanced surface plasmon resonance based on differential phase measurement, *Sens. Actuators B,* **176**, pp. 1128–1133.

Zenkevich, E. I., Blaudeck, T., Shulga, A. M., Cichos, F., and von Borczyskowski, C. (2007). Identification and assignment of porphyrin-CdSe heteronanoassemblies, *J. Luminesc.*, **122–123**, pp. 784–788.

Zenkevich, E., Cichos, F., Shulga, A., Petrov, E., Blaudeck, T., and von Borczyskowski, C. (2005). Nanoassemblies designed from semiconductor quantum dots and molecular arrays, *J. Phys. Chem. B,* **109**, pp. 8679–8692.

Zenkevich, E. I., Gaponenko, S. V., Sagun, E. I., and von Borczyskowski, C. (2013). Bioconjugates based on semiconductor quantum dots and porphyrin ligands: Properties, exciton relaxation pathways and singlet oxygen generation efficiency for PDT applications, *Rev. Nanosci. Nanotechnol.,* **2**, pp. 184–207.

Zenkevich, E. I., Sagun, E. I., Knyukshto, V. N., Shulga, A. M., Mironov, A. F., Efremova, O. A., Bonnett, R., Songca, S. P., and Kassem, M. M. (1996). Photophysical and photochemical properties of potential porphyrin and chlorin photosensitizers for PDT, *J. Photochem. Photobiol. B: Biol.*, **33**, pp. 171–180.

Zenkevich, E. I., Sagun, V. N., Knyukshto, A. S., Stasheuski, V. A., Galievsky, A. P., Stupak, Blaudeck, T., and Von Borczyskowski, C. (2011). Quantitative analysis of singlet oxygen ($^1O_2$) generation via energy transfer in bioconjugates based on semiconductor quantum dots and porphyrin ligands, *J. Phys. Chem. C.*, **115**, pp. 21535–21545.

Zenkevich, E., Stupak, A., Göhler, C., Krasselt, C., and von Borczyskowski, C. (2015). Tuning electronic states of a CdSe/ZnS quantum dot by only one functional dye molecule, *ACS Nano*, **9**, pp. 2886–2903.

Zenkevich, E. I., Stupak, A. P., Kowerko, D., and Von Borczyskowski, C. (2012). Influence of single dye molecules on temperature and time dependent optical properties of CdSe/ZnS quantum dots: Ensemble and single nanoassembly detection, *Chem. Phys.*, **406**, pp. 21–29.

Zenkevich, E. I., and Von Borczyskowski, C. (2009). Photoinduced relaxation processes in self-assembled nanostructures: Multiporphyrin complexes and composites CdSe/ZnS quantum dot-porphyrin, *Macroheterocycles*, **2**, pp. 206–220.

Zenkevich, E., and Von Borczyskowski, C. (2012). Formation principles and excited states relaxation in self-assembled complexes: Multiporphyrin arrays and "Semiconductor CdSe/ZnS Quantum Dot-Porphyrin" nanocomposites. In: *Handbook of Porphyrin Science with Application to Chemistry, Physics, Materials Science, Engineering, Biology and Medicine. Volume 22—Biophysical and Physicochemical Studies of Tetrapyrroles* (Kadish, K., Smith, K. M., and Guilard, R., eds.), Chapter 104, pp. 68–159, World Scientific Publishing, Singapore.

Zenkevich, E. I., and von Borczyskowski, C. (2014). Self-Organization principles in the formation of multiporphyrin complexes and "Semiconductor Quantum Dot-Porphyrin" nanoassemblies, *J. Porphyrins Phthalocyanines*, **18**, pp. 1–19.

Zenkevich, E., and Von Borczyskowski, C. (2015). Surface photochemistry of quantum dot-porphyrin nanoassemblies for singlet oxygen generation. In: *Photoinduced Processes at Surfaces and in Nanomaterials, ACS Symposium Series* (Kilin, D., ed.), vol. **1196**, Chapter 12, pp. 235–272.

Zhan, N., Palui, G., Safi, M., Ji, X., and Mattoussi, H. (2013). Multidentate zwitterionic ligands provide compact and highly biocompatible quantum dots, *J. Am. Chem. Soc.,* **135**, pp. 13786–13795.

Zhang, C. Y., Yeh, H. C., Kuroki, M. T., and Wang, T. H. (2005). Single-quantum-dot-based DNA nanosensor, *Nat. Mater.,* **4**, pp. 826–831.

Zhdanova, K. A., Bragina, N. A., Bagratashvili, V. N., Timasheva, P. S., and Mironov, A. F. (2014). Noncovalent assemblies of CdSe semiconductor quantum dots and an amphiphilic long-chain *meso*-arylporphyrin, *Mendeleev Commun.,* **24**, pp. 247–249.

Chapter 6

Nanolithography and Decoration of Generated Nanostructures by Dye Molecules

Harald Graaf and Thomas Baumgärtel

Institute of Physics, Centre for Nanostructured Materials and Analytics, Technische Universität Chemnitz, Reichenhainer Str. 70, D-09107 Chemnitz, Germany

borczyskowski@physik.tu-chemnitz.de

6.1 Introduction

Fluorescence microscopy and spectroscopy are versatile tools which have become standard techniques in many branches of physics, chemistry, and biology. One of the most advantageous qualities of these methods is that the measured photophysical properties of the investigated entities (e.g., dye molecules or nanocrystals) allow a deep insight into other fundamental physical or chemical processes within themselves or their direct environment. Thus fluorescent molecules or other particles can be utilized as probes which are reliable, comparably cheap, and relatively easily accessible by experiment. However, the interpretation of the measured data is in many cases not trivial.

Self-Assembled Organic-Inorganic Nanostructures: Optics and Dynamics
Edited by Eduard Zenkevich and Christian von Borczyskowski

ISBN 978-981-4745-43-7 (Hardcover), 978-981-4745-44-4 (eBook)
www.panstanford.com

Fluorescence is a quantum-mechanical process originating from the electronic structure of matter.

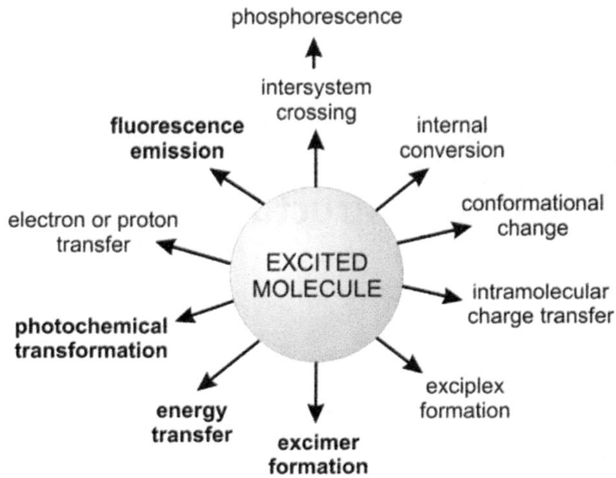

Scheme 6.1 Possible de-excitation pathways of excited molecules. Adapted from Valeur (2002), p. 10.

Especially for organic molecules, there is a strong interplay between their electronic structure, molecular geometry and the electronic structure of the environment. Once the molecule has been excited, there are different de-excitation pathways in which fluorescence is only one amongst many. Scheme 6.1 gives an overview of the involved processes. Of course, these pathways are not independent from each other but often connected, e.g., an energy transfer or an excimer formation may require a conformational change or rearrangement of the involved molecules (Kasha et al., 1965).

For a usage of molecules or nanoparticles as a fluorescent probe in several technical applications (e.g., Lab-On-A-Chip), it is necessary to immobilize the probe at a selectively defined location on the surface of the substrate. For this controlled attachment in a rather fast and reproducible way a top-down bottom-up approach is a very promising way. Here, in the first step, the substrate will be structured on the micrometer or nanometer scale leading to a two-dimensional texture. In the second step, the created structure can be selectively functionalized directly or via a linker by the probe molecule or nanoparticle.

In the following subsection, an effective method to create nanostructures on silicon, its functionalization by different binding mechanisms, and especially the optical properties of the probes will be presented.

Here it should be noted that the silicon substrates used in the first step were functionalized by self-assembled monolayers (SAMs). These SAM were formed through covalent binding of 1-alkenes to silicon (Sieval et al., 1998), which led to fully covered substrates. Such SAMs are very thin (about a few nanometers depending on the alkyl chain length) but dense packed and very stable (Yamada et al., 2004), thus preventing the oxidation of the silicon by atmospheric oxygen. Nevertheless these alkyl-terminated silicon can be structured, e.g., by local anodic oxidation (see Section 6.2).

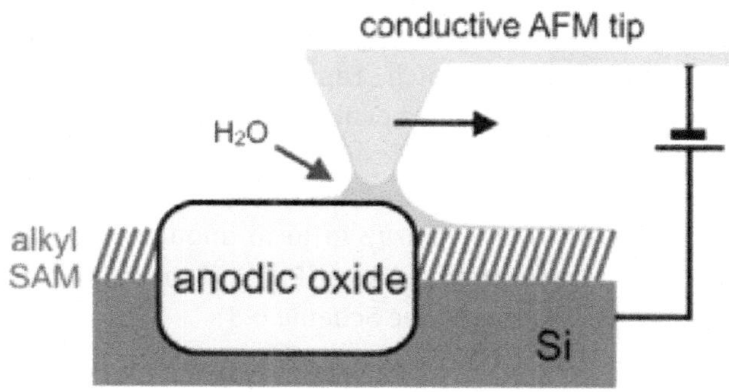

Scheme 6.2 Schematic view of the local anodic oxidation on an alkyl-terminated silicon substrate From Baumgärtel et al. (2012).

Not only monolayers are a suitable resist for lithographic purposes, but they also allow for tuning the surface properties of the substrate, e.g., surface hydrophobicity (Polster et al., 2010; Polster and Graaf, 2013). This can be used to influence the formed water meniscus and therefore the shape of the resulting oxide structures (Baumgärtel, 2011b).

6.2 Silicon Oxide Nanostructures on Alkyl-Terminated Silicon Surfaces

Local anodic oxidation (LAO) nanolithography by atomic force microscopy (AFM) is a reliable and versatile method to create nanoscaled patterns on metal or semiconductor surfaces (Garcia et al., 2006; Ara et al., 2002a; Yang et al., 2009; Baumgärtel and Graaf, 2011). From a chemical point of view LAO is an electrochemical reaction which is spatially confined by the AFM tip–sample geometry (for details about the oxidation process see (Yang et al., 2006)). Using a silicon substrate leads to silicon oxide which is characterized by volume expansion above the sample surface resulting in a topographic structure (see Scheme 6.2).

The structure geometry itself can be varied by controlling the oxidation parameters such as cantilever *velocity* and applied voltage. The formed structures can be used as templates for further functionalization (Baumgärtel and Graaf, 2011; Wouters and Schubert, 2003). The process of oxidation is self-limited and saturates after an oxide thickness of a few nanometers (Avouris et al., 1997). Dagata et al. and Dubois et al. developed models that explain the self-limitation through the build-up of space charge within the LAO oxide which reproduced the experimentally obtained data very well (Dagata et al., 1998; Dubois and Bubendorff, 2000). Chiesa and Garcia (Chiesa and Garcia, 2010) as well as Baumgärtel et al. (Baumgärtel et al., 2012) verified the existence and the sign of the space charge in local anodic oxide on silicon by Kelvin probe force microscopy (KPFM) (Nonnenmacher et al., 1991; Palermo et al., 2006). (See Scheme 6.3).

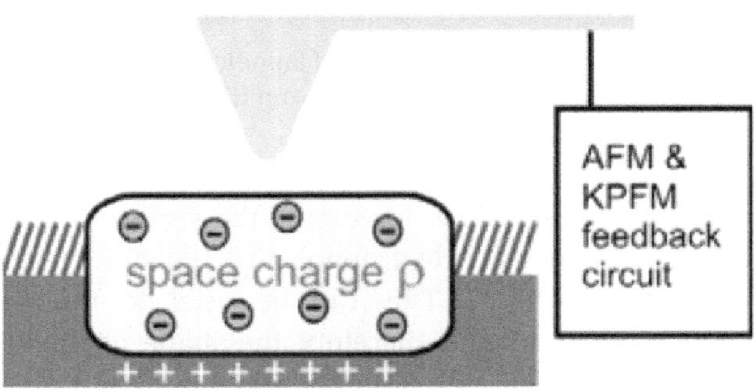

Scheme 6.3 Schematic view of an AFM and KPFM measurement of the space charge within the local oxide. From Baumgärtel et al. (2012).

Figure 6.1a shows the non-contact AFM image of a rectangular local silicon oxide pattern on a dodecyl-terminated silicon substrate. The measured oxide height above the surface is about 3.5 nm.

It has to be taken into account that about 40% of the generated oxide is below the silicon interface (Fontaine et al., 1998) and the dodecene monolayer (thickness ≈1.3 nm (Sieval et al., 1998)) decreases the apparent height further, so that the absolute thickness of the oxide can be estimated to about 8 nm.

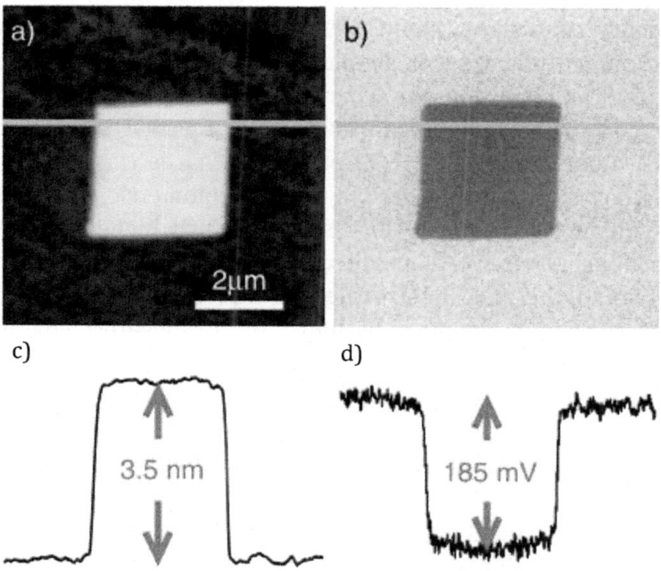

Figure 6.1 (a) AFM non-contact topography image of a nanolithography oxide structure on a dodecyl-terminated silicon surface. (b) KPFM image of the structure 47 min after generation at 51% RH. (c)–(d) Profile section along the green lines in the respective images in (a)–(b). From Baumgärtel et al. (2012).

The KPFM image (see Fig. 6.1b) indicates that the structure is charged negatively as the surface potential of the local oxide is about 185 mV lower than the surface potential of the surrounding dodecyl-terminated silicon. This surface potential difference (SPD) between oxide structures an alkyl-terminated silicon depends strongly on the surrounding humidity. (Baumgärtel et al., 2012) Investigations at dry conditions (around 4% r.H.) showed a three times higher SPD compared to the presented result obtained at about 51% r.H. Reason for this difference in the KPFM contrast is the shielding effect of adsorbed water on the surface of the oxide as it is known from previous KPFM studies on dielectric surfaces (Sugimura et al., 2002). On the other hand, the stored charges are quite stable (even over several hours and in contact with liquid water) (Baumgärtel et al., 2012) which is in clear contrast to charges introduced to silicon oxide layers by AFM-induced charging. Such charges dissipate rather quickly (within several minutes up to few hours) (Maedler and Graaf, 2013) even under dry conditions (about 20% r.H.) and especially in contact with liquid water. Here it should be noticed that in the case of AFM-induced charging the charges were deposited into more or less stoichiometric silicon oxide (e.g., thermal silicon oxide). On the other hand, in the LAO oxide the charge carriers are OH^- ions, which are trapped within a non-stoichiometric environment. Therefore it is reasonable to assume a much higher stability of the charges in the LAO oxide compared to electrons or holes inserted into thermal silicon oxide.

As a second independent indicator for negative charges in the LAO oxide optical methods can be used. For example, special dye molecules being spectroscopically sensitive to electric fields can be applied. Such organic molecules have been successfully used to measure surface polarity (Macquarrie et al., 1999), membrane potentials (Montana et al., 1989) or charge transfer processes in silicon nanoparticles (Martin et al., 2005).

Figure 6.2a shows a typical confocal microscope fluorescence image of a square LAO oxide structure after di-4-ANEPPS (aminonaphthyl-ethenylpyridinium, see Scheme 6.4 for the chemical structure) treatment. Di-4-ANEPPS is presumably deposited on the whole sample surface during the spin coating process but can only be optically detected on the local oxide. Reason for this is a strong quenching of the dye fluorescence by an effective energy transfer to the underlying silicon: Dye molecules on the oxide structure are separated from the silicon by the oxide (thickness $\approx$ 8 nm) while dye molecules on the surrounding surface are much closer to the silicon (thickness of alkyl monolayer in this case about 1.3 nm). As it is known that the energy transfer between the dye molecules and the silicon substrate depends strongly on their spacing (Sluch et al., 1995), it is easily understandable why molecules on the oxide structure show a detectable but weak fluorescence, while the fluorescence from molecules on the alkyl-terminated silicon is quenched and therefore below the detection limit.

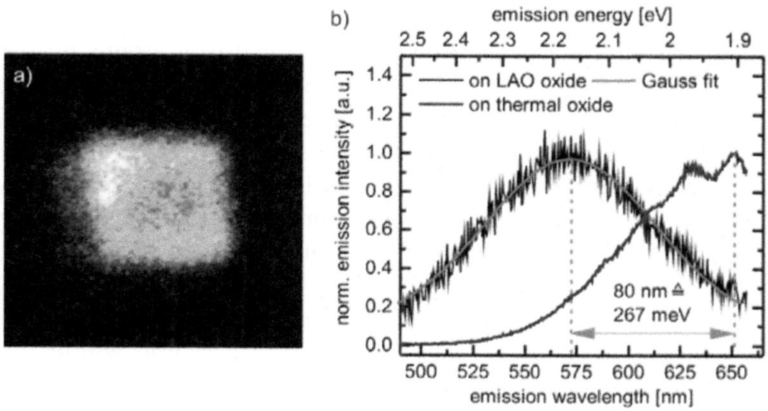

Figure 6.2 (a) Confocal microscope image of di-4-ANEPPS spin-coated onto a substrate containing a square LAO structure. The excitation power was about 2 W and the integration time 5 ms for each pixel. (b) Emission spectra of di-4-ANEPPS on the structure (black) and on an uncharged reference substrate with 4 nm thermal oxide (blue). Spectra were integrated over 5 and 30 s, respectively. The spectral range of the spectrometer is limited to 655 nm. Gaussian fits (red) are used to obtain the emission maximum. From Baumgärtel et al. (2012).

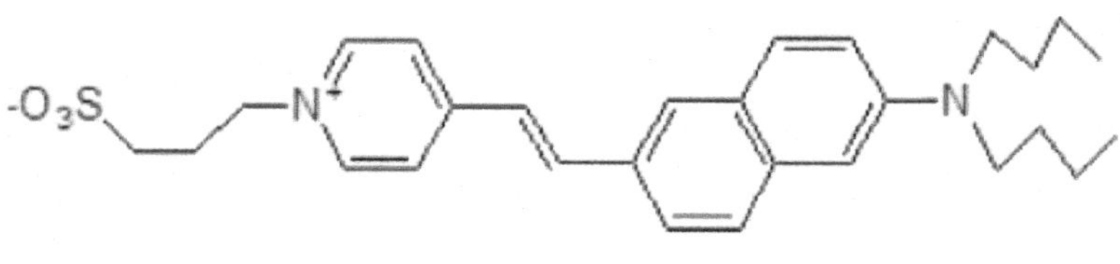

Scheme 6.4 Chemical structure of aminonaphthylethenylpyridinium (di-4-Anneps).

For deeper understanding of the influence of the charged silicon oxide on the optical properties of di-4-ANEPPS, emission spectra on a local anodic oxide structure (see Fig. 6.2b (black curve)) and on uncharged thermal oxide of similar thickness (blue curve) were compared to each other. In the case of di-4-ANEPPS on the LAO structure, a pronounced blue-shift of about 267 meV (80 nm) was found.

The emission of di-4-ANEPPS on thermal oxide is characterized by a double peak, which is related to the pre-treatment of the substrate (cleaning by "piranha" solution, H_2SO_4/H_2O_2) (Baumgärtel et al., 2012) which leads to hydroxyl groups on the oxide surface. Some of the di-4-ANEPPS molecules located closely to the hydroxyl groups may be influenced by their partial negative charge and therefore show a relatively small blue-shift. Other dye molecules which are not close to the hydroxyl groups possess a wavelength of maximum emission similar to that reported for di-4-ANNEPS on pure quartz glass (around 650 nm (Martin et al., 2005)).

The strong blue-shift of the emission on the LAO oxide structure is most likely caused by the space charge within the oxide. Reason for this shift is an electrochromic mechanism where the energy of the electronic transition is altered in the presence of an external electric field (Fluhler et al., 1985; Loew et al., 1978). From experiments with di-4-ANEPPS bound to hemispherical lipid bilayers it is known that the fluorescence emission maximum changes from 640 to 540 nm ($\Delta E \approx 359$ meV) upon a change of the transmembrane potential in the order of about 100 mV (Fluhler et al., 1985). A direct comparison of the blue shift of the emission maximum with the external electric field could not be made as the orientation of the molecule to the electric field plays an important role (Loew et al., 1978). In case of the lipid bilayers the di-4-ANEPPS is incorporated into the lipid layer and therefore nearly parallel to the electric field. In case of di-4-ANNEPS adsorbed on the LAO oxide the dye will be more or less parallel to the surface, so the influence of the electric field on the optical properties of the probe molecule is lower. Therefore the blue-shift is smaller in case of the LAO oxide.

Di-4-ANEPPS was also used to investigate the stability of the charge in the LAO oxide. Emission spectra have been taken 2, 5, and 42 h after generation of the LAO structure. Table 6.1 summarizes the results for different observation times after preparation as well as values from the literature. Even 42 h after the structure has been generated; the blue-shift of emission is still nearly the same indicating that the surface potential has not decreased. Further, the measurements indicate a slight enhancement as the blue-shift increases even further. It is not likely that the number of charges within the oxide increases, so this effect may be connected with a rearranging of either the dye molecules or the charges within the oxide. Interestingly in this optical investigation a shielding effect of adsorbing water as it has been detected by KPFM is not observable, so it can be assumed that water condenses on top of the di-4-ANEPPS film.

Finally it can be summarized that both methods, KPFM and optical spectroscopy, show clearly the existence and stability of the negative charges incorporated in the LAO oxide. Unfortunately, without further information and experimental control, a quantitative statement about the number of charges and the charge density is not possible.

Table 6.1 Fluorescence maximum position λ_{cm}^{max} and E_{em}^{max} of di-4-ANEPPS on different systems (values from (Baumgärtel et al., 2012))

Surface/condition	λ_{cm}^{max} (nm)	E_{em}^{max} (eV)
On quartz	655[a]	1.893
On thermal oxide ("piranha"-treated, 4 nm)	651	1.905
On LAO oxide		
after 2 h	571 ± 2	2.171
after 5 h	561 ± 2	2.210
after 42 h	559 ± 2	2.218
On lipid membrane	640[b]	1.937
On lipid membrane at −100 mV membrane potential	540[b]	2.296

[a]From Martin et al. (2005).
[b]From Sluch et al. (1995).

6.3 Attachment of CdSe-Nanocrystals on Alkyl-Terminated Silicon

In order to test the selectivity of specific interactions a TOPO-capped CdSe/ZnS nanocrystal (see Scheme 6.5) was chosen as a probe. Due to the TOPO surfactants the outer part was expected to be hydrophobic, which should lead to an interaction with the hydrophobic alkyl monolayer as has been shown in (Chen et al., 1999).

Immersion of the alkyl terminated and structured silicon substrates in a solution of TOPO-covered CdSe/ZnS nanoparticle in toluene led to a specific adsorption of these nanoparticles on the alkyl terminated silicon but not on the silicon oxide as demonstrated in Fig. 6.3. In this rather larger image it can be seen that the height of the adsorbed nanoparticles (nearly all adsorbed on the alkyl monolayer) is large compared to the LAO oxide with a height of about 3 nm. On top of the oxide structures nearly no adsorption of nanoparticles can be seen. As water soluble nanoparticles do not adsorb on the alkyl monolayer, it is reasonable to conclude that the driving force for the adsorption are hydrophobic interactions between the alkyl monolayer and the alkyl part of the TOPO shell around the nanoparticles.

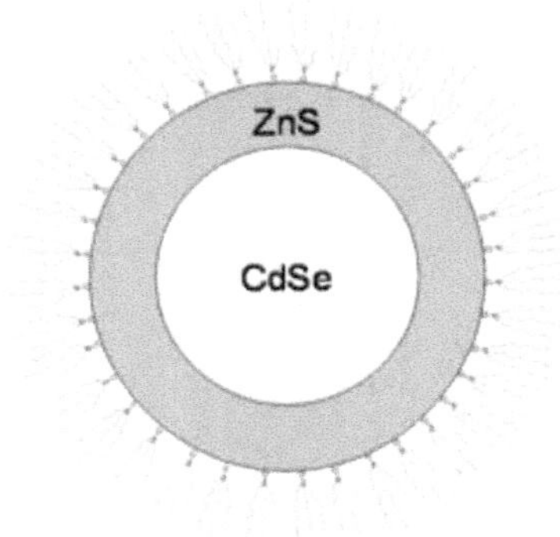

Scheme 6.5 Principle structure of a CdSe/ZnS nanocrystal with a TOPO shell.

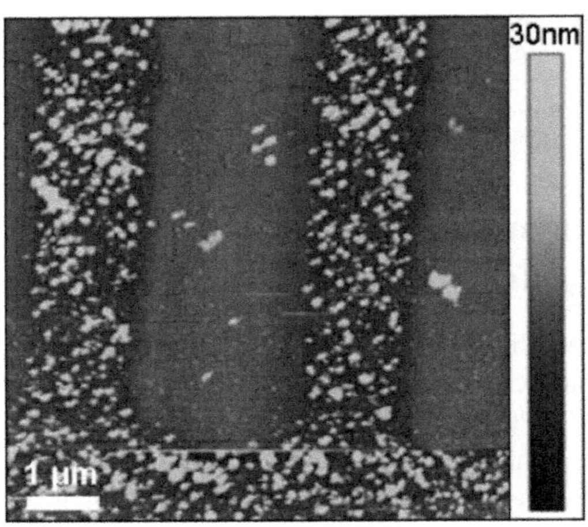

Figure 6.3 AFM height image after adsorption of CdSe/ZnS nanocrystals. Height maximum 30 nm. The bare parts of the image with nearly no nanoparticles are parts of three silicon oxide bars. From Graaf et al. (2007).

The adsorbed nanoparticles themselves did not form a close monolayer on the alkene monolayer as can also be seen in Fig. 6.3. Most of the nanoparticles were isolated from each other while some obviously formed larger aggregates consisting of at least a few nanoparticles. These aggregates dominate the AFM-Height-Image. Looking closer to the single particles a ring-like structure around each single particle and each aggregate was found, which was described in literature as a "moat" and a wall around a castle or an impact crater (Graaf et al., 2007). Such a structure is depicted in more detail in Fig. 6.4. The inset shows the section where the height profile has been measured.

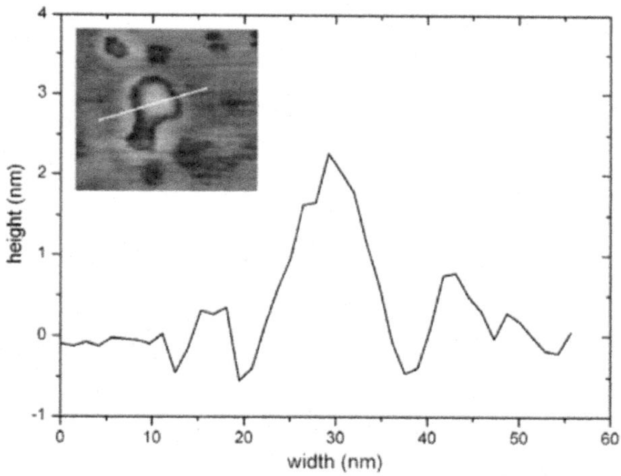

Figure 6.4 Height profile of a single CdSe/ZnS nanocrystal on an alkyl monolayer. The small inset shows the corresponding AFM image section with the profile line. From Graaf et al. (2007).

Inspecting the height profile reveals a central peak (height 2.8 nm as referred to the lowest level of about –0.5 nm), and symmetrically surrounding moats (–0.5 nm deep), which are themselves surrounded by walls of about 0.5 nm height (Graaf et al., 2007).

Such structures have not been observed in the case of a pure alkyl monolayer or after treatment with functional molecules as molecules. It is therefore obviously that the nanoparticles are responsible for this finding.

The area between the nanocrystals is found to be flat with statistically distributed holes of a medium depth of about 1.2 ±

0.2 nm and an average diameter of about 10–30 nm (see Fig. 6.5). These structures were only found on the alkyl monolayer (the hydrophobic part of the sample). Comparing this depth with the size of a TOPO molecule (around 1.1 nm length (Graaf et al., 2007)) reasonably suggests that these holes are due to missing TOPO molecules.

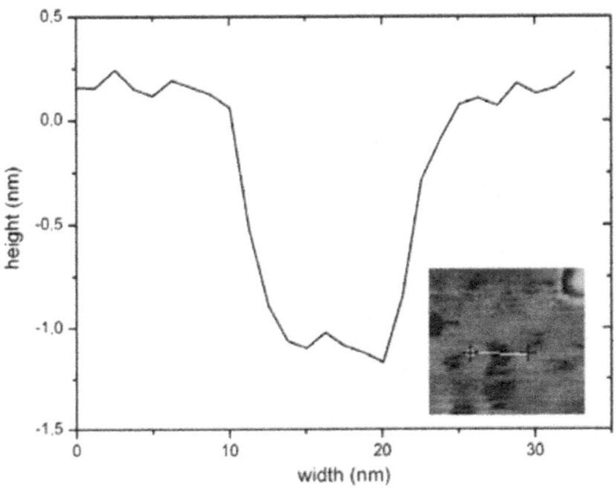

Figure 6.5 Height profile of a typical area of the hydrophobic part of the sample. The small inset shows the corresponding AFM image section. From Graaf et al. (2007).

The photoluminescence image supported the results obtained by AFM investigations. No photoluminescence signal (see Fig. 6.6) was detected from the LAO oxide structure which corresponds to the fact that nearly no CdSe-nanoparticle absorbed here. On the other hand, the surrounding (the alkyl monolayer regime) was characterized by a clear luminescence signal. This indicates that the luminescence of the CdSe/ZnS nanocrystals was not fully quenched by the silicon substrate below the dodecyl monolayer with a thickness of around 1.2 nm. Bright spots dominated the photoluminescence image. These higher intensities might be due to larger aggregates of CdSe/ZnS nanocrystals as shown in the AFM image in Fig. 6.3.

In conclusion, it has been demonstrated that hydrophobic interactions, e.g., between a alkyl terminated silicon substrate and hydrophobic nanoparticle (the hydrophobic character is here introduced by the TOPO capping) led to a selective binding of the nanoparticles. During (or after) the absorption of the Topo-capped Nanoparticles some (or all) of the TOPO molecules arrange in a ring-like structure around a (bare) crystal. The crystals themselves formed no dense-packed monolayer. Most of them are single ones while some form aggregates. The aggregates itself show the highest luminescence intensity. Additionally to the particle adsorption there is also an adsorption of abundant TOPO-molecules from solution on the alkyl monolayer. Here it should be noted that also hydrophilic interactions can be used for selective attachment of nanoparticles, as it has been reported recently for polymer beads suspended in aqueous solution (Belgardt et al., 2014).

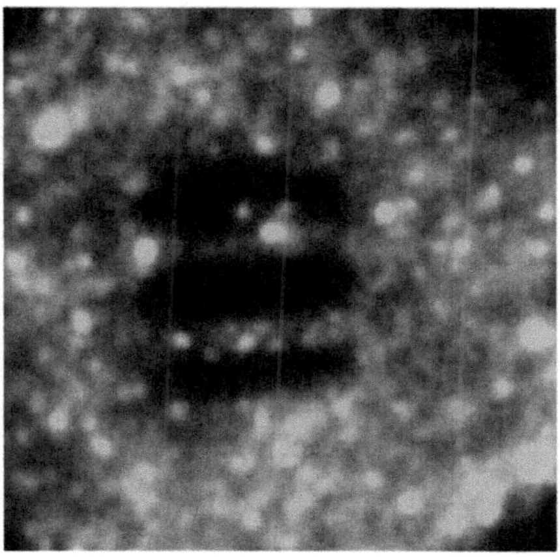

Figure 6.6 Luminescence image after CdSe/ZnS TOPO-covered nanoparticles adsorption. The dark areas in the middle are related to the oxide bars shown in Fig. 6.4. The image is taken with a wide field fluorescence microscope and the focus spot is about 20 μm in diameter. From Graaf et al. (2007).

6.4 Attachment of Charged Dye Molecules

The negative charges within the LAO oxide (see Section 6.2) can be used to immobilize cationic species like dye molecules (Graaf et al., 2007; Baumgärtel et al., 2010; Baumgärtel and Graaf, 2014) or organic semiconductors (Losilla et al., 2008; Losilla et al., 2009) selectively on the structured surface. The high stability of the incorporated charges in the oxide allows a selective adsorption of the dyes from aqueous solution due to electrostatic interactions. Different fluorescent dyes have been deposited to the structure: rhodamine 6G (R6G) (Scheme 6.6B), cresyl violet (CV) (Scheme 6.6A) and a spermine-functionalized perylene bisimide (sf-PBI) (Scheme 6.6C).

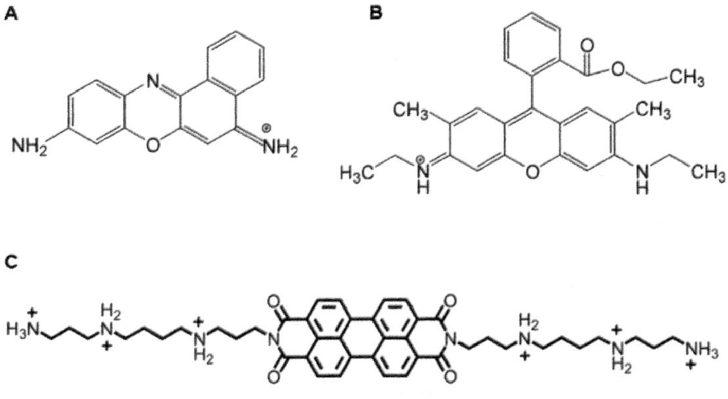

Scheme 6.6 Chemical structure of a cresyl violet cation (CV) (A), a rhodamine 6G cation (R6G) (B) and a spermine-functionalized perylene bisimide cation (sf-PBI) (C).

6.4.1 Attachment of Rhodamine 6G and Cresyl Violet to the LAO Oxide Structure

In the case of R6G and CV, the positive charge is mainly located at the nitrogen atoms of the molecules and is part of the chromophoric system of the molecule. After adsorption of the dye from solution, a height increase of the LAO oxide structure had been detected (see Fig. 6.7).

The height increase for R6G is about 0.55 nm, which corresponds to a monolayer of molecules on the LAO oxide. The two peaks in the height image could be explained by the formation of small R6G aggregates during the deposition.

Looking at the fluorescence of the R6G on the samples showed clearly that there is no emission from the alkyl-monolayer but a clear emission from the oxide structures (see Fig. 6.8).

In Fig. 6.8 several oxide lines generated via LAO using different voltages from −8 to −4 V are displayed. In Fig. 6.8 (left), the AFM height image of the structures before attachment of the R6G molecules are shown. The line width is below 100 nm and the line height is in the range from 1 to 4.5 nm (see Fig. 6.8, left).

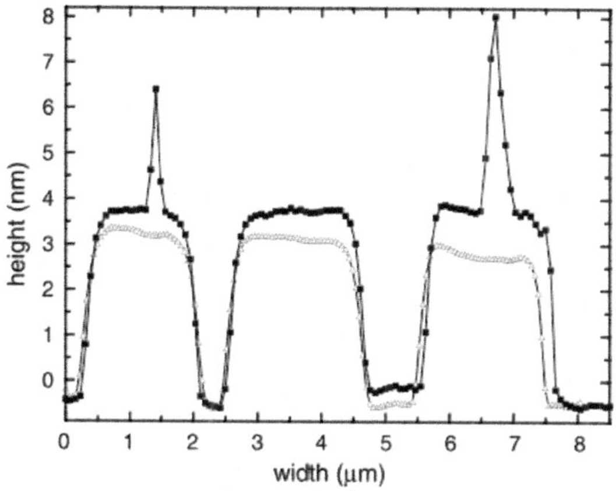

Figure 6.7 Height profile of bare LAO oxide structures (-Δ-) on a dodecyl terminated silicon substrate (applied voltage 12 V). Height profile of the same structure after R6G treatment (-■-). From Graaf et al. (2007).

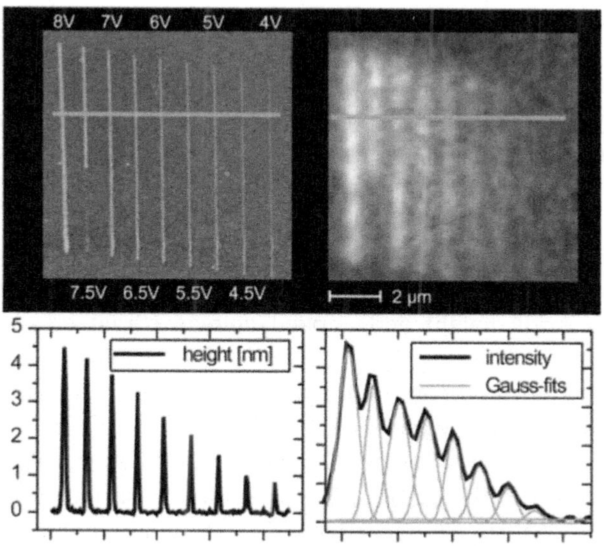

Figure 6.8 LAO-oxide line patterns of different size on dodecyl-terminated silicon before and after dye attachment. Left: AFM height image before dye attachment (the corresponding lithography voltages are given above and below the lines). Right: Optical wide field microscope images of the same spot on the surface after dye attachment. The diagrams below correspond to a cross section along the blue line in the above image. From Baumgärtel et al. (2010).

The optical investigation of the same area after R6G deposition using an optical wide field microscope shows a signal in the same general shape of the structures (Fig. 6.8, right). Thus it can be concluded also from the optical investigations that the dye molecules were attached to the LAO oxide. It is also evident that the dye molecules can still be excited optically and those they still emit fluorescent light. On the other hand, the resolution of the optical image is limited and therefore not as sharp as the AFM image. It has to be noted that also the emission intensity from the dye-functionalized nanostructures is relatively low, which can be attribute to a strong photoluminescence quenching. This quenching process lowers also the life time of the excited state from 3.7 ns (R6G in aqueous solution) down to about 0.6 ns (R6G on LAO oxide) (Baumgärtel et al., 2010).

For an evaluation of the cross sections in Fig. 6.8, each peak was fitted by a Gaussian function. From these functions, characteristic parameters of each line such as height, width, and area under the curve were determined. In the case of the optical image, the Gauss fits reveal that the FWHM of the detected intensity for each line is below the line separation so that there is only a negligible contribution to the neighboring line maximum. Figure 6.9 shows the height of the fitted Gaussian curves for both the topographical AFM image as well as the optical image for each line (every line corresponds to a different voltage used to generate the line).

In agreement with other groups, there is a linear correlation between oxide height and applied voltage (Graaf et al., 2008a; Yang et al., 2005; Teuschler et al., 1995; Cavallini et al., 2003). Interestingly the peak photoluminescence (PL) intensity also decreases linearly with the voltage. This means that the PL intensity depends linearly on the oxide height. This is in contradiction to the classical theory of dipole–surface interaction, which predicts a d^{-3} dependence of the fluorescence resonance energy transfer (FRET) rate on the distance d between dye and semiconductor is expected (Hayashi and Castner, 1983; Whitmore et al., 1983). On the other hand, different groups found also deviations from this d^{-3} behavior for $d < 10$ nm (Hayashi and Castner, 1983; Whitmore et al., 1983; Sluch et al., 1995; Barnes 1998). As d is also below 10 nm (oxide height) here, the findings confirm that for small distances a simple model of dipole–semiconductor energy transfer is not valid. This result is supported by the fact that the PL intensity shows also a linear relation to the oxide volume (Baumgärtel et al., 2010). Therefore it is assumed that the PL intensity is mainly attributed to the number of dye molecules and not to a distance-dependent PL quenching mechanism. The found quenching is therefore dominated by fast non-radiative decay channels from the excited state to trap states in the LAO oxide, as it was suggested in (Liang and Ponte Goncalves, 1985).

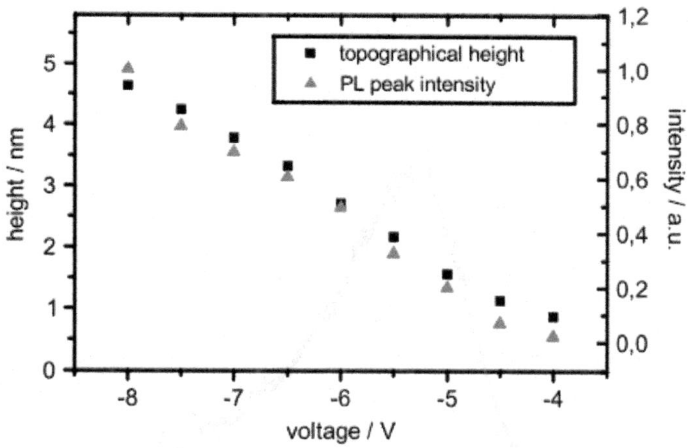

Figure 6.9 Topographical height (black squares) and peak PL intensity obtained from Gaussians (red triangles) against the voltage at which the corresponding line has been generated. From Baumgärtel et al. (2010).

For a deeper understanding, spectral information of the emission of R6G (Fig. 6.10) and CV (Fig. 6.11) has been obtained. For comparison spectra of the dye in aqueous solution (and for R6G on native SiO_x) are given.

For both dyes, the general shape of the emission spectrum of the dyes on LAO oxide is quite similar to that in solution. Yet a closer look reveals a clear spectral broadening especially towards blue wavelength. This broadening can be attributed to an additional emission. The maximum intensities for the emissions in solution as well as on LAO oxide are given in Table 6.2.

Table 6.2 Fluorescence maximum position E_{cm}^{max} and its energetic shift of R6G and CV on different systems

Dye	E_{em}^{max} (eV) [solution]	E_{em}^{max} (eV) [LAO-Oxide]	Shift (eV)
R6G[a]	2.25	2.25	0
		2.33	0.08
CV[b]	2.00	2.00	0
		2.032	0.032

[a]From Baumgärtel et al. (2010).
[b]From Baumgärtel and Graaf (2014).

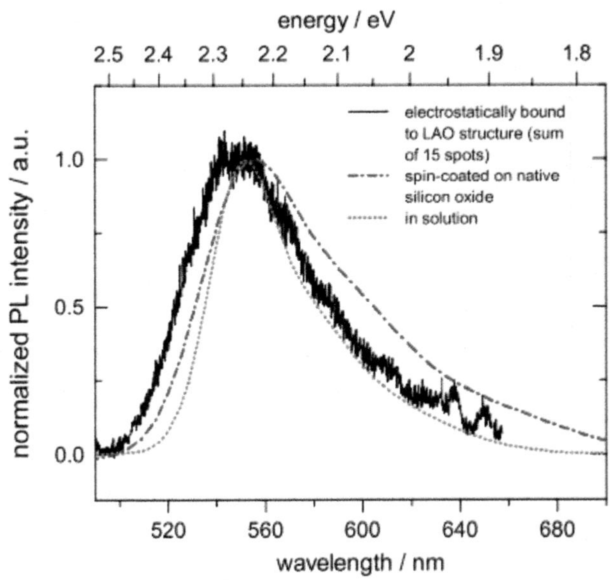

Figure 6.10 Emission spectra of R6G attached to an LAO-oxide structure (black), in diluted aqueous solution (red) and as a sub-monolayer on silicon with a native SiO_x layer (blue). All spectra are normalized; the excitation wavelength was 465 nm and the excitation power 4 μW for the spectra measured by confocal microscopy. From Baumgärtel et al. (2010).

For such a spectral shift, different reasons are discussed:

As the distance of the dye molecules on the LAO oxide is quite small, a formation of ordered aggregates of molecules with a coupling of transition dipoles is possible. Such a formation is known, e.g., for CV on SiO_2 particles, where dimers are formed. This dimerization causes a splitting of the dimer excitonic singlet state into two levels. In the case of CV only a transition to higher energetic level is quantum mechanically allowed (H-aggregates), so that the absorption is blue-shifted with respect to the monomer absorption (Martini et al., 1997; Liu and Kamat, 1996). These dimers are characterized by a very small emission (Liu et al., 1995), so that such an aggregation can be excluded as an explanation for the observed blue-shift of emission on LAO oxide. This assumption is supported by reports on R6G, where a clear red-shifted emission has been found if the intermolecular distance is close to the FÖRSTER radius (5 nm for R6G). The shift is in this case due to resonant energy transfer (Zondervan et al., 2003). As on the LAO oxide a monolayer of dye molecules can be assumed, the intermolecular distance should be smaller than the FÖRSTER radius. However, the emission of dyes on LAO oxide is blue shifted, energy splitting and energy transfer could be excluded as mechanisms for the spectral shift.

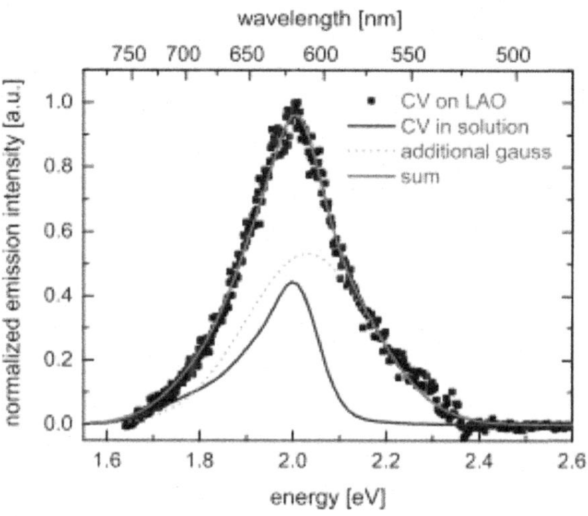

Figure 6.11 Emission spectra of CV attached to LAO-oxide (black) and in diluted aqueous solution (blue). The measured spectrum on LAO-oxide can be satisfactorily fitted by a superposition (red) of the emission spectrum in solution and an additional blue-shifted Gaussian peak (green). The excitation wavelength was 514 nm and the excitation power 200 μW. From Baumgärtel and Graaf (2014).

A second explanation may be a change in the dielectric environment of the dye molecules on the surface compared to that of dye molecules in solution. The dielectric surroundings influence the intermolecular energy levels and thus, if they are dissimilar, can cause a spectral shift. To understand the influence of silicon oxide on spectral properties of the emission a sub-monolayer of R6G was prepared on a silicon substrate with a native oxide layer at the surface and the emission spectrum was investigated. The emission from these R6G molecules, however, is shifted about 36 meV to the red with respect to the solution spectrum. Such an inhomogeneous broadening towards larger wavelengths has been reported before for R6G in close vicinity to metallic surfaces (Cnossen et al., 1993) and was explained as an effect of the reduced excited state life time on the homogeneous damping rate. In contrast, the emission from the dyes on LAO oxide is shifted to the blue, which can be explained by the difference between the LAO oxide and native or thermally grown oxide on silicon. One main difference between the native oxide and the LAO oxide are the negative charges incorporated into the LAO oxide.

From previous reports it is known that some cationic dyes in solution undergo a spectral shift in the presence of anions to shorter wavelengths. This phenomenon (known as metachromasia (Schubert and Hamerman, 1956)) is attributed to the formation of cation–anion charge-transfer complexes (Pal and Ash, 1973). Since the LAO oxide includes negative charges and dye molecules are in proximity to the outermost negative space charges within the oxide, it is reasonable to assume that there is a strong coupling of the positively charged dye molecules with the negative space charges. This could cause a comparable shift to higher energy for the absorption (and therefore also the emission) as it was found for charge-transfer complexes (Pal and Ash, 1973). In the case of R6G and CV, the interaction with a negative charge in the oxide can lead to localization of the positive charge in chromophoric system, which leads to a slight change of the energetic of the relevant energy levels.

As the other explanations (aggregation and change in dielectrics of the surroundings) can be excluded, it is suitable to assume that the formation of analogues to charge-transfer complexes (here localization of positive charge in the chromophoric system by the negative charges in LAO oxide) explains the blue shifted emission.

Here it should be taken into account that there is still an important fraction of adsorbed dye molecules show no blue-shifted emission, which suggested that only a part of the molecules are characterized by strong electrostatic interactions and therefore the distance between negative charge and molecule is quite small. From comparison of the integral intensities of monomer and additional Gaussian peak (see Fig. 6.11) it can be concluded that about two of three molecules are more or less located close to negative charges.

Besides the energetic blue-shift of the emission spectra, the LAO oxide influences also other optical properties. Depending on the power of the excitation light, the intensity of the emission decreases rapidly within seconds to a few minutes. This mechanism is known as photobleaching (a process which converts fluorescent particles into non-fluorescent entities (Zondervan et al., 2004)). The mechanism is still not fully understood but supposed that various pathways with different intermediate states lead finally to a non-fluorescent state (Schuster et al., 2007). In the case of the bound dyes, an irreversible and complete bleaching takes place for long exposure of exciting light, as even after long periods without excitation no recovery of the emission takes place.

The bleaching of the dyes (R6G and CV) with excitation time was investigated and compared to a dye (R6G) on thermal oxide of equivalent thickness. In all cases, the time dependence of the decrease in emission intensity can be fitted by a bi-exponential decay (see Eq. 6.1).

$$I_{\mathrm{pl}} = I_0 + A \cdot \exp\left(-\frac{t}{\tau_1}\right) + B \cdot \exp\left(-\frac{t}{\tau_2}\right) \qquad (6.1)$$

Such a bi-exponential decay has been reported for R6G before (Liang et al., 1984). The determined time constants as well as their relation and the excitation power are given in Table 6.3.

Table 6.3 Time constants extracted from bi-exponential fit of emission intensity decay for R6G and CV.

Dye/condition	P_{exc} (μW)	τ_1 (s)	τ_2 (s)	τ_2/τ_1
R6G on LAO-oxide (4 nm thick)[a]	6.5	5.6	33	5.89
R6G on thermal oxide (4 nm thick)[a]	6.5	7.7	52	6.75
CV on LAO-oxide[b]	200	0.44	2.70	6.13

[a]From Baumgärtel et al. (2010).
[b]From Baumgärtel and Graaf (2014).

Up to know it is still not clear why the bleaching process is governed by two different characteristic times. This could be a hint for the existence of two different species or states of the molecules, showing different photochemistry which leads to the transformation into the non-fluorescent state. As time constants for the decay of R6G differs only slightly (factor of about 1.6) for both substrates and same bi-exponential law explained both processes, an influence of the electrostatic binding mechanism can be excluded.

For CV the found time constants are about 12 times smaller than that for R6G, which is attributed to the higher excitation power. On the other hand, the relation between the two time constants is for both dyes roughly the same (about 6). Since the non-fluorescent state is reached by a photochemical process (possibly an oxidation) it could be expected that due to similarities in the chemical structure of the chromophoric system also similar chemical degradation takes place.

6.4.2 Attachment of a Spermine-Functionalized Perylene Diimide Derivative to the LAO Oxide

Compared to R6G and CV the spermine-functionalized perylene diimide derivative (sf-PBI) (for chemical structure see Scheme 6.6C and (Rehm et al., 2010)) contains six positive charges. Additionally the charges are located in the two side chains of the molecule (three positive charges in each chain) and are therefore not part of the chromophoric system. The chromophoric system is the well known perylene bisimide, which is characterized by a high photostability, chemical inert behavior and high fluorescence quantum yield. These properties make them an ideal system for single molecule spectroscopy (Kowerko et al., 2009; Krause et al., 2011), light induced energy transfer processes (Prathapan et al., 2001) and light collection (Würthner, 2004).

Furthermore, they are known to form highly fluorescent J-type aggregates under certain conditions by hydrogen bonding and π–π stacking interactions (Kaiser et al., 2009). In recent studies, evidence was found for an efficient long-range transport of energy by exciton migration in such aggregates (Lin et al., 2010) which could be very useful for photovoltaic devices or organic circuits.

As the spermine side chains are bound to the perylene core at imide positions, the planar structure of the perylene core is sustained and the optical properties of the PBI unit are not significantly altered. Also, an aggregation of the molecules via π–π stacking is still possible, although with decreased binding constants and a preference for the formation of small dimer aggregates instead of extended columnar aggregates due to the repulsion of the cationic side chains (Rehm et al., 2010).

AFM height measurements of LAO oxide nanostructures before and after sf-PBI adsorption show a small height increase between 0.3 nm for thin line structures and 0.7 nm for larger two-dimensional structures (Baumgärtel et al., 2014).

As the height growth is very homogeneous and also the roughness of the residual surface does not change, a nearly uniform deposition of sf-PBI onto the LAO structure was assumed. Furthermore, the height increase is rather small in comparison to the length of a single sf-PBI molecule (approximately 4 nm). Thus it is reasonable to assume that the molecules are lying almost flat on the surface. This is supported by the fact that the positive charges (they are attracted toward the negative charges in the LAO oxide) are located along both spermine chains. Hence a flat orientation of the molecules on the surface brings all charges of sf-PBI as close as possible to the charges in the oxide. The measured height increase (0.3 to 0.7 nm) is within the range of the diameter along the short in-plane axis of the sf-PBI molecule (0.5 nm) (Baumgärtel et al., 2014).

6.4.2.1 Optical emission at room temperature

In Fig. 6.12 the emission spectrum of an ethanol solution of sf-PBI and the emission spectrum of sf-PBI bound to a LAO oxide structure (see inset) is given. Compared to the sf-PBI fluorescence in solution (here monomeric molecules are present), there is a red shift and a broadening (up to around 800 nm) of the fluorescence spectrum of sf-PBI on LAO oxide. The peak at 540 nm can be attributed to the $S_1 \rightarrow S_0$ transition of monomeric sf-PBI species except of a red shift by about 28 meV (540 nm compared to 533 nm in solution). This shift might be caused by the different dielectric environment for molecules in solution and bound to LAO oxide. In the broad long wavelength part of the spectrum a substructure at 585 nm can be identified. This is attributed to the first vibronic progression of the $S_1 \rightarrow S_0$ transition as the energetic separation from the main peak is about $\Delta E \approx (175 \pm 5)$ meV (Baumgärtel et al., 2014), which is in good agreement with typical values for perylene tetra carboxylic acid derivatives in solution or thin films (Bulovic et al., 1996; Hennessy et al., 1999; Proehl et al., 2004).

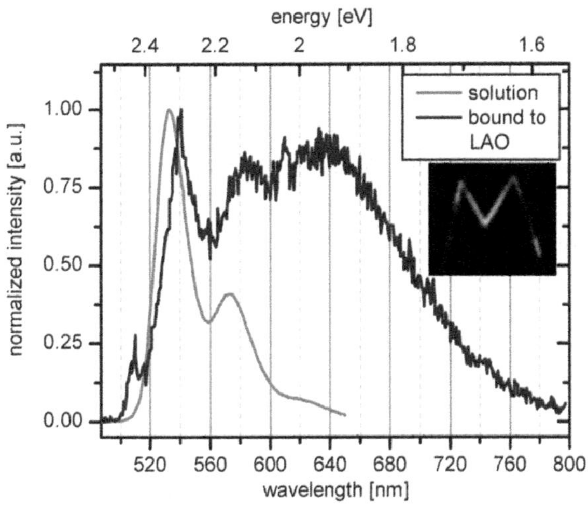

Figure 6.12 Emission spectra of sf-PBI in ethanol solution (red) and bound to a LAO-oxide nanostructure (blue). A confocal microscopic image of the functionalized structure is shown as inset. (The intensity below 520 nm (for emission spectrum of sf-PBI on LAO-oxide) is an artifact caused by used optical filter). From Baumgärtel et al. (2014).

The prominent broad peak in the lower energy regime with a maximum at about 640 nm could not be related to a monomer emission. It is known that the emission spectra of PBI dyes change dramatically if thin films of these molecules are prepared (Ahrens et al., 2004; Graaf et al., 2011). Here the formation of aggregates with predominant H-type coupling and the possibility of relaxation into excimers is given as an explanation.

Gao et al. (Gao et al., 2011) and Engel et al. (Schubert et al., 2013) were able to simulate recently the H-dimer spectrum in good agreement with experimental data using an elaborate model which mixes Frenkel excitons (FE), intermolecular charge-transfer excitons (CTE) and also includes electron-phonon coupling. They found that the broad and the strongly red-shifted emission of the H-dimers can be explained by a change of the optimal dimer geometry between ground state and excited state. Upon excitation, the H-dimer undergoes a relaxation onto a new geometry which possesses a lower energy of the exciton. As the exciton life time is much longer in the optimized geometry than in the excited state of the monomer and since the relaxation of the structure is slow, there is an emission from either states (or any intermediate state) which extends the emission significantly in wavelength (Baumgärtel et al., 2014).

For comparison the emission spectra of spin-coated films using sf-PBI solution of different concentration were prepared. In Fig. 6.13 two of these spin-coated films are shown. It should be noted that for a concentration of 1.6×10^{-6} M a spectrum of monomers are found, while at concentrations above 1.6×10^{-4} M no further change in spectra was found (see (Baumgärtel et al., 2014)). As clearly seen in Fig. 6.13 the emission is strongly red-shifted as well as broadened and similar to the fluorescence spectra of α-perylene crystals which are composed of π-stacked perylene dimers (in that case denoted as E-type emission) (Sumi, 1989). E-type-like emission has been found also for other perylene samples such as films (Graaf et al., 2011; Graaf et al., 2008b), designed H-dimers (Giaimo et al., 2008; Veldman et al., 2008), stacks (Ford and Kamat, 1987; Bisht et al., 1996; Akimoto et al., 1997) or polymers (Hernando et al., 2004). Especially interesting is that for sf-PBI films from solutions with a concentration of 1.6×10^{-5} M (blue in Fig. 6.13), the emission can be explained to be a combination of a monomer and an aggregate spectrum. Obviously there is a transition region of concentrations, which is around 1.6 × 10^{-5} M for the used spin-coating parameters, at which both species of the sf-PBI molecules (monomers and dimers) coexist on the surface.

Finally it should be noted that the relation of the intensities of the emission at 540 and 640 nm in the case of sf-PBI bound to LAO oxide showed variations. For sf-PBI bound to two-dimensional LAO oxide nanostructures the emission intensity of the broad peak at 640 nm is much higher compared to peak at 540 nm (see spectrum at 298 K in Fig. 6.14). This indicates an increase of dimers compared to monomeric entities on two-dimensional nanostructures compared to linear structures.

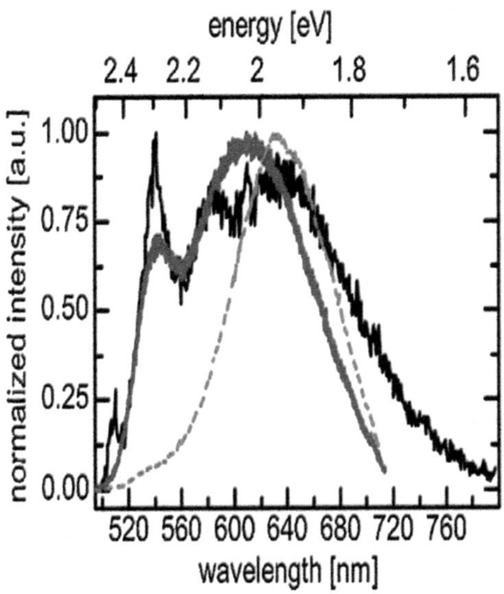

Figure 6.13 Comparison of Fluorescence spectra of spin coated films prepared by a 1.6·10^{-5} M (blue) and a 1.6·10^{-4} M (red dashed) solution of sf-PBI on 100 nm thick SiO_2 with the emission of sf-PBI on a LAO oxide nanostructure (black). From Baumgärtel et al. (2014).

6.4.2.2 Optical emission at low temperature

The intensity ratio between monomer-like and E-excimer emission (at about 640 nm) is clearly a function of temperature, whereby the relative excimer emission is decreased at low temperatures (see Fig. 6.14).

Comparable behavior has been reported for α-perylene crystals where emission from the E-excimer vanishes at low temperatures while an additional emission appears at higher energies. This emission has been denoted as Y-type emission and which shows a similar spectral shape as the monomer spectrum (Sumi, 1989; von Freydorf et al., 1978; Walker et al., 1985). A phenomenological model has been developed to describe the excimer formation in perylene crystals, see, e.g., Walker et al. (1985) and Mahrt et al. (1994). In this model, the fluorescent E-excimer state is populated only via thermally activated energy transfer from the Y state. This model could partly explain the found experimental observations for sf-PBI on LAO oxide: The E-type emission at high temperatures is a result of the formation of E-states following a thermally activated change in the relative position of two (coupled) sf-PBI molecules. Unfortunately the exact nature of this positional change is still not fully understood. Two models on molecular level are discussed in literature:

(i) The change is related to the distance of the dimer subunits (Sumi, 1989; Tanaka, 1963)
(ii) Orientational and conformational changes are responsible for the changes (Gao et al., 2011; Veldman et al., 2008).

In both cases at high temperatures (e.g., room temperature), the motional freedom of the molecules allows such a position change in the dimer which enables the formation of an excimer state. At lower temperatures, the barrier cannot be overcome and the formation of the E-state is not possible anymore.

Interestingly for aggregates of sf-PBI on thermal silicon oxide (not shown here, see (Baumgärtel et al., 2014)) no vanishing of the excimer emission was found when decreasing the temperature to 77 K. This is in good agreement with earlier reports for the temperature dependence of the fluorescence of α-perylene crystals where the Y-state does not occurred for temperatures above 60 K and the E-emission completely vanishes only for temperatures below 30 K (Walker et al., 1985). In case of highly concentrated perylenyldodecaoic acid in Langmuir–Blodgett (LB) films, the E-type emission does not disappear at all, even not at very low temperatures (2 K) (Mahrt et al., 1994; Weiss et al., 1992). This indicates that even at 77 K the thermal energy is high enough in order to allow for configuration changes of the sf-PBI molecules within the thin films (Baumgärtel et al., 2014).

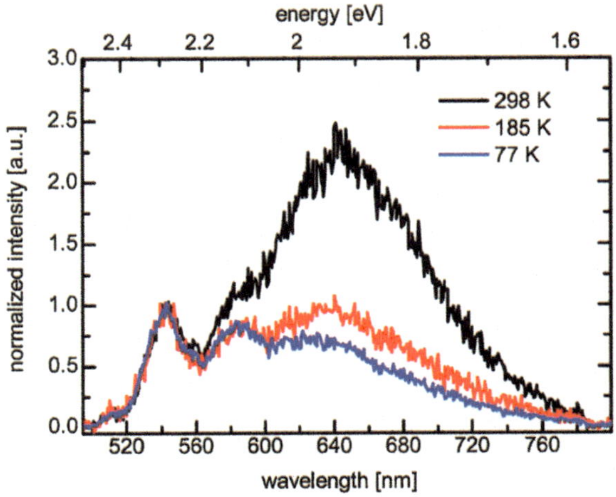

Figure 6.14 Emission spectra of sf-PBI on a (squared) LAO nanostructure measured in vacuum at 298 K (black), 185 K (red) and 77 K (blue) ($\lambda_{exc.}$ = 488 nm). Spectra have been measured during cooling down from room temperature within about 1 h and are normalized to the monomer peak at 545 nm. From Baumgärtel et al. (2014).

The difference in the temperature dependence is a clear indication for a major influence of the negative charges in the LAO oxide on the molecular re-arrangement of sf-PBI during cooling down. For a deeper understanding the samples were brought to room temperature again and the emission was investigated (see Fig. 6.15).

The increase of temperature did not lead to an instantaneous recovery of the E-emission. Even after 14 h at room temperature, only a part of the former E-excimers reappeared. This is quite surprising, as the system has been heated up to room temperature again and the thermal motion should allow a re-arrangement of the perylene units.

From these findings it was concluded that in the case of thin films of sf-PBI on thermal oxide the intermolecular interactions are dominated by the attractive π–π interaction, while the repulsive interaction of the positive charges in the side-chains are rather small due to conformation, which maximize the distance between the chains. For the sf-PBI on the LAO oxide, the situation is a bit more complex: The first molecule is lying on the oxide and slightly immobilized by the electrostatic attraction of the spermine side chains to the substrate. With a second molecule a dimer is formed. This dimer is obviously determined by an equilibrium between electrostatic attraction of the spermine chains to the substrate, on the one hand, as well as a repulsion between the positively charged spermine side chains (Fig. 6.16A, red arrows) and the tendency of the PBI cores to form stacks, on the other hand (Fig. 6.16A, blue arrows). The configuration of the sf-PBI molecules on the LAO structure is therefore a complex function of those interactions and the temperature-dependent three-dimensional degrees of freedom of the molecular entities (Baumgärtel et al., 2014).

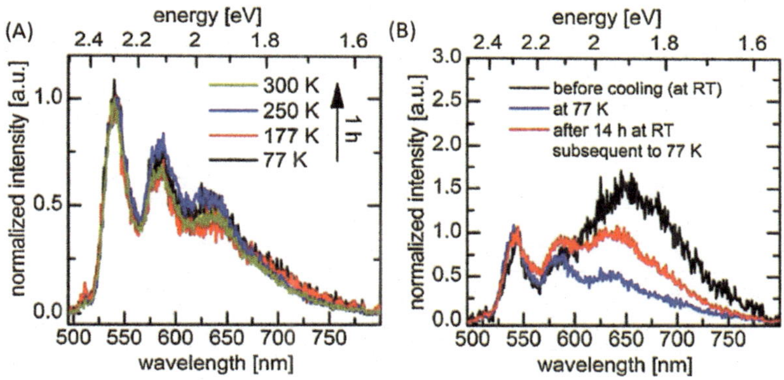

Figure 6.15 (A) Emission spectra of sf-PBI bound to a LAO oxide nanostructure measured in vacuum at 77 K (black), 177 K (red), 250 K (blue) and 300 K (yellow). The spectra have been measured during heating from 77 K to 300 K within 1 h. Spectra are normalized to the monomer peak. (B) Emission spectra of sf-PBI bound to a LAO nanostructure before cooling down (black), at 77 K (blue) and after 14 h at room temperature (RT) subsequent to cooling to 77 K (red). Spectra are normalized to the monomer peak. From Baumgärtel et al. (2014).

At room temperature, the thermal energy is high enough for the spermine side chains to minimize the electrostatic repulsion between each other by bending upwards (see Fig. 6.16B). This configuration also allows for a relaxation of the PBI cores into an optimal π–π-stacking (E-excimer geometry). If the temperature is lowered, the thermal energy becomes too small to keep the charged chains apart from each other. Thus the chains bend down and the electrostatic repulsion between them forces the molecule to rotate (see Fig. 6.16C). Finally, the molecules orient in a new minimal energy configuration (Fig. 6.16D) which is much more different from the optimal stacking for an E-type geometry but is also very stable because of electrostatic binding of now all four spermine side chains of the dimer with the LAO oxide. The stability of this configuration explains the very long recovery time of the E-emission once the sample has been cooled down and heated up again. In summary, the interaction energy depends strongly on the inter-fragment distance R (at least $\sim R^{-6}$) (Grimme, 2008), meaning that the stacking interaction is only strong for short distances between the perylene cores. The electrostatic forces between the spermine groups and the negatively charged substrate otherwise have a much longer ranged dependence ($\sim R^{-2}$) (Baumgärtel et al., 2014).

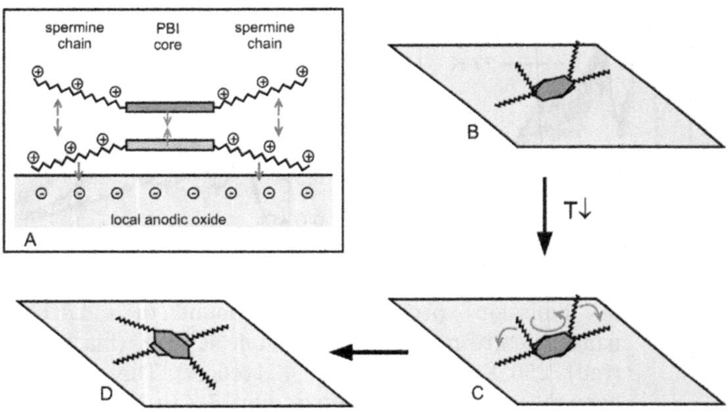

Figure 6.16 (A) Schematic representation of different interactions governing the arrangement of sf-PBI dimers on local anodic oxide. Blue arrows denote π–π-interactions and red arrows symbolize electrostatic interactions. B–D: Exemplary configurational re-orientation of a sf-PBI dimer upon lowering of the temperature. At room temperature, the spermine side chains can thermally activated bend upwards and allow for a relaxation of the PBI cores into an optimal E-type geometry (B). Lowering the temperature leads to a bending down of the chains and an accompanied reorientation of the monomeric units relative to each other (C). The final configuration (D) is much more different from the optimal E-type geometry and also more stable because of the electrostatic binding of all of the 4 spermine side chains. From Baumgärtel et al. (2014).

6.4.2.3 Fluorescence life time investigations

Determination of the fluorescence life time showed a mono-exponential decay for sf-PBI in aqueous solution (see Fig. 6.17). From this decay a life time of (4.1 ± 0.1) ns was obtained, which is in agreement with previous investigations of sf-PBI (Rehm et al., 2010). The decay of the bound sf-PBI is clearly multi-exponential and there are at least three life time components needed to deliver a satisfactory fit result for whole curve (see black dotted line in Fig. 6.17). Following fit-function (Eq. 6.2) has been used for analyzing the fluorescence life time results:

$$I(t) = A_1 \exp\left(-\frac{t}{\tau_1}\right) + A_2 \exp\left(-\frac{t}{\tau_2}\right) + A_3 \exp\left(-\frac{t}{\tau_3}\right) \quad (6.2)$$

The shortest time constant τ_1 is typically in the of in the order of 1 ns, the medium time constant τ_2 is around 4 ns and the long decay time τ_3 is in the order of 14 ns. To understand the multi-exponential nature of the fluorescence decay, three different mechanisms have to be considered:

(i) As it is evident from the spectral investigations, there are at least two different species of sf-PBI present on the LAO nanostructure: monomer-like emitting molecules as well as molecules that show E-type emission. Both species possess quite different fluorescence life times, for the monomer it should be very similar to that in solution (4 ns) whereas the fluorescence life time for the E-emission of perylene crystals is in the order of 100 ns (Walker et al., 1985).

(ii) An efficient excitation energy transfer is expected for dye molecules that are closed to each other (as it is the case for electrostatically bound sf-PBI) which leads to a decrease of the fluorescence life time. It has been shown that upon increase of the perylene dye concentration in LB films, there is a decrease of the life time as well as a transition from mono-exponential to a bi-exponential decay (Akimoto et al., 1997). This was explained by energy transfers between monomers, two different excimers as well as monomers and excimers. Also for chemically designed u-shaped PBI dimers, a double-exponential decay of the fluorescence has been found (Veldman et al., 2008). Here the authors attributed this to the existence of two excited states of the dimerized PBI.

(iii) As the sf-PBI molecules on the LAO oxide nanostructures are separated from the underlying silicon only by the oxide height (a few nm), a very strong energy transfer between the dyes and the silicon is expected by classical electromagnetic theory (Dexter, 1979; Alivisatos et al., 1987) that should decrease the measured fluorescence life time dramatically.

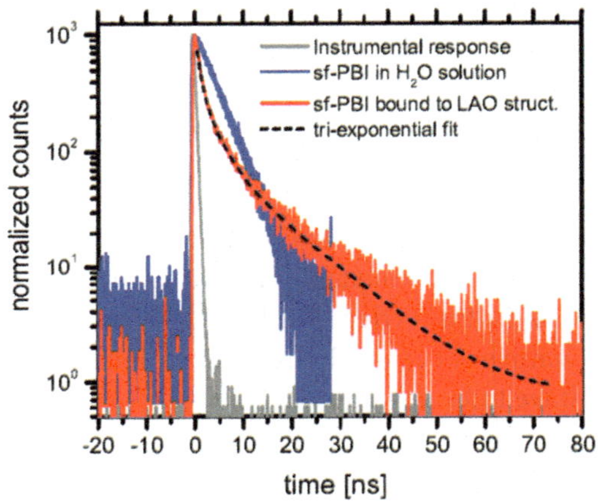

Figure 6.17 Log-lin plot of the normalized fluorescence decay curve of sf-PBI in a 10^{-7} M aqueous solution (blue) and bound to a LAO oxide nanostructure (red) including a tri-exponential fit (dotted black). The decay has been measured over the whole spectral range. The instrumental response function for the LAO measurement is displayed as a gray curve.

As a result of all three processes, the fluorescence decay of sf-PBI bound to the nanostructures is in average faster than in solution and possesses at least tri-exponential characteristics. In the following, the different aspects will be investigated and discussed in greater detail.

As a control experiment in order to exclude the quenching of the underlying silicon, again films of sf-PBI spin-coated on silicon wafers with a 100 nm thick thermal silicon oxide layer were investigated. Fluorescence decay curves of sf-PBI films prepared from solution of different concentrations (1.6×10^{-3} M to 1.6×10^{-8} M) of sf-PBI were prepared (see Fig. 6.18).

For μM concentrations and below, the curves are very similar to each other and can be attributed to the monomer emission. At a concentration of 10^{-5} M, there is a transition to a stronger multi-exponential behavior of the fluorescence decay (spectrally there is a conjunction of monomer and E-excimer spectrum) and at a concentration of 10^{-4} M and above the curves are again very similar to each other (predominant E-excimer spectrum). However, even for low concentrations the decay curves are not purely mono-exponential.

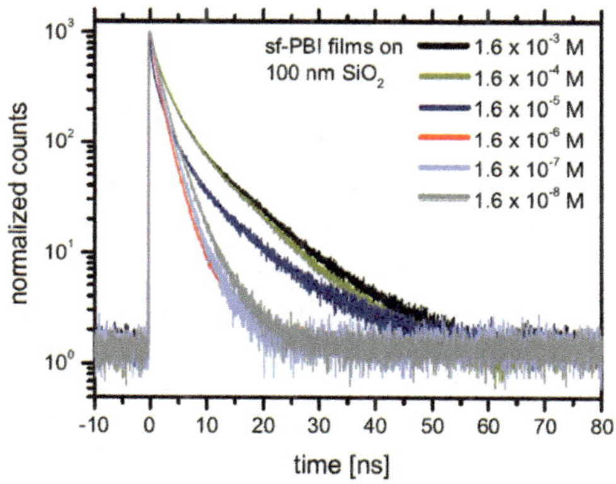

Figure 6.18 Log-lin plot of the normalized fluorescence decay of sf-PBI films on 100 nm thermal silicon oxide on silicon substrates for dye concentrations of 1.6 $\times 10^{-8}$ M (gray), 1.6 $\times$ 10^{-7} M (light blue), 1.6 $\times$ 10^{-6} M (red), 1.6 $\times$ 10^{-5} M (dark blue), 1.6 $\times$ 10^{-4} M (yellow) and 1.6 $\times$ 10^{-3} M (black) in the spin-coated solution. The decay has been measured over the whole spectral range.

The reason for this is still unclear but may be connected to differences of the dielectric environment of the dye molecules on the silicon oxide surface. Nevertheless, the curves have been evaluated by a tri-exponential fit according to Eq. 6.2 which has turned out to be sufficient in order to retrace the decay satisfactorily. The obtained life times and amplitudes from those fits are displayed in Fig. 6.19. The evaluation of the decay also confirms the conclusions from above.

For concentrations above 10^{-6} M, the longest time constant increases significantly (by more than a factor of two) and also the medium decay time become larger. Yet there is no clear change in the amplitudes of the different decay components. The medium time constant has the strongest weighting of the three components (roughly two-thirds) and the other decay times share the remaining third. This means that though the magnitude of the three time constants changes for different concentrations, their distribution remains rather similar. As the spectral properties are changing dramatically from low concentrations (monomer emission) to high concentrations (E-type emission), it can be concluded that the mere existence of three different decay times cannot be attributed directly to the three different emissive states (monomer, Y state, E state). The multi-exponential nature of the decay seems to be a result of the system itself (sf-PBI on silicon oxide) and independent of the origin of the emission (isolated molecules, dimers or excimers). Up to now this origin of the multi-exponential decay is still unclear. However, there is clearly a transition region between monomer and E-type emission (at around 10^{-5} M). In this concentration regime a decrease of the amplitude of the long time constant and an increase of the amplitude of the shortest decay time is observed (see Fig. 6.19 and Fig. 6.18). This may be a hint for a fast energy transfer between monomeric sf-PBI molecules and aggregates that coexist on the surface at this concentration.

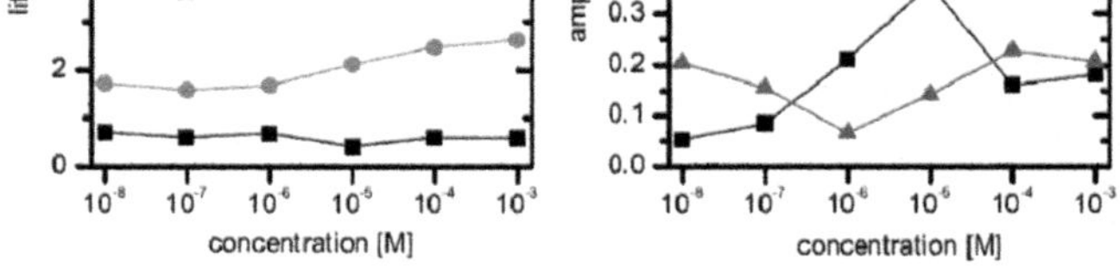

Figure 6.19 Obtained life times τ_n (left) and their respective amplitudes A_n (right) for tri-exponential fits of the fluorescence decay curves in Fig. 6.18 according to Eq. 6.2.

As a next step, the quenching of the underlying silicon has been the subject of investigation. Therefore, LAO line structures with a different height have been prepared followed by binding of sf-PBI and investigation by time-resolved fluorescence microscopy (see Fig. 6.20).

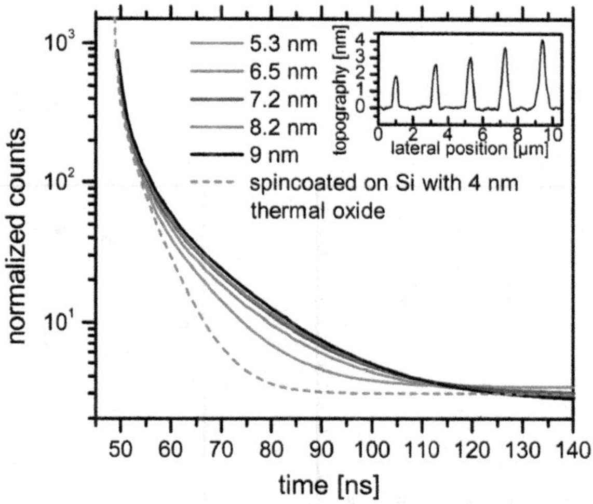

Figure 6.20 Log-lin plot of the normalized tri-exponential fluorescence decay fits of the decay curves of sf-PBI bound to LAO nanostructures with different thickness ranging from 5.3 to 9 nm (solid curves) and spin-coated from a 1.6×10^{-5} M solution on silicon with a 4 nm thermal oxide layer (dashed curve). The decay has been measured over the whole spectral range. The inset shows an AFM height profile of the used LAO line structures. Note that the oxide height is larger than the detected topographic height as 40% of the LAO oxide is below the silicon surface (Fontaine et al., 1998) and the self-assembled monolayer leads to an additional virtual decrease of the height (Sieval et al., 1998).

As the measurement exhibits a rather low signal-to-noise-ratio, differences between the curves are small (the oxide height can be adjusted only in a rather narrow window). Therefore, not the measured data is displayed but the tri-exponential fits of the curves in Fig. 6.20. To exclude local difference within the oxide lines, several measurements on different spots on each line have been performed and the fluorescence decay curves have been averaged for each line. The obtained parameters from the tri-exponential fits for different oxide thickness are displayed in Fig. 6.21. All three life time components are becoming larger for an increasing oxide thickness. This is just as expected because the energy transfer between excited molecules and the underlying silicon depends strongly on their distance (Alivisatos et al., 1987). Therefore also the shortest life time component possesses the largest amplitude in contrast to the sf-PBI films on 100 nm oxide. To get a more quantitative analysis of the quenching, the evaluated data was compared with predictions from a theoretical model.

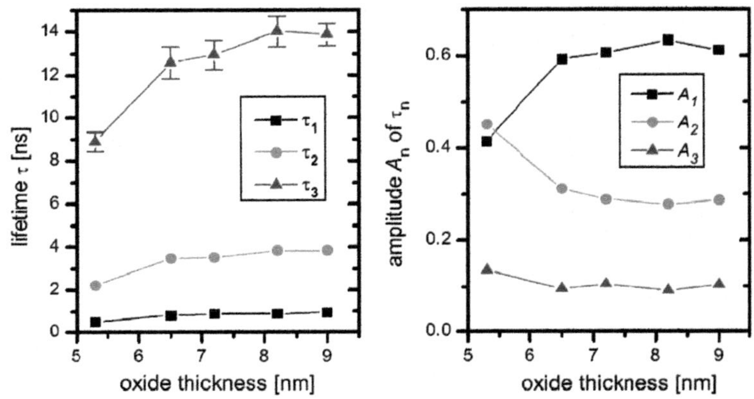

Figure 6.21 Obtained excited state life times τ_n and their respective amplitudes An for tri-exponential fits according to Eq. 6.2 of the fluorescence decay curves of bound sf-PBI depending on the LAO-oxide thickness.

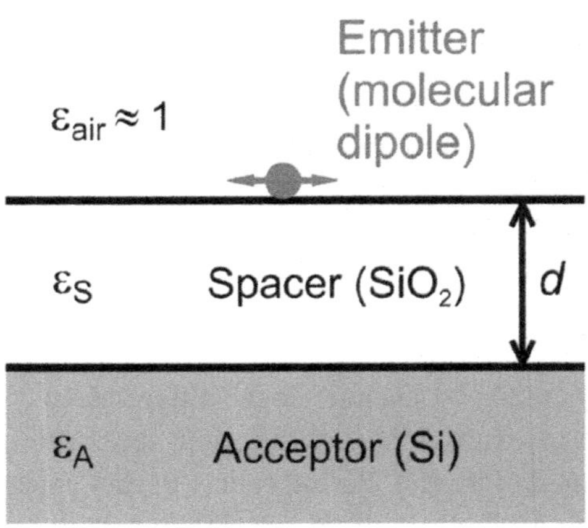

Scheme 6.7 Schematic geometry relevant for molecule-semiconductor or metal energy transfer. The ε_s denote the complex dielectric constants of the involved media.

The energy transfer rate between excited molecules and a metal or semiconductor surfaces and the consequential lowering of the fluorescence life time has been described by several classical and quantum mechanical approaches (Chance et al., 1976; Chance et al., 2007; Persson and Lang, 1982; Stavola et al., 1985). Briefly, those models generally regard an electronically excited molecule as oscillating point dipole, separated by a spacer with a certain thickness d from the acceptor medium (see Scheme 6.7).

The dipole field incident on the acceptor is partially reflected and partially absorbed in a manner governed by the dielectric constant ε of the involved media. Thus the imaginary part of the electric field at the dipole can be related to the life time of the molecule. In general, such models yield to the result that the fluorescence life time is related to the spacer thickness d by a d^{-4} dependency in the case of metal acceptor substrates or a d^{-3} relation in the case of a semiconductor respectively. Several experimental studies of distance dependent lowering of the fluorescence life time through molecule semiconductor energy transfer have found a good agreement with a simple cubic relation in the form of Eq. 6.3 (Sluch et al., 1995; Hayashi et al., 1983; Danos et al., 2008):

$$\tau_d = \frac{\tau_0}{1 + \left(\frac{d_0}{d}\right)^3} \quad (6.3)$$

Here τ_d is the distance dependent life time, τ_0 is the "free" life time of the emitter without the presence of the acceptor and d_0 is the critical transfer distance where the rate constant for the energy transfer is equal to that for fluorescence of the molecule in absence of the acceptor. However, the d^{-3} relation in this form is too arbitrary in order to be compared to the obtained data points as there is no information about the value of d_0 which would remain a fitting parameter which is hardly assessable. Therefore, a more fundamental model has been used to compare the experimental results with the theoretically expected behavior. The employed theory has been established by Persson and co-workers (Persson and Lang, 1982; Persson, 1978; Persson and Persson, 1980; Avouris and Persson, 1984) and was already used by other groups to evaluate the distance dependency of the fluorescence life time of R6G on aluminum (Cnossen et al., 1993) and rhodamine B on silicon substrates (Ishikawa et al., 1998). In this theory, the excited state life time $\tau_d(d)$ of an oscillating dipole at a small (k_d << 1) distance above a metal surface is given by

$$\tau_d^{-1}(1) = \tau_\infty^{-1}\left\{1+\frac{\eta}{8}(dk)^{-3}\left[2\mathrm{Im}\left(\frac{\varepsilon_\mathrm{A}(\omega)-\varepsilon_1}{\varepsilon_\mathrm{A}(\omega)+\varepsilon_1}\right)+6\xi\frac{1}{k_\mathrm{F}d}\frac{\omega}{\omega_\mathrm{P}}+18\frac{\omega_\mathrm{F}}{\omega_\mathrm{P}}\frac{\omega}{\omega_\mathrm{P}}\frac{1}{k_\mathrm{F}d}\right]\right\}, \tag{6.4}$$

where τ_∞ is the fluorescence life time at an infinite distance from the acceptor, k is the magnitude of the wave vector at the emitted frequency ω, $\varepsilon_\mathrm{A}(\omega)$ is the complex dielectric constant of the acceptor medium, ε_1 is the dielectric constant of the medium in which the dipole is embedded, η is an orientational parameter, k_F is the Fermi wave vector of the electron gas, ω_F the Fermi frequency, ω_P is the plasma frequency and ξ a constant which depends on the electron gas density. The last two terms in the square brackets contain electron-gas parameters and for very small molecule-surface separations (d < 5 nm) those terms are dominant and cause the d^{-4}-dependency in the case of nearly free-electron-like metals (Cnossen et al., 1993). The first term is describes the bulk scattering contribution to the dampening of the emissive rate and shows a d^{-3} dependence. A derivation and detailed discussion of the different terms in Eq. 6.4 is beyond the scope of this chapter and can be found in the mentioned literature (Persson and Lang, 1982; Persson, 1978; Persson and Persson, 1980; Avouris and Persson, 1984). As the thickness of the LAO nanostructures is above 5 nm and the substrate is not a metal with a free electron gas but a semiconductor, only the first term is important for the following analysis and consequently the last two terms in Eq. 6.4 can be omitted (Ishikawa et al., 1998). Thus Eq. 6.4 simplifies to

$$\tau_d^{-1}(1) = \tau_\infty^{-1}\left\{1+\frac{\eta}{8}(dk)^{-3}\left[2\mathrm{Im}\left(\frac{\varepsilon_\mathrm{A}(\omega)-\varepsilon_1}{\varepsilon_\mathrm{A}(\omega)+\varepsilon_1}\right)\right]\right\}. \tag{6.5}$$

For the involved parameters, the following values can be assumed: As the sf-PBI molecules are most likely oriented in parallel fashion to the substrate, the orientational parameter η is 0.75 (Cnossen et al., 1993). The complex dielectric constant of the acceptor is a function of the energy of the dipole radiation. Unfortunately, the spectrum of the sf-PBI emission is rather wide, so there is no discrete dipole energy. For a coarse approximation, an energy of 1.91 eV (≈650 nm) was assumed, as this is roughly the maximum position of the dominant E-type emission. The imaginary part of ε_A of weakly doped silicon is about 0.126 for this energy (Aspnes and Studna, 1983). In Eqs. 6.4 and 6.5, ε_1 denotes the dielectric constant of the medium that surrounds the dipole. As the dipole is situated at the interface between silicon oxide ($\varepsilon_{Si} \approx 2.1$ (Philipp, 1971)) and air ($\varepsilon_{air} \approx 1$), theoretical curves have been calculated for both values as limiting cases of ε_1. Values for τ_∞ have been taken from the investigations of sf-PBI films on 100 nm oxide at a concentration of 10^{-5} M (see Fig. 6.19). For these films the quenching by the underlying silicon is negligible because for spacer thicknesses above 40 nm, there is nearly no influence of the substrate (Alivisatos et al., 1987; Sluch et al., 1995)).

The determined theoretical curves for the long life time (τ_∞ = 8.46 ns) and the medium life time (τ_∞ = 2.12 ns) are displayed in Fig. 6.22 together with the measured values from Fig. 6.21. Evidently, the distance dependence of the middle time constant τ_2 and the short time constant τ_1 of sf-PBI on the nanostructures corresponds to the long (8.46 ns) and the middle (2.12 ns) time constant of the emission in absence of the quenching substrate. The shortest decay component (0.4 ns) of the sf-PBI film emission on 100 nm oxide is likely to be quenched below the detection limit and is not observable anymore. However, the existence of the third time constant τ_3 is completely unexpected. As it is even larger than the long decay component of the films far away from the silicon, it would be related to a substantially slower decay which is only present for sf-PBI bound to the LAO structures. However, the relatively low amplitude (≈0.1; see Fig. 6.21) of this long component makes such a conclusion arguable. Although the theoretical curves show a reasonable agreement with the measured decay times, an examination of the decay times for oxide thicknesses over several magnitudes would be necessary in order to provide a verifiable survey of the model. Since the controlled LAO oxide thickness is limited to one order of magnitude at the most, such an investigation is unfortunately not possible with the investigated systems. Furthermore, there are many factors that provide a certain scope for uncertainties.

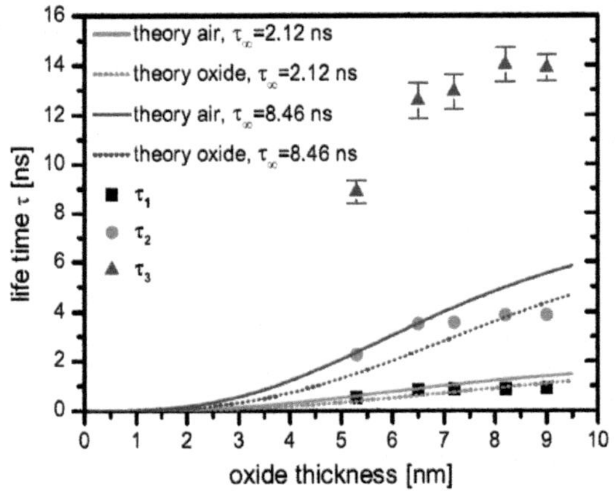

Figure 6.22 Fluorescence life times of sf-PBI for different LAO-oxide thickness from Fig. 6.20 compared to the theoretically expected curves according to Eq. 13.5 for different free life times τ_∞ and surrounding media.

First of all, the theoretical model was derived for a single emitter on a flat and infinite dielectric surface. The LAO oxide surface, on the other hand, is laterally confined and the molecules are packed densely on it.

The oligomerization of the sf-PBI molecules, as evident from the spectral investigations, and a possible energy transfer between them may further influence the true distance dependency of the quenching by the silicon substrate. As mentioned above, the model has been derived for a dipole which is isotropically surrounded by the dielectric medium with ε_1. In the present experimental case, the sf-PBI molecules are located at the silicon oxide/air interface. This has been tried to be accounted for by the calculation for both ideal cases and the measured values are indeed mostly within the intermediate region of both curves. Nevertheless, possibly occurring surface effects that are not included in the calculations may lead to a deviation from the pure d^{-3} dependence.

Last but not least, Eq. 6.5 is only valid for a dipole with a discrete energy. The emission spectrum of sf-PBI on the LAO nanostructures is rather wide though. This means that the influence of the underlying silicon on the dampening of the emissive rate should also have a rather broad distribution. In Fig. 6.20, also the fluorescence decay fit of a sf-PBI film spin-coated from a $1.6 \cdot 10^{-5}$ M solution on silicon with a 4 nm thermal oxide layer is displayed (dashed curve). As expected, the fluorescence decay is faster than for the 5.3 nm LAO oxide nanostructure (especially the long time component has smaller amplitude and is shorter).

A spectrally correlated life time measurement would be very helpful in order to assign the different life time components to different spectral features or emissive states. Therefore a different sample has been investigated time-resolved while separating the fluorescence emission into two regions by utilizing a 570 nm short pass (SP) filter and a 585 nm long pass (LP) filter (see Fig. 6.23). Figure 6.24 shows the normalized fluorescence decay curves for both spectral regimes. Again, the decay can only be fitted satisfactorily by using a tri-exponential function in the form of Eq. 6.2. The obtained fit parameters are given in Table 6.4.

Table 6.4 Life times τ_n and relative amplitudes A_n of the decay components of the tri-exponential fits in Fig. 6.24

	Short component		**Medium component**		**Long component**	
Spectral regime	τ_1 / ns	A_1	τ_2 / ns	A_3	τ_3/ns	A_3
<570 nm (monomer)	0.11	16.5%	0.32	80%	1.68	3.5%
>585 nm (E-excimer)	0.43	43.7%	1.22	42.7%	4.23	13.6%

The fluorescence emission below 570 nm, which is attributed mostly to the monomer-like emission, is dominated mainly by a very short decay time constant of 0.32 ns. This can be understood in terms of a quenching by an energy transfer to the underlying silicon as well as to the close-by dye aggregates or monomers. Consequently, the already short life time component of the monomers is further decreased. It should be mentioned that the very short life time for this regime (τ_1 = 0.11 ns) is below the temporal resolution of the detector (IRF-width ≈200 ps) and therefore should not be regarded as "real" value.

The emission above 585 nm consists to nearly equal parts of the two decay components 0.43 ns and 1.22 ns as well as a smaller but clear observable decay component with a life time of 4.23 ns. From that as well as the visual impression of the decay curves in Fig. 6.24, it can be concluded that even for a close proximity to the silicon, the E-emissive states are significantly longer-lived than the monomer-like sf-PBI molecules as it was already indicated by the temporal investigations of sf-PBI films on 100 nm silicon oxide. Furthermore, it seems that the fluorescence decay of the monomers possess a rather mono-exponential character whereas the E-type excimers show a broader distribution of life times.

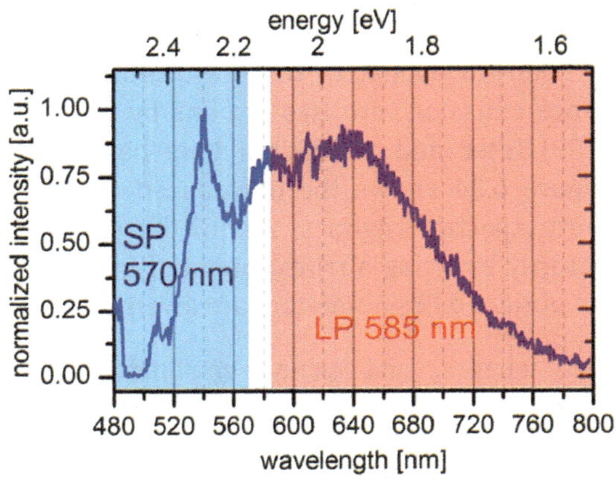

Figure 6.23 Spectral partition of the sf-PBI emission on the Lao-oxide nanostructures by a 570 nm short pass (SP) and a 585 nm long pass (LP) filter.

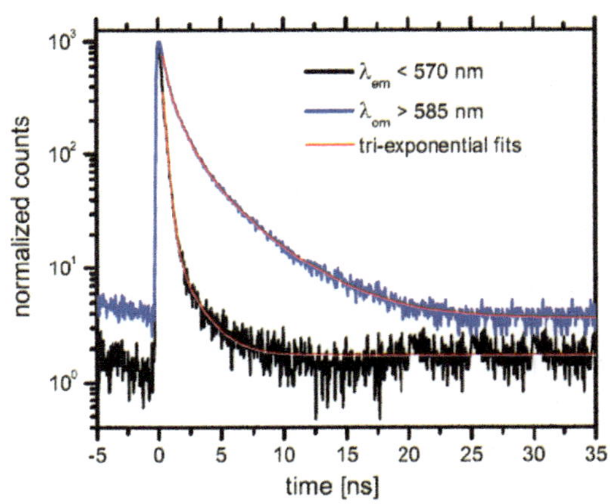

Figure 6.24 Decay curves and tri-exponential fits (red) of the fluorescence emission below 570 nm (black) and above 585 nm (blue) of sf-PBI bound to LAO-oxide nanostructures.

In summary, all measurements show that the fluorescence decay of sf-PBI bound to LAO nanostructures is clearly multi-exponential. This may be partly caused by energy transfer between the different species and/or states on the structure (monomers, dimers). A theoretically expected dependency of the emissive rate dampening on the dipole energy further enhances the multi-exponential character. The broad spectral and decay time distributions indicate that there is a multitude of different involved molecular states which may be understood in terms of a configurational disorder within the electrostatically assembled sf-PBI films. Such an assumption is also substantiated by the observation that for different samples with comparable LAO oxide height but different preparation conditions (LAO oxidation parameters, immersion time in the dye solution) the measured decay times differ slightly. The investigation of the influence of different preparation and system parameters on the decay characteristics with a high spectral resolution would be very beneficial for a deeper understanding of the true nature behind the assembly of sf-PBI on local anodic oxide nanostructures. Time resolved experiments at low temperatures may also help to further unravel the aggregation behavior.

6.5 Covalent Attachment of FITC

Besides a selective binding of dye molecules via electrostatic interactions to LAO oxide nanostructures, also a chemical route of functionalization can be chosen. Here at first a linker molecule is bound to the formed oxide (e.g., a silane molecule) and in a second step a dye is covalently attached to the linker. This approach leads to higher stability of the bond (covalent bonds possess comparable high dissociation energies [several 100 kJ/mol]).

A direct binding of the functional material to the oxide nanostructure is possible if the material possesses an appropriate anchoring group that binds selectively to silicon oxide surfaces. Silane compounds such as trichlorosilanes or ethoxysilanes have been used as standard substances for chemical modification of silicon oxide surfaces (Sagiv, 1980; Howarter and Youngblood, 2006). In order to provide the possibility for further chemical modification, functional silanes are of major interest. These molecules possess a functional head-group (e.g., carboxyl, amino or thiol group) in addition to the silane tail group. This functional head group can interact in a second step with other molecules resulting in an immobilization of the desired material on the structure with the silane molecule as linker. The binding of diverse silanes to silicon oxide surfaces has been the subject of several investigations (Ulman, 1996; Fadeev and McCarthy, 2000; Desbief et al., 2011). Though a previous study demonstrated the general feasibility to bind silane molecules to LAO nanostructures, there was strong evidence for only a partial coverage of the structure and no complete monolayer formation (Ara et al., 2002b).

For a reliable and controlled silanization yielding in perfect monolayer formation on LAO nanostructures, suitable conditions are required. A successful route for a homogeneous binding of functional silane molecules to silicon oxide nanostructures prepared by LAO lithography of alkyl-terminated silicon has been developed recently (Baumgärtel et al., 2013). In addition, the dye fluorescein-5-isothiocyanate (FITC, for chemical structure see Scheme 6.10, second step) was subsequently bound to the amino-terminated silane layer. FITC is a fluorescein derivative with an N=C=S functional group. This group is reactive towards nucleophiles such as amine or thiol groups. A successful large-scale binding of FITC to amino-terminated silicon surfaces (Ara et al., 2007) has been demonstrated previously by fluorescence measurements (Tada et al., 2002). The attachment of FITC on the oxide nanostructures is realized in two steps, as depicted schematically in Scheme 6.8. First, the functional silane aminopropyltriethoxysilane (APTES) is bound to the LAO oxide, which leads to an amino-functionalization of the structure. In a second step, FITC is bound to the amino group on the structure through its N=C=S group.

Figure 6.25 shows the results of comparative AFM height measurements after different functionalization steps: After binding of APTES, there is a height increase of approximately 0.3 ± 0.1 nm (blue dashed curve), which is in good agreement with the length of a hydrolyzed APTES molecule (0.3 nm). The subsequent binding of FITC to the amino-functionalized nanostructure leads to a further increase in height of about 0.7 ± 0.1 nm. As the length of a single FITC molecule is about 1 nm, it can be concluded that the molecules are not standing upright but are tilted from the surface normal.

A clear luminescence signal was found, which originates from the nanostructure only, The spectral shape of the fluorescence light resembles that of FITC in acetone solution and is shifted about 36 nm to the red due to the different dielectric environment (the small signal close to the filter edge at 485 nm is not part of the FITC fluorescence but probably scattered light from the environment) (see Fig. 6.26).

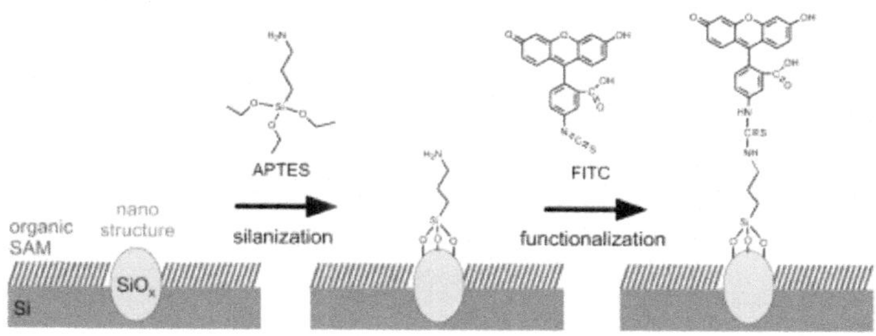

Scheme 6.8 Schematic route for covalent binding of FITC to a silicon oxide nanostructure on alkyl-terminated silicon. After local anodic oxidation, a layer of APTES molecules is bound to the silicon oxide as linker. In a second step, FITC is bound to the amino-functionalized oxide structure. From Baumgärtel et al. (2013).

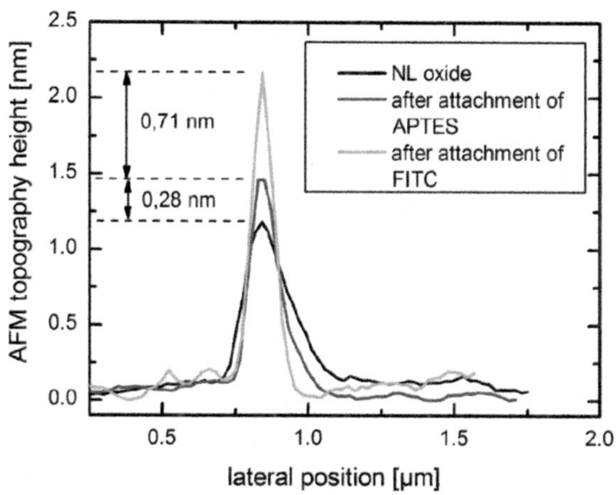

Figure 6.25 Height profile of an LAO oxide structure (black curve) after 6 h in 10 mM APTES solution dry toluene under dry conditions (blue curve) and after subsequent binding of FITC in aqueous solution (green curve). From Baumgärtel et al. (2013).

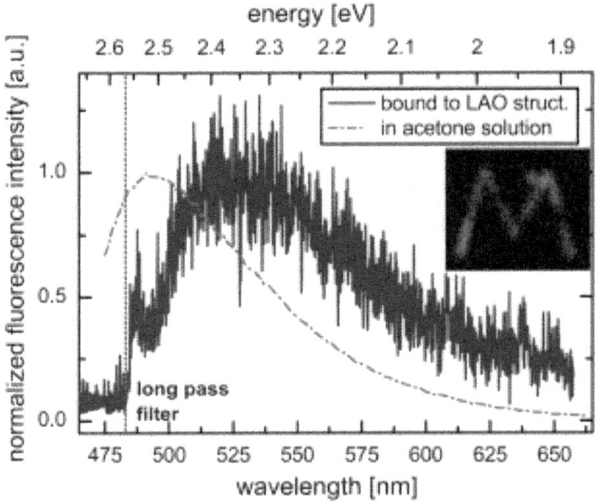

Figure 6.26 Fluorescence spectrum from the FITC-functionalized LAO oxide nanostructure shown in Fig. 6.25 (dark gray curve) compared to FITC fluorescence in acetone solution (red dashed curve). The inset shows a confocal microscope image of the structure (excitation wavelength: 465 nm, excitation power: 11 μW, integration time: 1 ms/px). From Baumgärtel et al. (2013).

It should be noted that he fluorescence signal is comparably weak (roughly a factor of 10 above the background noise level). This is surprising, because other high-quantum-yield xanthene dyes (e.g., R6G) bound to the LAO oxide nanostructures by electrostatic interactions, show a much higher signal-to-noise ratio for similar structure heights and excitation powers. Therefore quenching by the underlying silicon could not explain the low fluorescence intensities.

As a second parameter the orientation of the transition dipole moment plays an important role in the absorption and emission of radiation by the bound molecules it is known. From literature that xanthene derivatives, such as fluorescein or rhodamine, have a transition dipole moment for the $S_1 \rightarrow S_0$ transition that is typically oriented along the xanthene unit (Penzkofer and Wiedmann, 1980; Arbeloa and Martinez, 2006). As this unit is perpendicular to the binding axis of the FITC molecule, it should be oriented rather parallel to the surface (tilt angle α = 0°). As in the prepared samples the emission is rather low, here a very larger tilt angle

(α close to 90°) of the xanthenes unit would be necessary to explain the low emission intensity. Due to the AFM investigations as well as steric reasons such a strong tilting is very unlikely.

As a third possibility to explain the low emission intensity a quenching by an efficient energy transfer between quite densely packed FITC molecules on the structure. This effect has been described by Imhof et al. (Imhof et al., 1999) for FITC bound to silica spheres. Here a quenching of the fluorescence intensity for increasing dye concentration was found, which was attributed to an interaction between neighboring molecules. Additionally this interaction also causes a lowering of the excited state energy, which led to a 10 nm red shift of the emission spectrum from 531 nm to 541 nm with increasing dye concentration. Although it is difficult to exactly determine the peak position in Fig. 6.26 due to the low signal-to-noise ratio of the spectrum, the maximum of the fluorescence of FITC on LA-oxide clearly seems to be rather in the range of 530 nm than 540 nm. The absence of the further red-shift as well as the agreement of the approximate peak position and spectral shape with the results of Imhof et al. for low dye concentrations, leads to the conclusion that the observed low fluorescence intensity is likely not caused by a dense packing of the FITC molecules on the oxide structure.

Excluding all other explanations for the low fluorescence intensity (comparing to the intensity of electrostatically bound dyes) only a much lower surface density in the case of covalently bound FITC could explain the low emission intensity. Such a low dye density on the surface could be explained by a non-ideal (not well-ordered) and closed monolayer of APTES on the LAO oxide nanostructure (below a scale of 20 nm, as AFM height investigations showed a thickness of about one monolayer of APTES, see Fig. 6.25). It should be noted that further investigations are necessary to support the given explanation.

In conclusion, a route for a controlled covalent functionalization of silicon oxide nanostructures with an amino-terminated silane and FITC dye molecules has been successfully realized. Optical investigations of the bound FITC dye, on the other hand, seem to indicate that the forming APTES monolayer is not well-ordered on a scale below 20 nm. Here further optimizations of the monolayer formation conditions have to be done.

6.6 Conclusion

In the following, general and specific new findings presented within this chapter will be summarized briefly. Partly similar features with respect to the decoration of heterogeneous surfaces with fluorescent probes will be presented in Chapters 7 and 8.

(i) The recently verified negative space charge within the local oxide structures is stable at least over several days. A found decrease of the Kelvin probe force microscopy contrast over time at ambient conditions could be clearly attributed to an adsorption of ambient humidity on the structure which leads to a screening of the charge. The high stability of the charge was also confirmed by fluorescence measurements of a charge sensitive dye. Though both methods showed a good qualitative agreement, special attention should be paid to their quantitative interpretation which cannot be easily derived for such systems.

(ii) The surrounding alkyl monolayer can be used for selective attachments of nanoparticles with hydrophobic shell. On the other hand, particles with hydrophilic shell will mostly attached to hydrophilic oxide nanostructures.

(iii) The negative space charge allows for a reliable and easily applicable binding of different classes of cationic dye molecules. For the first time, a perylene derivative could be selectively immobilized on LAO nanostructures by using spermine-functionalized PBI molecules that have been synthesized by the Würthner group.

(iv) The spectral emission characteristics of electrostatically bound R6G and CV is very similar to that in solution or a physisorbed thin film on silicon oxide. However, there is a small yet clearly verified blue shift which was attributed to the interaction between the R6G molecules and the negative charges. No distinct influence of the electrostatic binding on the photophysical degradation (bleaching) of rhodamine 6G and the cresyl violet could be found.

(v) Emission spectra of sf-PBI functionalized LAO nanostructures can be interpreted as a combination of monomer-like emission and E-type excimer luminescence. Thus it can be concluded that some of the emitters exist in a state which is similar to the electronic configuration of monomeric sf-PBI molecules whereas others are able to relax into an E-type excimer geometry. The spectral findings for the electrostatically attached PBIs could also be compared to those of thin films spin-coated on silicon with a 100 nm silicon oxide layer.

(vi) The relative strength of the E-emission from the nanostructures decreases gradually by cooling the sample down to 77 K. This may be understood in terms of a motional constriction due to the electrostatic forces that bind the sf-PBI molecules to the local anodic oxide surface.

(vii) The decay process of the excited state has been found to be at least tri-exponential, both for the dense spin-coated films as well as for the dye bound to LAO oxide. This indicates a multitude of different involved molecular states which may be understood in terms of a configurational disorder within the electrostatically assembled sf-PBI films. The E-excimers possess a larger life time than the sf-PBI monomers which is also in agreement with previous findings as well as the investigations of the spin-coated films.

(viii) The influence of the underlying close-by silicon on the dampening of the emissive rate shows a reasonable agreement with an existing theoretical model.

(ix) A covalent attachment of an optically active material (here the functional fluorescein derivative FITC) to the LAO nanostructures via an amino-terminated silane linker has been successfully demonstrated. The binding was confirmed by AFM height measurements and fluorescence microscopy.

(x) Photoluminescence intensity and spectral investigations, however, indicate that the density of the bound FITC molecules is rather low compared to the electrostatic binding mechanisms of rhodamine 6G or sf-PBI. This may be connected to a relatively low order within the formed silane layer on a nanometer scale.

References

Ahrens, M. J., Sinks, L. E., Rybtchinski, B., Liu, W., Jones, B. A., Giaimo, J. M., Gusev, A. V., Goshe, A. J., Tiede, D. M. and Wasielewski, M. R. (2004)

Self-assembly of supramolecular light-harvesting arrays from covalent multi-chromophore perylene-3,4:9,10-bis(dicarboximide) building blocks, *J. Am. Chem. Soc.* 126, 8284–8294.

Akimoto, S., Ohmori, A., and Yamazaki, I. (1997) Dimer Formation and Excitation Relaxation of Perylene in Langmuir–Blodgett Films, *J. Phys. Chem. B* 101, 3753–3758.

Alivisatos, A., Arndt, M., Efrima, S., Waldeck, D. and Harris, C. (1987) Electronic energy transfer at semiconductor interfaces. I. Energy transfer from two-dimensional molecular films to Si(111), *J. Chem. Phys.* 86, 6540–6549.

Ara, M., Graaf, H. and Tada, H. (2002) Nanopatterning of alkyl monolayers covalently bound to Si (111) with an atomic force microscope, *Appl. Phys. Lett.* 80, 2565–2567.

Ara, M., Graaf, H. and Tada, H. (2002) Atomic force microscope anodization of Si(111) covered with alkyl monolayers, *Jpn. J. Appl. Phys.* 41, 4894–4897.

Ara, M., Tsuji, M.and Tada, H. (2007) Preparation of amino-terminated monolayers via hydrolysis of phthalimide anchored to Si(111), *Surf. Sci.* 601, 5098–5102.

Arbeloa, F. L. and Martínez, V. M. (2006) Orientation of adsorbed dyes in the interlayer space of clays. 2 fluorescence polarization of rhodamine 6G in laponite films, *Chem. Mater.* 18, 1407–1416.

Aspnes, D. and Studna, A. (1983) Dielectric functions and optical parameters of Si, Ge, GaP, GaAs, GaSb, InP, InAs, and InSb from 1.5 to 6.0 eV, *Phys. Rev. B* 27, 985–1009.

Avouris, P. and Persson, B. (1984) Excited states at metal surfaces and their non-radiative relaxation, *J. Phys. Chem.* 88, 837–848.

Avouris, P., Hertel, T. and Martel, R. (1997) Atomic force microscope tip-induced local oxidation of silicon: kinetics, mechanism, and nanofabrication, *Appl. Phys. Lett.* 71, 285–287.

Barnes, W. L. (1998) Fluorescence near interfaces: the role of photonic mode density, *J. Mod. Optic* 45, 661–669.

Baumgärtel, T., von Borczyskowski, C. and Graaf, H. (2010) Fluorescence studies of Rhodamine 6G functionalized silicon oxide nanostructures, *Nanotechnology*, 21, 475205.

Baumgärtel, T. and Graaf, H. (2011) In *Local Anodic Oxidation Nanolithography on Silicon: Chemical Routes to Functional Nanostructures* Lithography: principles, processes and materials ed. by Hennessy, T. C., Nova Science publishers, Hauppauge, NY pp. 175–193.

Baumgärtel, T., von Borczyskowski, C. and Graaf, H. (2012) Detection and stability of nanoscale space charges in local oxidation nanolithography *Nanotechnology*, 23, 095707.

Baumgärtel, T. (2012) in *Fluorescence life time investigations of bound sf-PBI dyes*, Dissertation, Technische Universität Chemnitz, pp. 83–96.

Baumgärtel, T., von Borczyskowski, C. and Graaf, H. (2013) Selective surface modification of lithographic silicon oxide nanostructures by organofunctional silanes, *Beilstein J. Nanotechnol.* 4, 218–226.

Baumgärtel, T. and Graaf, H. (2014) Spectral shift of cresyl violet luminescence on charged silicon oxide nanostructures, *Phys. Status Solidi A*, 211, 905–909.

Baumgärtel, T., Rehm, S., Würthner, F., von Borczyskowski, C. and Graaf, H. (2014) Functional bisimide dyes bound via electrostatic interactions to oxide nanostructures generated by AFM lithography, *Appl. Surf. Sci.*, 318, 51–58.

Belgardt, C., Blaudeck, T. von Borczyskowski, C. and Graaf, H. (2014) Self-Assembly of Ordered Colloidal Nanoparticle Films in Few-Micron Wide Laser-Desorbed Lines of Octadecylsiloxane Monolayers on Silicon Oxide Surfaces, *Adv. Eng. Mater.* 16, 1090–1097.

Bisht, P., Fukuda, K. and Hirayama, S. (1996) Ripple structures in the fluorescence spectra of *N,N'*-bis(2,5-di-tert-butylphenyl)-3,4,9,10-perylenebis(dicarboximide) (DBPI) in spherical micrometer-sized polystyrene beads, *Chem. Phys. Lett.* 258, 71–79.

Bulovic, V., Burrows, P., Forres, J. C. S. R. and Thompson, M. (1996) Study of localized and extended excitons in 3,4,9,10-perylenetetracarboxylic dianhydride (PTCDA) I. Spectroscopic properties of thin films and solutions, *Chem. Phys.* 210, 1–12.

Cavallini, M., Mei, P., Biscarini, F. and Garcia, R. (2003) Parallel writing by local oxidation nanolithography with submicrometer resolution, *Appl. Phys. Lett.*, 83, 5286–5288.

Chance, R. R., Prock, A. and Silbey, R. (1976) Comments on the classical theory of energy transfer. II. Extension to higher multipoles and anisotropic media, *J. Chem. Phys.* 65, 2527–2531.

Chance, R. R., Prock, A. and Silbey, R. (2007) in *Molecular Fluorescence and Energy Transfer Near Interfaces,* Advances in Chemical Physics Vol. 37 ed. by Prigogine, I. and Rice, S. A., Wiley-Interscience, New York, pp. 1–65.

Chen, C. C., Yet, C. P., Wang, H. N. and Chao, C. Y. (1999) Self-Assembly of Monolayers of Cadmium Selenide Nanocrystals with Dual Color Emission, *Langmuir* 15, 6845–6850.

Chiesa, M. and Garcia, R. (2010) Nanoscale space charge generation in local oxidation nanolithography, *Appl. Phys. Lett.* 96, 263112.

Cnossen, G., Drabe, K. E. and Wiersma, D. A. (1993) Fluorescence properties of submonolayers of rhodamine 6G in front of a mirror, *J. Chem. Phys.* 98, 5276–5280.

Danos, L., Greef, R. and Markvart, T. (2008) Efficient fluorescence quenching near crystalline silicon from Langmuir–Blodgett dye films, *Thin Solid Films* 516, 7251–7255.

Dagata, J. A., Inoue, T. I., Itho, J., Matsumoto, K. and Yokoyama, H. (1998) Role of space charge in scanned probe oxidation, *J. Appl. Phys.* 84, 6891–6899.

Desbief, S., Patrone, L., Goguenheim, D., Guérin, D. and Vuillaume, D. (2011) Impact of chain length, temperature, and humidity on the growth of long alkyltrichlorosilane self-assembled monolayers, *Phys. Chem. Chem. Phys*, 13, 2870–2879.

Dexter, D. (1979) Two ideas on energy transfer phenomena: Ion-pair effects involving the OH stretching mode, and sensitization of photovoltaic cells, *J. Luminescence*, 18/19, 779–784.

Dubois, E. and Bubendorff, J.-L. (2000) Kinetics of scanned probe oxidation: space-charge limited growth, *J. Appl. Phys.* 87, 8148–8154.

Fadeev, A. Y. and McCarthy, T. J. (2000) Self-Assembly Is Not the Only Reaction Possible between Alkyltrichlorosilanes and Surfaces: Monomolecular and Oligomeric Covalently Attached Layers of Dichloro- and Trichloroalkylsilanes on Silicon, *Langmuir*, 16, 7268–7274.

Fluhler, E., Burnham, V. G. and Loew, L. M. (1985) Spectra, membrane binding, and potentiometric responses of new charge shift probes, *Biochemistry* 24, 5749–5755.

Fontaine, P. A., Dubois, E. and Stievenard, D. (1998) Characterization of scanning tunneling microscopy and atomic force microscopy-based techniques for nanolithography on hydrogen-passivated silicon, *J. Appl. Phys.* 84, 1776–1781.

Ford, W. and Kamat, P. V. (1987) Photochemistry of 3,4,9,10-perylenetetracarboxylic dianhydride dyes. 3. Singlet and triplet excited-state properties of the bis(2,5-di-tert-butylphenyl)imide derivative, *J. Phys. Chem.* 91, 6373–6380.

Gao, F., Zhao, Y. and Liang, W. (2011) Vibronic Spectra of Perylene Bisimide Oligomers: Effects of Intermolecular Charge-Transfer Excitation and Conformational Flexibility, *J. Phys. Chem. B* 115, 2699–2708.

Garcia, R., Martinez, R. V. and Martinez, J. (2006) Nano-chemistry and scanning probe nanolithographies, *Chem. Soc. Rev.* 35, 29–38.

Giaimo, J., Lockard, J., Sinks, L., Scott, A., Wilson, T. and Wasieliewski, M. R. (2008) Excited Singlet States of Covalently Bound, Cofacial Dimers and Trimers of Perylene-3,4,9,10-bis(dicarboximide)s, *J. Phys. Chem. A* 112, 2322–2330.

Graaf, H., Vieluf, M. and von Borczyskowski, C. (2007) *Selective binding of dye molecules and CdSe nanocrystals on nanostructures generated by AFM lithography of silicon surfaces, Nanotechnology*, 18, 265306.

Graaf, H., Baumgärtel, T., Vieluf, M. and von Borczyskowski, C. (2008) Local anodic oxidation on dodecyl terminated silicon(100) *Superlatt. Microstruct.* 44, 402–410.

Graaf, H., Unold, T., Mattheus,C. and Schlettwein, D. (2008) *Self-organization of crystalline domains in originally amorphous perylene diimide films J. Phys. D: Appl. Phys.* 41, 105112.

Graaf, H., Friedriszik, F., Wagner, C. and von Borczyskowski, C. (2011) *Optical spectroscopy of trap states in amorphous perylene derivative films J. Phys. Chem. C* 115, 8150–8154.

Grimme, S. (2008) Do Special Noncovalent π–π Stacking Interactions Really Exist?, *Angew. Chem. Int. Ed.* 47, 3430–3434.

Hayashi, T., Castner, T. and Boyd, R. (1983) Quenching of molecular fluorescence near the surface of a semiconductor, *Chem. Phys. Lett.* 94, 461–466.

Hennessy, M., Soos, Z.G., Pascal Jr. Z. R. and Girlando, A. (1999) Vibronic structure of PTCDA stacks: the exciton–phonon-charge-transfer dimer, *Chem. Phys.* 245, 199–212.

Hernando, J., de Witte, P., van Dijk, E., Korterik, J., Nolte, R., Rowan, A., Garcia-Parajo, M. and van Hulst, N. (2004) Investigation of Perylene Photonic Wires by Combined Single-Molecule Fluorescence and Atomic Force Microscopy, *Angew. Chem. Int. Ed.* 43, 4045–4049.

Howarter, J. A. and Youngblood, J. P. (2006) Optimization of Silica Silanization by 3-Aminopropyltriethoxysilane, *Langmuir* 22, 11142–11147.

Imhof, A., Megens, M., Engelberts, J. J., de Lang, D. T. N., Sprik, R. and Vos, W. L. (1999) Spectroscopy of Fluorescein (FITC) Dyed Colloidal Silica Spheres, *J. Phys. Chem. B* 103, 1408–1415.

Ishikawa, M., Yogi, O., Ye, J., Yasuda, T. and Maruyama, Y. (1998) Grouping of Independent Single Molecules on Silicon Surfaces, *Anal. Chem.* 70, 5198–5208.

Kaiser, T., Stephanenko, V. and Würthner, F. (2009) Fluorescent J-Aggregates of Core-Substituted Perylene Bisimides: Studies on Structure–Property Relationship, Nucleation–Elongation Mechanism, and Sergeants-and-Soldiers Principle, *J. Am. Chem. Soc.* 131, 6719–6732.

Kasha, M., Rawls, H. and El-Bayoumi, M. (1965) The exciton model in molecular spectroscopy, *Pure Appl. Chem.* 11, 371–392.

Kowerko, D., Schuster, J. and von Borczyskowski, C. (2009) *R*estricted conformation dynamics of single functionalized perylene bisimide molecules on SiO_2 surfaces and in thin polymer films, *Mol. Phys.* 107, 1911–1921.

Krause, S., Kowerko, D., Börner, R., Hübner, C. and von Borczyskowski, C. (2011) *S*pectral Diffusion of Single Molecules in a Hierarchical Energy Landscape, *ChemPhysChem*, 12, 303–312.

Liang, Y., Moy, P. F., Poole, J. A. and Ponte Goncalves, A. M. (1984) Fluorescence of rhodamine B on semiconductor and insulator surfaces: dependence of the quantum yield on surface coverage, *J. Phys. Chem.* 88, 2451–2455.

Liang, Y. and Ponte Goncalves, A. M. (1985) Time-resolved measurements of the fluorescence of Rhodamine B on semiconductor and glass surfaces, *J. Phys. Chem.* 89, 3290–3294.

Lin, H., Camacho, R., Tian, Y., Kaiser, T., Würthner, F. and Scheblykin, I. (2010) Collective Fluorescence Blinking in Linear J-Aggregates Assisted by Long-Distance Exciton Migration, *Nano Lett.*10, 620–626.

Liu, D., Hug, G. L. and Kamat, P. V. (1995) Photochemistry on Surfaces. Intermolecular Energy and Electron Transfer Processes between Excited Ru(bpy)32+ and H-Aggregates of Cresyl Violet on SiO_2 and SnO_2 Colloids, *J. Phys. Chem.* 99, 16768–16775.

Liu, D. and Kamat, P. V. (1996) Dye-Capped Semiconductor Nanoclusters. One-Electron Reduction and Oxidation of Thionine and Cresyl Violet H-Aggregates Electrostatically Bound to SnO_2 Colloids, *Langmuir* 12, 2190–2195.

Loew, L. M., Bonneville, G. W. and Surow, J. (1978) Charge shift optical probes of membrane potential. Theory, *Biochemistry* 17, 4065–4071.

Losilla, N. S., Oxtoby, N. S., Martinez, J., Garcia, F., Garcia, R., Mas-Torrent, M., Veciana, J. and Rovira, C. (2008) Sub-50 nm positioning of organic compounds onto silicon oxide patterns fabricated by local oxidation nanolithography, *Nanotechnology* 19, 455308.

Losilla, N. S., Martinez, J. and Garcia, R. (2009) Large area nanoscale patterning of silicon surfaces by parallel local oxidation, *Nanotechnology* 20, 475304.

Macquarrie, D. J., Taverner, S. J., Gray, G. W., Heath, P. A., Rafelt, J. S., Saulzet, S. I., Hardy, J. J. E., Clark, J. H., Sutra, P, Brunel, D., di Renzo, F. and Fajula, F. (1999) The use of Reichardt's dye as an indicator of surface polarity, *New J. Chem.*, 23, 725–731

Maedler, C. and Graaf, H. (2013) Local Charge Storage in Thin Silicon Oxide Films: Mechanisms and Possible Applications, *J. Phys. Chem. C*, 117 5358–5363.

Mahrt, J., Willig, F., Storck, W., Weiss, D., Kietzmann, V, Schwarzburg, K., Tufts, B. and Troesken, B. (1994) Luminescence and Configurations of Perylene Dimers in a Langmuir-Blodgett Film, *J. Phys. Chem.* 98, 1888–1894.

Martin, J., Cichos, F. and von Borczyskowski, C. (2005) Confocal microscopy of electrostatic properties of Si-quantum dots and silica surfaces by charge sensitive dye molecules, *Opt. Spectrosc.*, 99, 281–288.

Martini, I., Hartland, G. V. and Kamat, P. V. (1997) Ultrafast Photophysical Investigation of Cresyl Violet Aggregates Adsorbed onto Nanometer-Sized Particles of SnO_2 and SiO_2, *J. Phys. Chem. B* 101, 4826–4830.

Montana, V., Farkas, D. L. and Loew, L. M. (1989) Dual-wavelength ratiometric fluorescence measurements of membrane potential, *Biochemistry,* 28, 4536–4539.

Nonnenmacher, M., O'Boyle, M. P. and Wickranmashinghe, H. K. (1991) Kelvin probe force microscopy, *Appl. Phys. Lett.* 58, 2921–2923.

Pal, M. K. and Ash, S. K. (1973) Metachromasia of basic dyes induced by mercuric chloride. II, *J. Phys. Chem.* 78, 536–540.

Palermo, V., Palma, M. and Samori, P. (2006) Electronic characterization of organic thin films by Kelvin probe force microscopy, *Adv. Mater.* 18, 145–164.

Penzkofer, A. and Wiedmann (1980) Orientation of transition dipole moments of rhodamine 6G determined by excited state absorption, *Opt. Commun.* 35, 81–86.

Persson, B. (1978) Theory of the damping of excited molecules located above a metal surface, *J. Phys. C*, 11, 4251–4269.

Persson, B. and Persson, M. (1980) Damping of vibrations in molecules adsorbed on a metal surface, *Surf. Sci.* 97, 609–624.

Persson, B. and Lang, N. (1982) Electron-hole-pair quenching of excited states near a metal, *Phys. Rev. B*, 26, 5409–5415.

Philipp, H. (1971) Optical properties of non-crystalline Si, SiO, SiO_x and SiO_2, *J. Phys. Chem. Solids*, 32, 1935–1945.

Polster, D., Graaf, H., Baumgärtel, T., von Borczyskowski, C., Benedikt, U. and Auer, A. A. (2010) Beyond Cassie's Law: A Theoretical and Experimental Study of Mixed Alkyl Monolayers, *Langmuir* 26, 8301–8308.

Polster, P. and Graaf, H. (2013) Advancing and receding angles – Dynamic contact angle measurements on mixed alkyl monolayers, *Appl. Surf. Sci.* 265, 88–93.

Prathapan, S., Yang, S., Seth, J., Miller, M., Bocian, D., Holten, D. and Lindsey, J.(2001) Synthesis and Excited-State Photodynamics of Perylene–Porphyrin Dyads. 1. Parallel Energy and Charge Transfer via a Diphenylethyne Linker, *J. Phys. Chem. B* 105, 8237–8248.

Proehl, H., Dienel, T., Nitsche, R. and Fritz, T. (2004) Formation of Solid-State Excitons in Ultrathin Crystalline Films of PTCDA: From Single Molecules to Molecular Stacks, *Phys. Rev. Lett.* 93, 097403.

Rehm, S., Stepanenko, V., Zhang, X., Rehm, T. and Würthner, F. (2010) Spermine-Functionalized Perylene Bisimide Dyes—Highly Fluorescent Bola-Amphiphiles in Water, *Chem. Eur. J.* 16, 3372–3382.

Sagiv, J. (1980) Organized monolayers by adsorption. 1. Formation and structure of oleophobic mixed monolayers on solid surfaces, *J. Am. Chem. Soc.* 102, 92–98.

Schubert, M. and Hamerman, D. (1956) Metachromasia; chemical theory and histochemical use, *J. Histochem. Cytochem.* 4, 159–189.

Schubert, A., Settels, V., Liu, W., Würthner, F., Meier, C., Fink, R. F., Schindlbeck, S., Lochbrunner, S., Engel, B. and Engel, V. (2013) Ultrafast Exciton Self-Trapping upon Geometry Deformation in Perylene-Based Molecular Aggregates, *J. Phys. Chem. Lett.* 4, 792–796.

Schuster, J., Brabandt, J. and von Borczyskowski, C. (2007) Discrimination of Photoblinking and Photobleaching on the Single Molecule Level *J. Lumin.* 127, 224–229.

Sieval, A. B., Demirel, A. L., Nissink, J. W. M., Linford, M. R., van der Maas, J. H., de Jeu, W. H., Zuilhof, H. and Sudhoelter, E. J. R. (1998) Highly stable Si-C linked functionalized monolayers on the silicon (100) surface, *Langmuir* 14, 1759–1768.

Sluch, M. I., Vitukhnovsky, A. G. and Petty, M. C. (1995) Anomalous distance dependence of fluorescence lifetime quenched by a semiconductor, *Phys. Lett. A* 200, 61–64.

Stavola, M., Dexter, D. and Knox, R. (1985) Electron-hole pair excitation in semiconductors via energy transfer from an external sensitizer, *Phys. Rev. B*, 31, 2277–2289.

Sugimura, H., Ishida, Y., Hayashi, K. and Takai, O. (2002) Potential shielding by the surface water layer in Kelvin probe force microscopy, *Appl. Phys. Lett.* 80 1459–1461.

Sumi, H. (1989) Two kinds of excimers in α-perylene and pyrene crystals: Origin of Y and V emissions, *Chem. Phys.* 130, 433–449.

Tada, H., Ara, M. and Tanaka, (2002) Wet-Process Molecular Planting in A Specific Site on Silicon with Si-C Covalent Bonds, MRS Proceedings, 739, H7.37 doi:10.1557/PROC-739-H7.37.

Tanaka, J. (1963) The Electronic Spectra of Aromatic Molecular Crystals. II. The Crystal Structure and Spectra of Perylene, *Bull. Chem. Soc. Jpn.* 36. 1237–1249.

Teuschler, T., Mahr, K., Miyazaki, S., Hundhausen, M. and Ley, L. (1995) *Appl. Phys. Lett.* 67, 3144–3146.

Ulman, A. (1996) Formation and Structure of Self-Assembled Monolayers, *Chem. Rev.,* 96, 1533–1554.

Valeur. B. (2002) *Molecular Fluorescence Principles and Applications.* WILEY-VCH, Weinheim.

Veldman, D., Chopin, S. Meskers, S., Groeneveld, M., Williams, R. and Janssen, R. (2008) Triplet Formation Involving a Polar Transition State in a Well-Defined Intramolecular Perylenediimide Dimeric Aggregate, *J. Phys. Chem. A* 112, 5846–5857.

von Freydorf, E., Kinder, J. and Michel-Beyerle, M. (1978) On low temperature fluorescence of perylene crystals, *Chem. Phys.* 27, 199–209.

Walker, B., Port, H. and Wolf, H. (1985) The two-step excimer formation in perylene crystals, *Chem. Phys.* 92, 177–185.

Weiss, D., Kietzmann, R., Mahrt, J., Tufts, B., Storck, W. and Willig, F. (1992) E-excimer and Y-type luminescence of perylene dimers in a Langmuir-Blodgett film at 1.5 K, *J. Phys. Chem.* 96, 5320–5325.

Whitmore. P. M., Alivisatos, A. P. and Harris, C. B. (1983) Distance Dependence of Electronic Energy Transfer to Semiconductor Surfaces: $^3n\pi'$ Pyrazine/GaAs(110), *Phys. Rev. Lett,.* 50, 1092–1094.

Wouters, D. and Schubert, U. (2003) Constructive nanolithography and nanochemistry: local probe oxidation and chemical modification, *Langmuir* 19, 9033–9038.

Würthner, F. (2004) Perylene bisimide dyes as versatile building blocks for functional supramolecular architectures, *Chem. Comm.* 1564–1579.

Yang, M., Zheng, Z., Liu, Y. and Zhang, B. (2005) Scanned probe oxidation on an octadecyl-terminated silicon (111) surface with an atomic force microscope: kinetic investigations in line patterning, *Nanotechnology,* 17, 330–337.

Yang, M., Zheng, Z., Liu, Y. and Zhang, B. (2006) Kinetics of atomic force microscope-based scanned probe oxidation on an octadecylated silicon (111) surface, *J. Phys. Chem. B* 110 10365–10373.

Yang, M., Wouters, D., Giesbers, M., Schubert, U. S. and Zuilhof, H. (2009) Local Probe Oxidation of Self-Assembled Monolayers on Hydrogen-Terminated Silicon, *ACS Nano* 3, 2887–2890.

Zondervan, R., Kulzer, F., Orlinskii, S. B. and Orrit, M. (2003) Photoblinking of Rhodamine 6G in Poly(vinyl alcohol): Radical Dark State Formed through the Triplet, *J. Phys. Chem. A* 107, 6770–6776.

Zondervan, R., Kulzer, F., Kol'chenko, M. and Orrit, M. (2004) Photobleaching of rhodamine 6G in poly (vinyl alcohol) at the ensemble and single-molecule levels, *J. Phys. Chem. A* 108, 1657–1665.

Chapter 7

Identification of Heterogeneous Surface Properties via Fluorescent Probes

Daniela Täuber[a] and Christian von Borczyskowski[b]

[a]*Chemical Physics, Lund University, Getingevägen 60, S-22241 Lund, Sweden*
[b]*Institute of Physics, Center for Nanostructured Materials and Analytics, Technische Universität Chemnitz,Reichenhainer Str. 70, D-09107 Chemnitz, Germany*

dantaube@gmx.de, borczyskowski@physik.tu-chemnitz.de

7.1 Introduction

Silicon wafers with native or thermal oxide are widely used in technical applications, for example, in solar cells or for chip fabrication (Venema, 2011). The performance of such devices is influenced by surface chemistry (Shemella and Nayak, 2009) and by local interface properties (Lüttich et al., 2012). In particular, silica surfaces are covered by hydroxyl groups (silanols) and water, depending on fabrication and temperature treatment (Zhuravlev, 2000; Dugas and Chevalier, 2000). Owing to the high fabrication temperature (1000°C) of the thermal oxide, the silanol coverage is significantly lower compared to fumed silica (Dugas and Chevalier, 2000). Yet, the amount and distribution of surface

Self-Assembled Organic-Inorganic Nanostructures: Optics and Dynamics
Edited by Eduard Zenkevich and Christian von Borczyskowski

ISBN 978-981-4745-43-7 (Hardcover), 978-981-4745-44-4 (eBook)
www.panstanford.com

silanols is known to affect nanotechnical fabrication routines, for example, nanolithography (see also Chapter 6) (Baumgärtel and Graaf, 2011) and coatings with self-assembled monolayers (Belgardt et al., 2010). However, up to now only average silanol coverages were known from ensemble measurements on macroscopic scales (Zhuravlev, 2000). Here we use laser scanning confocal microscopy to investigate the distribution of fluorescent rhodamine 6G (R6G) adsorbed on Si wafers with 100 nm thermal oxide. According to Iler (1979), charged dye molecules preferentially physisorb to surface silanols. Thus, the derived spatial distribution of adsorbed R6G molecules unravels the distribution of surface silanols on a submicron range.

Heterogeneity might be even more complex since charges or photogenerated charges might also be present on heterogeneous substrates such as semiconductors or aggregates of them. Photogeneration of charges might occur in case that aggregates of semiconductors, single semiconductor nanocrystals or organic dye molecules are deposited onto a surface, which can be exploited for photoactive devices such as solar cells (O'Regan and Grätzel, 1991). In most of these cases, a microscopic description of the electronic interaction is missing and is difficult to approach since the corresponding interface between, e.g., semiconductors and dyes is also heterogeneous with respect to its electronic, chemical, and structural properties. A typical example is nanoporous TiO_2 covered with organic dyes (Nazeeruddin et al., 1993). Another example of semiconductor/dye heterogeneous interactions is nanoporous silicon, which is subject to quantum confinement with respect to its optical properties (Canham, 1990; Kovalev et al., 1999). To detect charges optically one needs charge-sensitive molecular species. In the present chapter, we report on such an approach by coating single Si–nanocrystals with aminonaphthylethenylpyridinium dye (ANEPPS) molecules, which are known to exhibit large spectral shifts upon creation of static or time-dependent electric fields (Fluhler et al., 1985). A similar approach has been reported in Chapter 6. Making use of scanning confocal microscopy, we report on spatially resolved luminescence spectra down to the level of single silicon nanoparticles.

7.2 Impact of Chemical Inhomogeneities of SiO_2 Surfaces

7.2.1 Optical Decoration of Silanol Groups

It is known that the surface of freshly cut Si wafers is covered by native oxide. Native oxides as well as thermally grown ones are chemically inhomogeneous with respect to the amount and distribution of terminating surface groups. This is schematically shown in Fig. 7.1. Isolated external silanol (OH–) groups can easily bind (electrostatically couple) to charged dye molecules (Albers et al., 2015) or to the surface of Ag nanoclusters (Krause et al., 2013), while geminal and vicinal groups show less effective coupling.

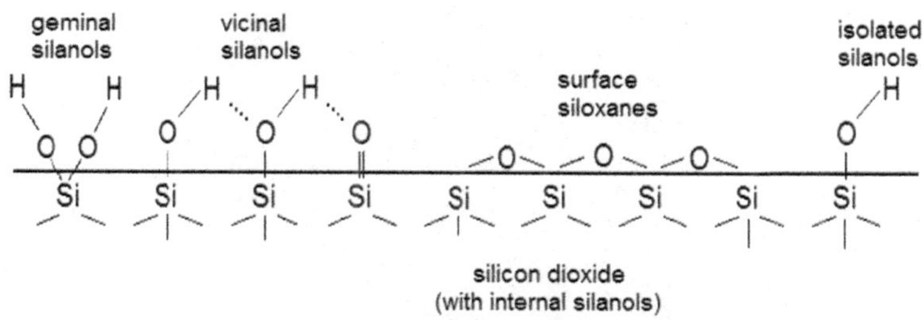

Figure 7.1 Schematic presentation of various terminating OH-groups on a SiO_2 surface.

Decoration of silanol groups with fluorescent markers is achieved by dip-coating them into a solvent (e.g., ethanol) containing a low concentration of the respective markers. After evaporation of the solvent markers remain attached to the respective reactive surface sites which can be inspected via optical microscopy. A typical microscopic image of fluorescent rhodamine 6G (R6G) dye molecules is shown in Fig. 7.2 at two different spatial resolutions. Comparison indicates that the distribution of fluorescent molecules is strongly inhomogeneous but similar on different length scales, which indicates a fractal cluster distribution. Naturally, spatial resolution is limited by diffraction to about 250 nm for optical microscopy. However, due to the influence from processing conditions and further treatments, islands of terminating silanol surface groups might vary from single isolated silanols (<1 nm) to patches containing various types of silanols (in the range of μm). Thereby, the distribution is, on the one hand, influenced by the density of initial silanols/silanol clusters and, on the other hand, by the conditions for growth or removal of silanols, which preferentially will occur at the edge of existing silanol clusters (Zhuralev, 2000).

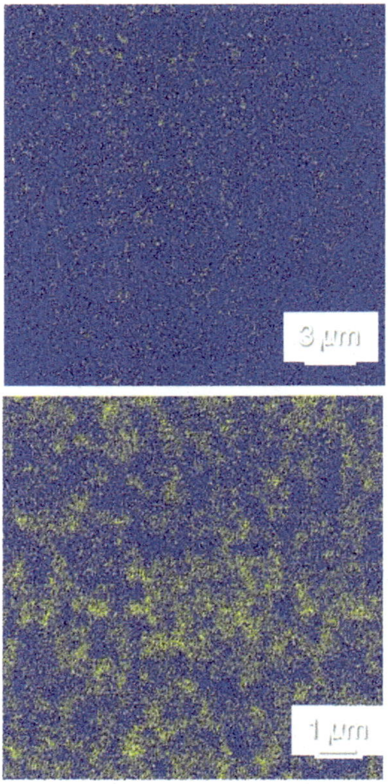

Figure 7.2 Confocal scan of rhodamine 6G fluorescence at two different spatial resolutions. Substrate is a Si(100) wafer with 100 nm thermal oxide (dry O_2/HCl processed at 1000°C) obtained from the Center for Microtechnologies (ZfM, Chemnitz, Germany) (Schubert, 2010).

The concentration and distribution of silanol groups can be controlled by physico-chemical procedures. Si substrates were boiled in ultraclean water for hydroxylation for 20 h ("hydroxylated"), while others were de-hydroxylated by tempering in air for 0.5 h at 800°C ("tempered"). Substrates which underwent no additional treatment besides cleaning by a 1:1 mixture of H_2SO_4 and H_2O_2 were used for comparison ("piranha"). The degree of hydroxylation can be macroscopically investigated by contact angle measurements as shown in Fig. 7.3.

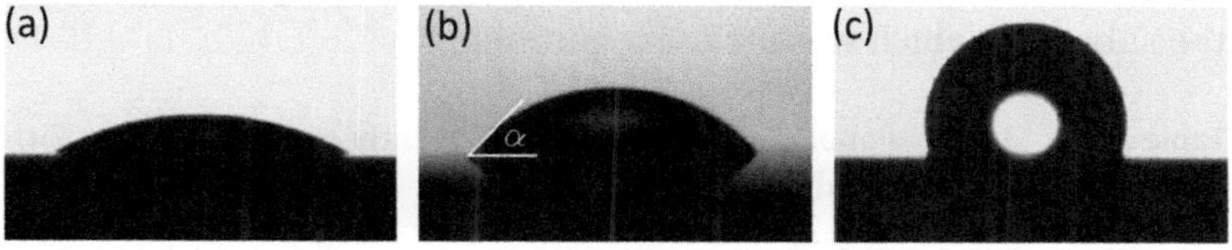

Figure 7.3 Water contact angle α for three substrates with different silanol densities. (a) As-grown SiO_2 (α = 28°), (b) decreased silanol density: tempered at 1000°C (α = 50°), (c) (by an octadecylsilan layer fully) silanized surface (α = 110°) (Krause et al., 2014).

Fluorescence intensities from R6G after evaporation of the solvent show an inhomogeneous distribution. In Fig. 7.4 fluorescence intensities from three differently treated substrates are shown after evaporation of ethanol. Raw images have been processed with WSxM 5.0 (Horcas et al., 2007). Similar results were obtained using toluene and n-hexane as solvents (not shown). Thus, the distribution of R6G molecules on the thermal oxide is not influenced by the kind of solvent, but is determined by the properties of the substrate surface.

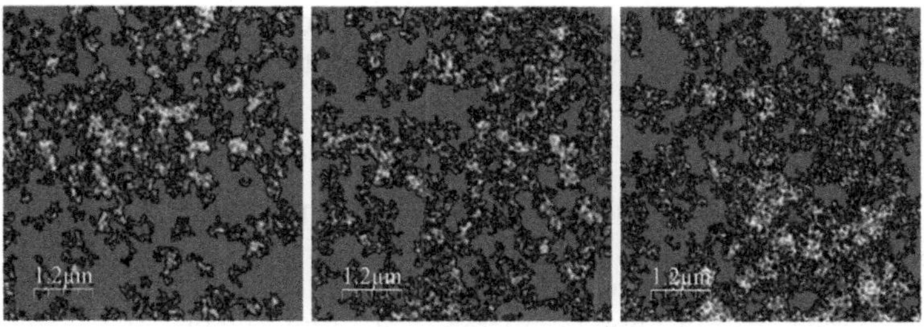

Figure 7.4 Distribution of R6G fluorescence intensities on thermal SiO_2 after (left) tempering at 800°C in air, (middle) cleaning in piranha with no further treatment, and (right) hydroxylation in water (Albers et al. 2015).

The results from image processing are given in Table 7.1. As can be seen, there is only a slight modification of the silanol coverage from 46.4% for the tempered substrate to 53.5% for the hydroxylated substrate. The largest average cluster size (1.97 ± 0.5 μm²) is observed for the hydroxylated substrate. The thermal oxide tempered for 30 min at 800°C shows only a slightly smaller average cluster size (1.42 ± 0.3 μm²), while the smallest average cluster size (0.64 ± 0.2 μm²) is obtained for the substrate which was etched in piranha.

Table 7.1 Local silanol distributions on differently treated thermal SiO_2 (Albers et al. 2015)

Substrate treatment	Tempered	Piranha	Hydroxylated
Investigated area [μm²]	110	110	81
# of clusters in area	36	85	22
Coverage [%]	46.4	49.4	53.5
Average cluster size and standard deviation [μm²]	1.42 ± 0.3	0.64 ± 0.2	1.97 ± 0.5
Nearest neighbor distance [μm]	1.75 ± 0.5	0.64 ± 0.2	0.66 ± 0.3

These results show that various surface treatments do not only change the coverage by dye molecules (depending on the silanol density) but also modify the clustering features. Larger average cluster sizes at similar surface coverage yield less homogeneous and more compact silanol clusters (see Fig. 7.4, left and right). In contrast, a smaller average cluster size represents a more homogeneous, but less compact silanol distribution, as can be seen in Fig. 7.4 (middle), which was obtained from etching in Piranha without further treatment.

7.2.2 Influence of Surface Inhomogeneities on Diffusion Dynamics in Ultrathin Liquid Films

Inhomogeneous solid-liquid interfaces influence dynamic properties of fluid films (Schuster et al., 2000; Fleury et al., 2010) and spatial diffusion of low concentrated (fluorescent) probes in such films. Optical investigation of single molecule diffusion in ultrathin liquid TEHOS (tetrakis-2ethyhexoxy-silane) films on quartz substrates revealed anisotropic diffusion of organic probe molecules (Schuster et al., 2003). This has been related to liquid layering at solid-liquid interfaces (Forcada and Mate, 2003; Yu et al., 1993). Similar effects have been observed using very small Ag clusters as optical probes via single-particle tracking (Krause et al., 2013). Additionally, also heterogeneous interface chemistry and morphology causes anisotropies in probe diffusion. An illustrative example is given in Fig. 7.5 showing the distribution of integrated fluorescence intensities from diffusing RhB molecules in an ultrathin TEHOS film on thermal oxide, which had been etched in Piranha solution. Bright white spots correspond to RhB molecules immobilized during fluorescence acquisition, while less bright areas stem from only temporary local presence of diffusing dye molecules. Detected trajectory paths of mobile RhB are overlayed in color. As can be seen, the trajectory paths are mostly restricted to the bright areas. However, one has to take into account that the experimental time resolution is limited to about 10 ms per diffusion step. This implies that fast diffusing tracer molecules escape the detection by tracking analysis (Täuber, 2011). The speculation therefore is that diffusion close to the substrate and close to silanol clusters is considerably slowed down, while diffusion between different silanol clusters is less slowed and thus may escape the detection by tracking analysis.

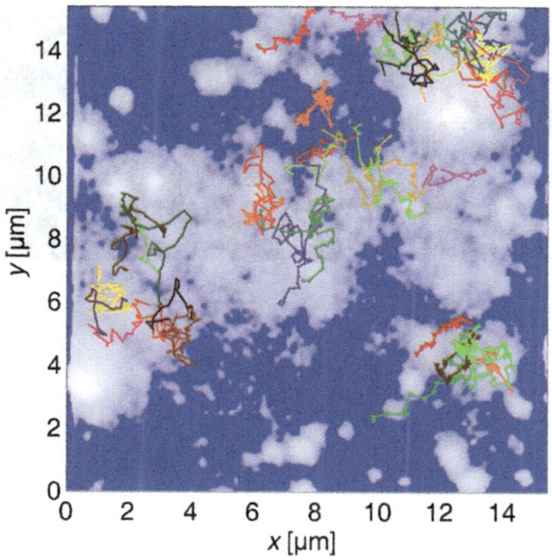

Figure 7.5 Time-integrated fluorescence intensities from single RhB molecules in an ultrathin TEHOS film on silicon with 100 nm thermally grown SiO_2. Detected diffusion trajectories of several mobile single RhB molecules are overlayed as colored paths. Bright areas correspond to high densities of RhB immobilized on silanol clusters.

To further elucidate this speculation, dye diffusion and subsequent detection was simulated for diffusion on substrates with varied silanol distributions as determined in the decoration experiment (Fig. 7.4). Figure 7.6 (top) shows the derived heterogeneous substrates used within these simulations, which employed a 100-times slowed diffusion close to the substrate on the silanol clusters (black areas) in respect to the diffusion in the upper film layer and on the white areas close to the substrate (Albers et al., 2015).

The integrated fluorescence intensity distributions obtained from the simulations are shown in Fig. 7.6 (bottom). As can be seen, the distributions of bright areas achieved from simulated diffusing dye molecules match the related silanol cluster distributions (top). Thereby, the brighter areas correspond to the silanol clusters themselves, while less bright areas (resembling faster diffusing dyes) are found in gaps between the silanol clusters. No dye molecules were detected on larger sized silanol-free areas. This result agrees with the above speculation that diffusion is slowed down close to the substrate on silanol clusters, while dye molecules diffuse faster in between the clusters.

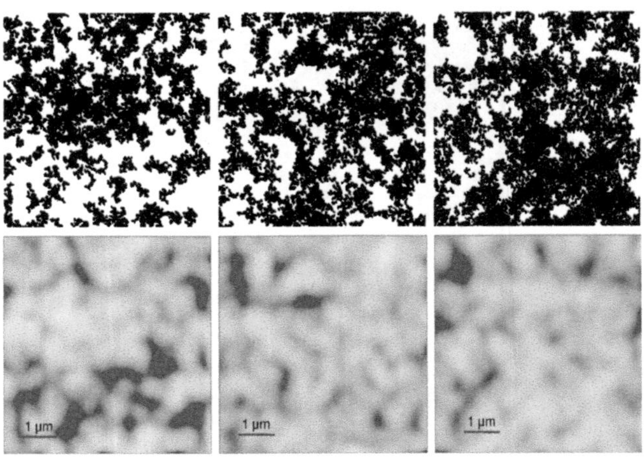

Figure 7.6 Different experimentally determined silanol distributions such as shown in Fig. 7.4: (Left) Tempered, (middle) piranha treated and (right) hydroxylated substrates. (Top) Heterogeneous substrate coverage used for the simulation, areas for slow diffusion are black. (Bottom) Time-integrated fluorescence intensities obtained from tracking analysis with tracking.sh (Albers et al., 2015).

Similar tracking experiments have been performed with Ag clusters with a diameter < 1 nm (Krause et al., 2013). In that case diffusion occurred in water films with n ($0 \leq n < 5$) monolayers of water and at different degrees of hydroxylation of the SiO_2 surface. The qualitative result is shown in Fig. 7.7. Coupling of Ag nanoparticles and silanol groups shows up spectroscopically as OH-vibrational sidebands of the fluorescence of the Ag particles. Diffusion becomes faster with increasing concentration of silanol groups. This is attributed to an increased wettability of the SiO_2 surface as shown on top of Fig. 7.7. This finding is opposite to the one for tracer molecules in thin TEHOS films for which we found a slowing down of diffusion in the areas of silanol clusters as discussed above. The reason for this is that the wettability with very thin water films drastically depends on the silanol concentration, whereas the wettability with non-polar TEHOS does not. Additionally, diffusion of Ag nanoparticles is slower in very thin water films, probably caused by two cooperating effects, namely the increased decoupling from weak binding of Ag particles to silanol groups and the softening of the water layers with increasing film thickness. Layering of liquids in thin films on surfaces is an often-observed phenomenon (Yu et al., 2000). In fact, careful analysis of diffusion reveals a kind of phase transition between frozen water layers and softened ones (see Fig. 7.7, bottom).

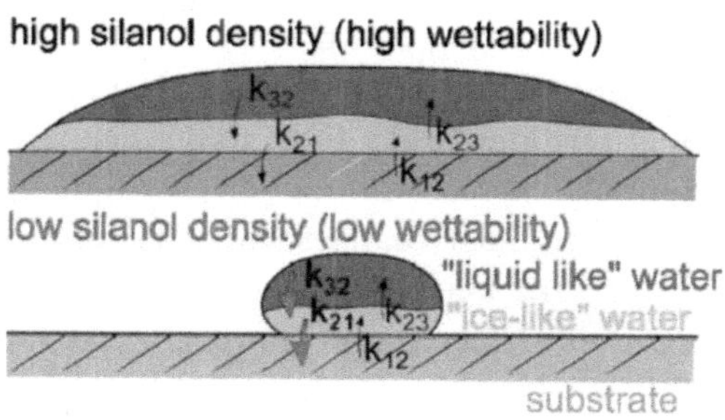

Figure 7.7 Scheme of at least two different heterogeneities influencing spatial diffusion in thin water layers on SiO_2 substrates. One of the heterogeneities is related to the distribution of silanols which determines the size and shape of water islands. The other one is related to the water layer film thickness. Layers close to the substrate are "frozen." We show exchange rates among the vertical arrangement of layers. Adapted from Krause et al. (2014).

7.3 Decorating Charged Si Nanoparticles with Charge-Sensitive Dye Molecules

Photoinduced charge generation has been suggested for colloidal semiconductor quantum dots such as CdSe (Nirmal et al., 1996) or Si nanocrystals (Si–NC) (Canham, 1990). The investigation of single quantum dots has revealed that the photoluminescence exhibits a luminescence intermittency (blinking) (Nirmal et al., 1990), a statistical sequence of dark and bright periods during which the luminescence is "off" or "on" (Cichos et al., 2007). It is often assumed (Efros and Rosen, 1996; Cichos et al., 2007) that the generation of dark periods is due to a photoinduced charge ejection resulting in the population of a broad distribution of (self-trapped) states in an amorphous or disordered environment (Cichos et al., 2007). Previously it has been shown that similar processes can also be observed for Si quantum dots (Cichos et al., 2007). The dependence of blinking phenomena on the embedding environment or surface attached molecules has been discussed extensively by von Borczyskowski (2016). If such chargeable semiconductor nanocrystals are deposited onto a Si-SiO_2 substrate (wafer) or are embedded in a polymer matrix the photogenerated charges might be transferred to the local environment. This will influence the electric/dielectric features of the material.

That intermittency might be related to charge separation processes has been confirmed on the nanometer scale by electric force measurements (Krauss et al., 2001). Charge separation and self-trapping on time scales from milliseconds to hours will be accompanied by changes in the electrostatic properties of the nanoenvironment. Such changes might be probed by special dye molecules that are known to be spectroscopically sensitive to electrostatic fields. Such an approach has already been discussed in Chapter 6 of this book. In the following we present results on coating single Si–NC nanocrystals dots deposited onto a Si-SiO_2 wafer with ANEPPS molecules, which are known to exhibit large spectral shifts upon creation of static or time-dependent electric fields (Fuhler et al., 1985). Making use of scanning confocal microscopy, it has been shown for spatially resolved luminescence spectra down to the level of single silicon nanocrystals (Si-NC) that photoinduced charging of Si-NC results in considerable spectral shifts of ANNEPPS molecules (Martin et al., 2005).

The experiment has been performed by depositing Si-NC surrounded by a SiO_x shell onto the substrate. Thereafter the sample has been spin coated by solutions containing ANEPPS molecules. After evaporation of the solvent, the sample is covered by a sub-monolayer fluorescent film of ANEPPS molecules (Martin et al., 2004), which is shown schematically in Fig. 7.8.

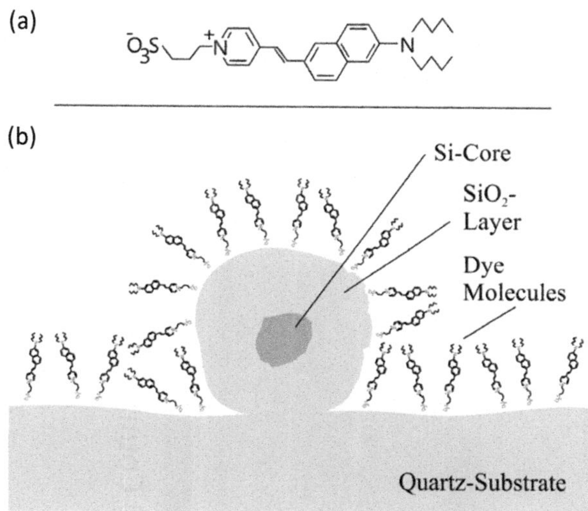

Figure 7.8 (a) Chemical structure of di-4-ANEPPS molecules and (b) schematic view of a silicon nanocrystal on a quartz substrate with a layer of di-4-ANEPPS molecules. It is assumed that the SO_3^- group of ANEPPS attaches to the surface causing ANEPPS to be nearly upright with respect to the Si-NC surface (Martin et al., 2005).

Figure 7.9 shows that the fluorescence intensity of ANEPPS is on average increased in the presence of Si-NCs. This can be followed in Fig. 7.10, which shows confocal images across a Si-NC. At the position of the Si-NC the ANEPPS fluorescence is increased as is indicated by the white areas. The spatial resolution is limited by the optical resolution of about 400 nm. Confocal images have been obtained at 2 different laser intensities. After a high laser intensity scan (see the horizontal line) the ANEPPS fluorescence is partly bleached as can be seen in the scan on the right.

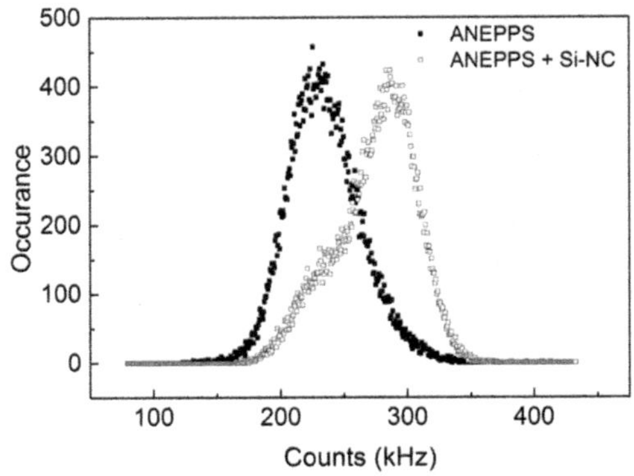

Figure 7.9 Histograms of fluorescence confocal images of a pure di-4-ANEPPS film on quartz (left) and a di-4-ANEPPS layer on a sample containing single silicon nanocrystals (right) (Martin et al., 2005).

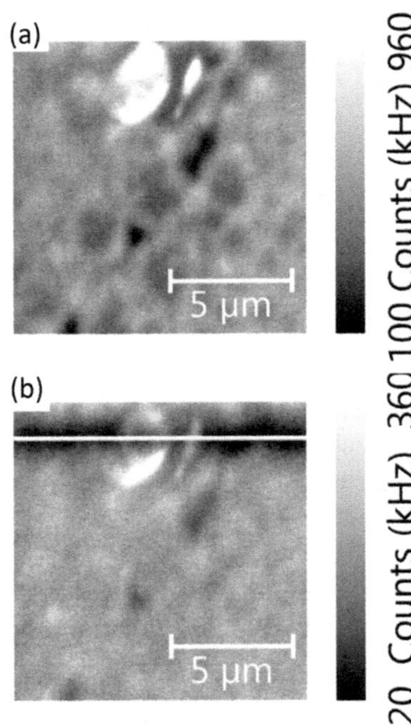

Figure 7.10 Confocal fluorescence images of the di-4-ANEPPS layer of a sample with a single Si–NC on a Si-SiO_2 wafer recorded before (a) and after (b) a laser scan with increased laser power across the bright spot. The scan is marked in (b) by a white horizontal line. The intensity scale is in counts/s (Martin et al., 2005).

Upon generation of a charge in a Si-NC, it might be ejected, e.g., to the SiO_2 layer or the substrate, the charge becomes self-trapped. Thus the surface of the substrate will be charged upon continuous laser excitation. This charged surface acts in reverse upon the electron distribution in ANEPPS molecules. The negatively charged SiO_2 surface will repel the excess charge of ANEPPS initially located at the SO_3^- head group towards the aromatic center of ANEPPS, thus screening and reducing the interaction with the (polar) environment (surfaces). This will result in an effective blue shift of the fluorescence, as can be seen in Fig. 7.11, which shows the energetic position of the maximum of the ANEPPS fluorescence spectrum across the Si-NC along the scan shown in Fig. 7.10a.

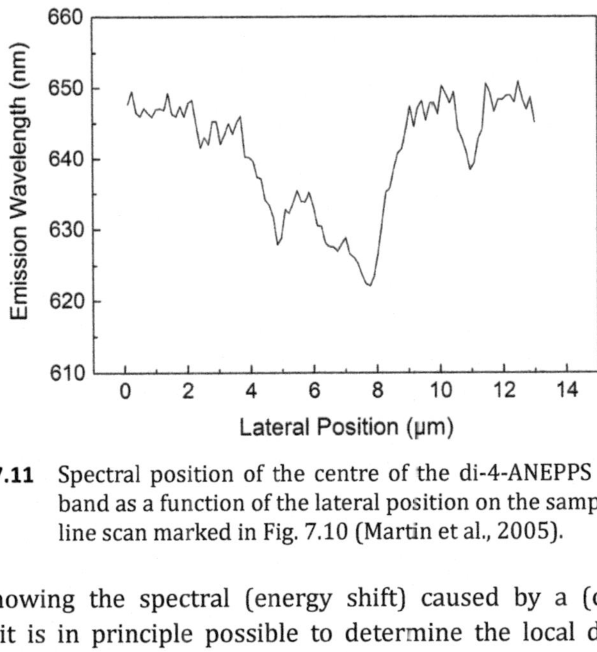

Figure 7.11 Spectral position of the centre of the di-4-ANEPPS emission band as a function of the lateral position on the sample for the line scan marked in Fig. 7.10 (Martin et al., 2005).

Knowing the spectral (energy shift) caused by a (charged) Si-NC it is in principle possible to determine the local dielectric constant. Figure 7.12 shows a relation between the energy of the ANEPPS fluorescence with the dielectric constant from a variety of experiments (Martin et al., 2005). Taking the spectroscopic data from Fig. 7.11, we can calculate a dielectric constant $\varepsilon \approx 3.3$ for the Si/SiO_2 substrate and $\varepsilon \approx 2.4$ above a (charged) Si-NC. The decrease of ε upon charging is in accordance with recent experiments on charged SiO_2 substrates (Martin et al., 2005).

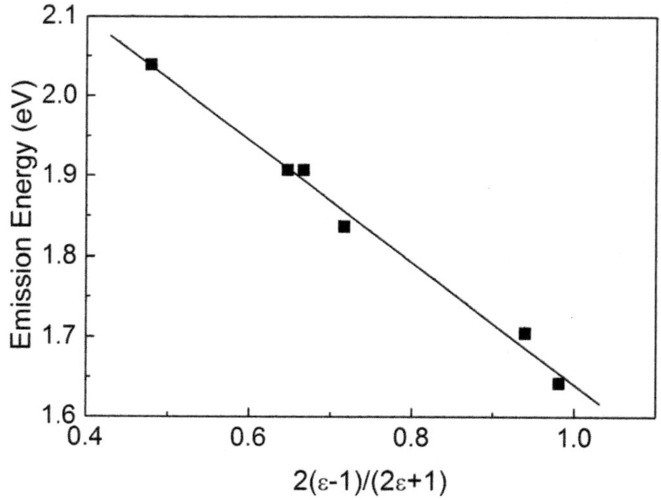

Figure 7.12 Di-4-ANEPPS fluorescence energy as a function of the dielectric constant ε of various solvents or environments. Data are according to Martin et al. (2005).

7.4 Conclusion

The examples of fluorescent probes discussed in this chapter (and in Chapter 6) show that an inhomogeneous distribution of chemical groups or charges on a dielectric surface can be identified by depositing adequately functionalized probe molecules or nanocrystals to the surface. Two limiting scenarios are possible: (1) Optical probes become immobilized at certain active binding centers or (2) are subject to "hindered" diffusion. The latter needs an abundant liquid film of at least a few monolayers thick.

The analysis follows the spatial distribution of luminescence intensities via microscopic scans or relies on tracking the diffusion of single probes. Depending on the types of probes, even single emitters can be detected. According to the size of such emitters, this results, in principle, in a spatial resolution of <1 nm in case of dye molecules or <3 nm in case of semiconductor nanocrystals. Unfortunately, a diffraction limited optical microscopy will limit the actual resolution to at most 250 nm. However, recent developments in super-resolution microscopy will circumvent this limitation in the future.

Naturally, other scanning microscope techniques such as STM, AFM, and electric force microscopy will provide much higher spatial resolutions. However, these techniques are not as readily applied as non-invasive optical techniques. Moreover, the optical approach provides additional spectroscopic information such as emission and absorption energies, polarization, and luminescence dynamics. Further on, optical techniques can easily follow mobilities or conformational changes on time scales down to a few ns using single-particle tracking or fluorescence correlation spectroscopy. Additionally, inner surfaces such as pores or channels can be accessed by optical means, revealing, e.g., the exact number of adsorbed molecules in pores of silica nanoparticles (Trenkmann et al., 2012).

Concluding, optical decoration techniques are a versatile tool to investigate chemically and physically heterogeneous (dielectric) surfaces on the nanoscale with extremely high sensitivity. This is an important approach for analytics of surfaces important, e.g., in sensoric applications via functionalized dyes or nanoparticles as has been also discussed in Chapter 6. A further example that includes a background for applications will be given in Chapter 8.

References

Albers, T., Bauer, M., von Borczyskowski, C., Gerlach F., Heidernätsch M., Kärger J., Kondrashova D., Radons G., Schubert S., Shakhov A., Täuber D., Valiullin R., and Zeigermann P. (2015) Driven diffusion in nanoscaled materials, *Diffusion Fundamentals*, 23(3), 1–25.

Baumgärtel, T., and Graaf, H. (2011) Local anodic oxidation nanolithography on silicon: Chemical routes to functional nanostructures, in: *Lithography: Principles, Processes and Materials* (ed. T. C. Hennessy), Nova Science Publishers, pp. 175–193.

Belgardt, C., Graaf, H., Baumgärtel, T., and von Borczyskowski, C. (2010) Self-assembled monolayers on silicon oxide, *Phys. Stat. Sol. C*, 7, 227–230.

Canham, L. T. (1990) Silicon quantum wire array fabrication by electrochemical and chemical dissolution of wafers, *Appl. Phys. Lett.*, 57, 1046.

Cichos, F., Martin, J., and von Borczyskowski, C. (2004a) Characterizing the non-stationary blinking of silicon nanocrystals, *J. Luminesc.*, 107, 160.

Cichos, F., Martin, J., and von Borczyskowski, C. (2004b) Emission intermittency in silicon nanocrystals, *Phys. Rev. B,* 70, 153141.

Cichos, F., von Borczyskowski, C., and Orrit, M. (2007) Power-law intermittency of single emitters, *Curr. Opin. Coll. Interf. Sci.,* 12, 272–284.

Dugas, V., and Chevalier, Y. (2000) Surface hydroxylation and silane grafting on fumed and thermal silica, *J. Coll. Interf. Sci.,* 264, 354–361.

Efros, A. L., and Rosen, M. (1997) Random telegraph signal in the photoluminescence intensity of a single quantum dot, *Phys. Rev. Lett.,* 78, 1110.

Fleury, J. B., Claussen, O., Herminghaus, S., Brinkmann, M., and Seemann, R. (2010) Mechanical stability of ordered droplet packings in microfluidic channels, *Appl. Phys. Lett.,* 99, 244104.

Fluhler, E., Burnham, V. G., and Loew, L. M. (1985) Spectra, membrane binding, and potentiometric responses of new charge shift probes, *Biochemistry,* 24, 5749.

Forcada, M. L., and Mate, C. M. (1993) Molecular layering during evaporation of ultrathin liquid films, *Nature,* 363, 527–529.

Hartmann, N., Klingebiel, B., Balgar, T., Franzka, S., and Hasselbrink, E. (2009) Laser-induced local dehydroxylation on surface-oxidized silicon substrates: Mechanistic aspects and prospects in nanofabrication, *Appl. Phys. A,* 94, 95–103.

Horcas, I., Fernandez, R., Gomez-Rodriguez, J. M., Colchero, J., Gomez-Herrero, J., and Baro, A. M. (2007) WSXM: A software for scanning probe microscopy and a tool for nanotechnology, *Rev. Sci. Instr.,* 78, 013705.

Iler, R. K. (1979) The surface chemistry of silica, in: *The Chemistry of Silica,* Wiley.

Kovalev, D., Heckler, H., Polisski, G., and Koch, F. (1999) Optical properties of Si nanocrystals, *Phys. Status Solidi B,* 215, 871.

Krause S., Hartmann, M., Kahle, I., Neuman, M., Heidernätsch, M., Spange, S., and von Borczyskowski, C. (2013) Optical tracking of single Ag clusters in nanostructured water films, *J. Phys. Chem. C,* 117, 24822–24829.

Krause, S., Heidernätsch, Hartmann, M., and von Borczyskowski, C. (2014). Optical Investigation of diffusion of single Ag markers in confined water films, *J. Nanoparticle Res.,* 16, 2700.

Krauss, T. D., O'Brien, S., and Brus, L. E. (2001) Charge and photoionization properties of single semiconductor nanocrystals, *J. Phys. Chem. B,* 105, 1725.

Lüttich, F., Lehmann, D., Friedrich, M., Chen, Z., Facchetti, A., von Borczyskowski, C., Zahn, D. R. T., and Graaf, H. (2012). Interface properties of OFETs based on an air-stable n-channel perylene tetracarboxylic diimide semiconductor, *Phys. Stat. Sol. A,* 209, 585.

Martin, J., Cichos, F., and von Borczyskowski, C. (2005) Confocal microscopy of electrostatic properties of Si quantum dots and silica surfaces by charge-sensitive dye molecules, *Opt. Spectrosc.,* 99, 281–288.

Nazeeruddin, M. K., Kay, A., Rodicio, I., Humphry-Baker, R., Mueller, E., Liska, P. Vlachopoulos, N., and Graetzel, M. (1993) Conversion of light to electricity by cis-X_2bis(2,2′-bipyridyl-4,4′-dicarboxylate) ruthenium(II) charge-transfer sensitizers (X = Cl^-, Br^-, I^-, CN^-, and SCN^-) on nanocrystalline titanium dioxide electrodes, *J. Am. Chem. Soc.,* 115, 6382.

Nirmal, M., Dabbousi, B. O., Bawendi, M. G., Macklin, J. J., Trautman, J. K., Harris, T. D., and Brus, L. E. (1996) Fluorescence intermittency in single cadmium selenide nanocrystals, *Nature,* 383, 802.

O'Regan, B., and Grätzel, M. (1991) A low-cost, high-efficiency solar cell based on dye-sensitized colloidal TiO_2 films, *Nature,* 353, 737.

Shemella, P., and Nayak, S. K. (2009) Electronic structure and band-gap modulation of graphene via substrate surface chemistry, *App. Phys. Lett.,* 94, 032101.

Schubert, S. (2010) Untersuchung der Diffusion und der Anbindung von Farbstoffen an nanostrukturierten Grenzflächen *Diploma Thesis,* TU Chemnitz.

Schuster, J., Cichos, F., and von Borczyskowski, C. (2003) Anisotropic diffusion of single molecules in thin liquid films, *Europ. Phys. J. E,* 12, 75–80.

Schuster, J., Cichos, F., Wrachtrup, J., and von Borczyskowski, C. (2000) Diffusion of single molecules close to interfaces, *Single Mol.,* 1, 299–305.

Täuber, D. (2011) Characterization of heterogeneous diffusion in confined soft matter *PhD Thesis,* TU Chemnitz, Chemnitz.

Trenkmann, I., Bok, S., Korampally, V., Gangopadhyay, S., Graaf, H., and von Borczyskowski, C. (2012) Counting single rhodamine 6G dye molecules in organosilicate nanoparticles, *Chem. Phys.,* 406, 41–46.

Venema, L. (2011) Silicon electronics and beyond, *Nature,* 479, 309.

Von Borczyskowski, C. (2016) Tuning luminescence intermittency, in: *Tuning Semiconducting and Metallic Quantum Dots: Spectroscopy and Dynamics* (eds. von Borczyskowski, C., and Zenkevich, E.) Pan Stanford Publishing, Singapore.

Yu, C. J., Richter, A. G., Kmetko, J., Datta, A., and Dutta, P. (2000) X-ray diffraction evidence of ordering in a normal liquid near the solid-liquid interface, *Europhys. Lett.*, 50, 487–493.

Zhuravlev L. T. (2000) The surface chemistry of amorphous silica. Zhuravlev model, *Colloids Surf. A,* 173, 1–38.

Chapter 8

Selective Surface Binding of Dye Molecules on Hybrid Humidity Sensors

Ines Trenkmann

Institute of Physics, Experimental Sensor Science,
Technische Universität Chemnitz, Reichenhainer Str. 70,
D-09107 Chemnitz, Germany

ines.trenkmann@physikphysik.tu-chemnitz.de

8.1 Introduction

Ideal sensors have to fulfill a wide range of expectations. On the one hand, the fabrication of sensors should be cost-efficient, forego the use of toxic chemicals and easily be variable in terms of the used (flexible) substrates and sizes. On the other hand, the manufactured devices should have a high sensitivity with a short response time and wide operation range, and simultaneously a high physical and chemical stability and lifetime (Wang et al., 2013).

For about three decades, relative humidity sensors based on polymer films have been widely used. The change of the relative humidity leads to the adsorption and desorption of water molecules between polymer chains and therefore to a detectable variation of the dielectric function of the device. The

Self-Assembled Organic-Inorganic Nanostructures: Optics and Dynamics
Edited by Eduard Zenkevich and Christian von Borczyskowski

ISBN 978-981-4745-43-7 (Hardcover), 978-981-4745-44-4 (eBook)
www.panstanford.com

main disadvantage of these devices results from the formation of water clusters at high humidity values, which lead to hysteresis effects (Chen and Lu, 2005). It was shown by Lin et al. (2004) that these effects can be minimized by integration of porous oxide particles into the polymer film and additionally the response time is decreased.

This approach will be continued here. For this purpose, hybrid structures formed by a poly(methylmethacrylate) (PMMA) film with a variable fraction of embedded porous silicon oxide (SiO_x) particles are investigated. Due to the fabrication process of the silica, the surface of the particles is covered by a high density of silanol groups. However, the number of free (≡Si-OH), geminal free (=$Si(OH)_2$), and bridged silanol groups depend on the used manufacturing method of the nanoparticles (Zhuravlev, 2000). At increasing relative humidity, water molecules can penetrated through the pores of the silica particles in the film and bound to the particles due to the formation of hydrogen bounds between silanol groups at the particle surface and water molecules (see Fig. 8.1). By this, the embedded silica particles improve the intake and outtake of water molecules at variation of the relative humidity.

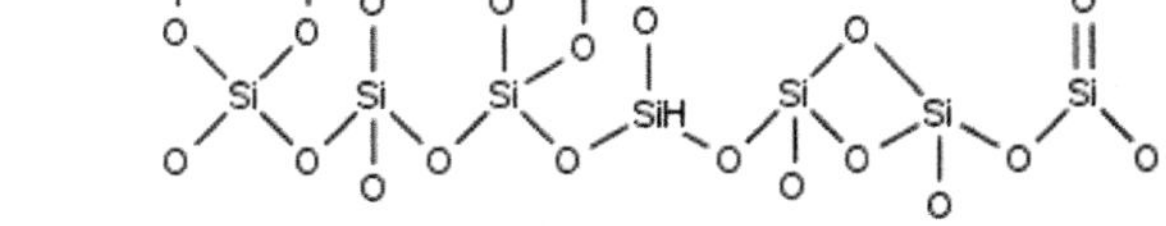

Figure 8.1 Illustration of adsorbed water on SiO_x particles. The binding of water arises from the formation of (strong) hydrogen bonds (dotted lines) between water molecules and silanol groups at the SiO_x surface which is for the sake of convenience indicated by the first atomic layer. Additionally, further hydrogen bonds develop between water molecules and lead to the formation of a water film.

To enable a free diffusion of water molecules within the particle pore and to prevent re-adsorption effects, it is important to use silica particles with a pore size larger than the water molecules (Zhuravlev, 2000). It is also required that a fraction of the embedded silica particles is not covered with the hydrophilic polymer film and has direct contact to the environment. However, due to solvent-based fabrication of the samples (see the next section) it is likely that the particles are almost completely embedded in the film and only a small part of the particles stands out of the polymer, which is schematically illustrated in Fig. 8.2. Therefore the control and variation of the fraction of the uncovered surface of the silica particles provides a way to improve the function of a relative humidity sensor.

Figure 8.2 Schematic illustration of a PMMA polymer film with embedded SiO_x particles.

For the analysis of the fraction of the particle standing out of the polymer film, a selective functionalization of the silica surface is used. Several studies have shown that such a selective binding of fluorescence dye molecules to nanostructured substrate can be used for optical analysis via fluorescence imaging (Graaf et al., 2007; Nafday and Lenhert, 2011; Zhuang et al., 2012). Baumgärtel et al. (2010, 2014) and Baumgärtel and Graaf (2014), for example, have demonstrated the selective binding of various cationic dyes from an aqueous solution on silica nanostructures on hydrophobic substrates. Details of the related procedures have been discussed in Chapter 6. Fluorescence microscope and atomic force microscope images of samples before and after functionalization illustrate that the cationic dye molecules only bind to (hydrophilic) silica nanostructures (Baumgärtel et al., 2010; Baumgärtel et al., 2014: Baumgärtel and Graaf, 2014). The adsorption of the cationic dye on silica arises, one the one hand, from the formation of hydrogen bounds between the dye molecules and silanol groups on the silica surface and, on the other hand, from strong ionic bounds between the positive charge of the dye and negatively charged de-protonated silanol groups (see Fig. 8.3, right) (Kikteva et al., 1999; Jal et al., 2004; Tleugabulova et al., 2005).

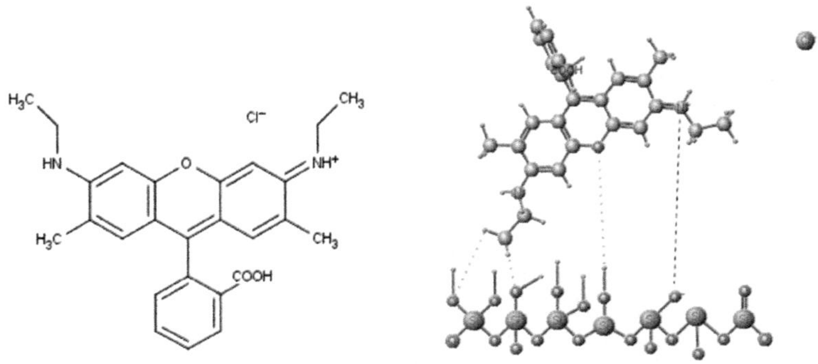

Figure 8.3 Chemical structure of rhodamine 6G (R6G) (left) and illustration of the adsorption of R6G on SiO_x particle surface (right) which is for the sake of convenience indicated by the first atomic layer. The binding of the cationic dye arises from the formation of hydrogen bonds (dotted lines) between the R6G molecule and silanol groups at the SiO_x surface and ionic bonds (dashed line) between the positive charge of the dye, mainly located at the nitrogen, and negative charged de-protonated silanol groups at the SiO_x surface.

Since both PMMA and silica are optical transparent, a wide range of water-soluble fluorescent dyes are available. In this study, rhodamine 6G (R6G, for chemical structure see Fig. 8.3, left) is used to selective functionalize the uncovered surface of the silica particles. This has the advantage that only a small number of dye molecules bind to the hydrophobic polymer. Additionally, the high quantum yield of R6G is ideal for high signal-to-noise-ratio in fluorescence images.

8.2 Experimental

We carried out emission and reflection measurements on hybrid structures based on polymer films with embedded inorganic particles, which are used as active layer for relative humidity sensors to analyze the fraction of the uncovered particle surface and to correlate the results with measurements of the sensor performance.

We prepared polymer films of poly(methylmethacrylate) (average molecular weight 120,000, Sigma-Aldrich, Germany) by spin-coating a PMMA solution onto cleaned glass substrates. An undoped PMMA solution was used as reference sample. For the analysis of the fraction of the uncovered inorganic particle surface, we prepared two hybrid structures, one with 80% PMMA and 20% particles and the second one with 40% PMMA and 60% particles, by adding the required amount of silicon oxide particles (particle size 2–25 μm, pore diameter 6 nm, Sigma-Aldrich, Germany) to the polymer film.

As first step transmission spectra of all three films were recorded with a UV-VIS photospectrometer (UV-PC-3100, Shimadzu, Germany) and an integration sphere (ISR-3100, Shimadzu, Germany).

For the functionalization of the uncovered fraction of the silica particle surface all samples are dip-coated in an aqueous rhodamine 6G (R6G, Radiant Dyes & Accessories GmbH, Germany) solution. Afterwards the samples were carefully rinsed with (purified) water (Milli-Q, Germany) and dried for 30 min on air. Emission and reflection images before and after functionalization with R6G were recorded with a confocal microscope LSM 510 (Zeiss, Germany). For direct excitation of R6G a 514 nm laser line was used and focused with a 10-fold objective (PlanNeofluar, NA 0.9) on the sample. The excitation and emission light was separated by a longpass dichroic mirror. Additionally, a 530 nm long pass filter and a 685 nm short pass filter were used to record the fluorescence and reflection images, respectively.

The analysis of the uncovered silica surface was performed on the fluorescence images of the dye-functionalized samples using the program WSxM (Horcas et al., 2007).

8.3 Results

8.3.1 Analysis of the As-Prepared Hybrid Structures

The transmission spectra of all three prepared structures are shown in Fig. 8.4. It is clearly that the transmission of the hybrid structures decreases with the addition of SiO_x particles and further deceases with increasing fraction of the SiO_x particles. This is caused by diffusive reflection of the irradiated light at the irregular shaped SiO_x particles (see Fig. 8.6) and the pores within the particles.

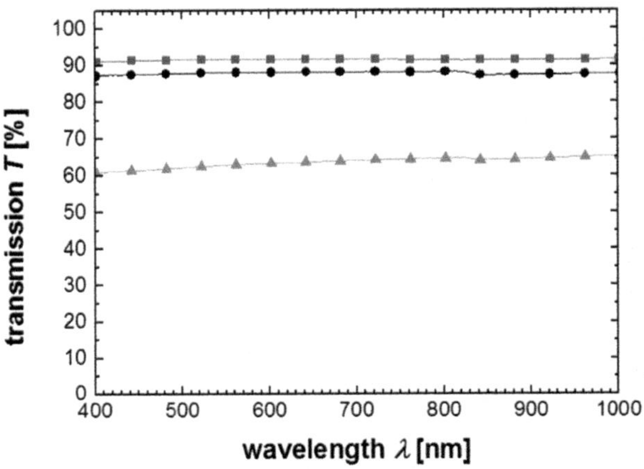

Figure 8.4 Transmission spectra of a pure PMMA film (squares), a hybrid structure with 80% PMMA and 20% SiO_x particles (circles) and with 40% PMMA and 60% SiO_x particles (triangles).

The reflection image (Fig. 8.5 left) of an undoped PMMA film shows, although PMMA is optical transparent, low emission intensities caused by a small amount of homogeneously distributed fluorescent impurities. Due to their low absorption and scattering cross section this impurities are not visible in the reflection image (Fig. 8.5 middle). Furthermore, no other (non-fluorescent) impurities are visible. Therefore it can be excluded that by binding of R6G to such impurities the fraction of the uncovered surface of SiO_x particles will be overestimated.

Figure 8.6 (left column) shows the emission and reflection images as well as the overlay of both signals of a hybrid structure with 80% PMMA and 20% SiO_x particles. The embedded oxide particles are clearly visible in the reflection image and differ considerably in size and shape. This becomes particular clear from the enlarged view shown in Fig. 8.7. The analysis of the image results in a size distribution of oxide particles from 2–35 μm, which is up to 40% larger than according to manufacturer's specification. The emission image of the hybrid structure with 20% oxide particles shows in contrast to the undoped PMMA film no homogenous distribution of emission. However, areas with smaller or even zero emission intensities can be identified. The overlay of the emission and reflection image shows that these areas coincide with the position of the embedded SiO_x particles. Moreover, ring-like structures are visible around the oxide particles (see Fig. 8.7). This suggests that the most oxide particles are covered by a thin polymer film on the surface as suggested in Fig. 8.2.

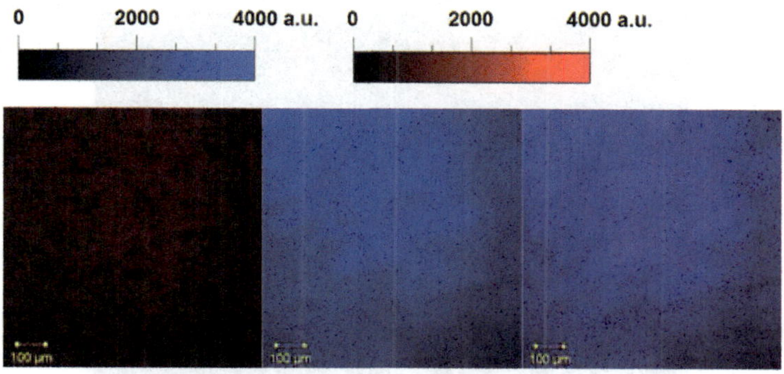

Figure 8.5 Typical emission (red, left) and reflection (blue, middle) image of PMMA polymer films and the superposition of both channels (right).

With increasing the fraction of oxide particle to 60% no individual particles can be identified (Fig. 8.6, right). Moreover, the concentration of the particles seems to be higher than 60%. This has two different reasons. First, the denoted fraction of oxide particles refers to the prepared solution. However, sample preparation via spin-coating the solution onto a substrate can lead to a higher concentration of oxide particles in the hybrid film. Second, due to the two-dimensional image of the film, the real spatial distribution of the oxide particles within the film is lost.

Therefore the question whether the individual particles within the film are covered by a polymer layer remains open. The image in Fig. 8.6 shows several areas with high emission intensities which cannot be correlated to features within the reflection image. A series of emission and reflection images of the sample were recorded while stepwise shifting the plane of the laser focus in the polymer film. In Fig. 8.8 two sections through the sample show the variation of the emission and reflection intensity as a function of the focal plane. The figure shows that the areas of high emission intensities correspond to areas with a small reflection signal. Apparently, in these areas more polymer (or impurities) is accumulated compared to the other areas of the sample.

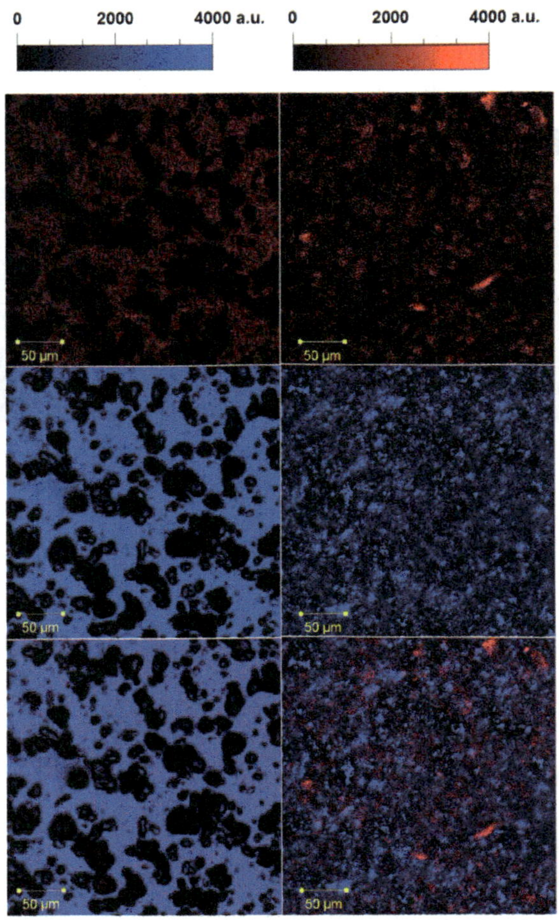

Figure 8.6 Typical emission (red, top) and reflection (blue, middle) images and the superposition of both channels (bottom) of films with 80% PMMA and 20% SiO_x particles (left column) and with 40% PMMA and 60% SiO_x particles (right column).

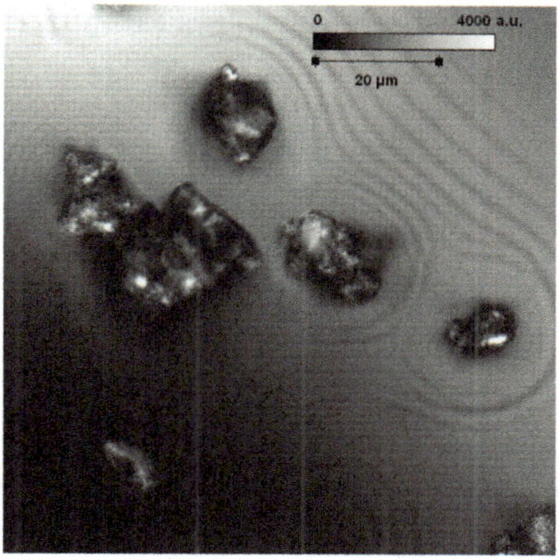

Figure 8.7 Reflection image of an enlarged area of the films with 80% PMMA and 20% SiO_x particles in Fig. 8.6. Besides the irregular shape of the SiO_x particles, (ring like) structures are visible inside the polymer film around the particles.

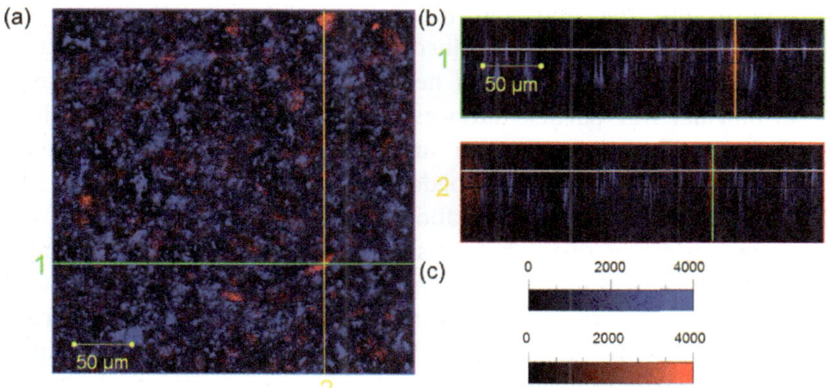

Figure 8.8 Overlay of the emission and reflection image of a film with 40% PMMA and 60% SiO_x particles (a). The intensity scale of the emission (red) and reflection (blue) is given in (c) in a.u. The focal plane was stepwise shifted by 1 μm and the emission and reflection image were recorded and overlaid onto each other. Lines 1 and 2 mark the respective sections of the film and show the variation of the emission and reflection intensities while shifting the focal plane. The white lines in both sections mark the normal focal plane which corresponds to the sample surface (b).

8.3.2 Analysis of Rhodamine-Functionalized Hybrid Structures

The emission and reflection images of both analyzed hybrid structures after functionalization of the uncovered oxide surface with the fluorescent dye R6G are shown in Fig. 8.9. The emission images show a clear increase of the detected emission intensities in several areas of sample due to binding of R6G. Especially the comparison of the reflection and emission images of the hybrid structure with 40% PMMA and 60% SiO_x particles clearly shows that the positions of the increased emission intensity correspond to positions of the embedded oxide particles. This becomes even more obvious upon recording emission and reflection along a single line through the film with variation of the focal plane (see Fig. 8.10). Furthermore, this focal cross section reveals that the areas of high emission intensity are restricted to the sample surface. The time during which the sample is immersed in the dye solution is obviously not long enough to permit the dye to diffuse into the film and to bind to SiO_x particles. This is important to avoid an incorrect determination of areas functionalized with R6G and hereby also of those surface areas of polymer free particle surfaces. The comparison of both hybrid structures clearly shows an increase of the dye-covered area with increased fraction of embedded oxide particles.

For the exact determination of the dye-covered area, the emission images of the hybrid structure are analyzed before and after functionalization of the uncovered oxide surface with R6G. At first, the intensity distributions of as-prepared films are determined. The sum of the highest occurring intensity plus twice the standard deviation is used as threshold for the analysis of the functionalized hybrid films. In turn, the intensity distributions are determined for the emission images. Intensity values below the threshold belong to the emission of impurities embedded in the polymer, values above the threshold to the emission of R6G. The program WSxM divides the emission images according to the chosen threshold value in areas with lower and higher emission intensity (Horcas et al., 2007). For the hybrid film with 80% PMMA and 20% SiO_x particles the area covered with R6G is $(2.7 \pm 0.2)\%$ of the film surface and for the hybrid structure with three-fold higher oxide particle fraction (40% PMMA and 60% SiO_x particles) the fraction of the uncovered oxide surface is with

(6.4 ± 0.3)% about 2.4 times larger. That the fraction of the uncovered surface does not increase to the same amount as the increased fraction of SiO_x particles might be due to the preparation procedure of the hybrid structures. Both films are prepared by the same spin-coating process, which leads for hybrid structures with 60% oxide particles to a higher average film thickness. In case of a constant film thickness, the increase of the fraction of embedded oxide particle should lead to the same increase of the uncovered oxide surface.

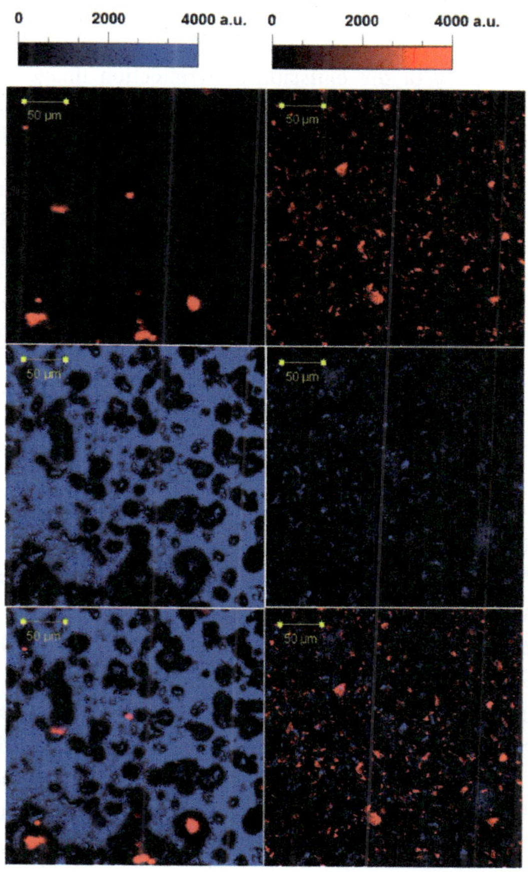

Figure 8.9 Typical emission (red, top) and reflection (blue, middle) images and the superposition of both channels (bottom) of films with 80% PMMA and 20% SiO_x particles (left column) and with 40% PMMA and 60% SiO_x particles (right column) after functionalization of the uncovered oxide surface with R6G.

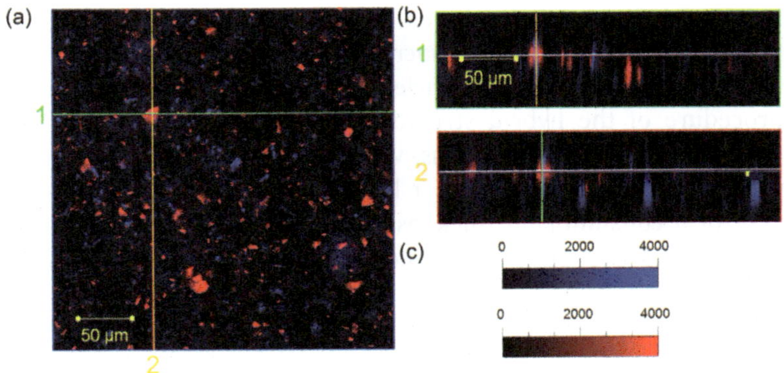

Figure 8.10 Overlay of the emission and reflection image of a polymer film with 40% PMMA and 60% SiO_x particles after functionalization of the uncovered oxide surface with R6G (a). The intensity scale of the emission (red) and reflection (blue) is given in (c) in a.u. The focal plane was stepwise shifted by 1 µm while recording the emission and reflection images. Line 1 and 2 mark the respective sections of the film and show the variation of the emission and reflection intensities upon shifting the focal plane. The white lines in both sections mark the normal focal plane which corresponds to the sample surface (b).

8.4 Discussion

The previous sections showed that it is possible to determine the uncovered oxide surface of SiO_x particles in hybrid structures by selective functionalization with a suitable fluorescent dye. For the uncovered surface of the hybrid structures with 80% PMMA and 20% SiO_x, an area of 2.7 ± 0.2% and with 40% PMMA and 60% SiO_x particles an area of 6.4 ± 0.3% is found.

Measurements with humidity sensors based on the presented hybrid structures show that the increase of the fraction of SiO_x particles from 20% to 60% results in a decrease of the response time from 17.5 to 12.0 s (−30%) upon an increase of the relative humidity from 33% to 95%. In parallel, the relative capacity increases in comparison to a sensor with an undoped PMMA film by about 51% (115%) upon embedding 20% (60%) SiO_x particles in the polymer film at 25°C and a relative humidity of 80% (Martin et al., 2010). This variation corresponds to an increase of the relative capacity by a factor of 2.3, which corresponds to the increase of the uncovered oxide surface when increasing the fraction of oxide particles.

Therefore both improvements of the relative humidity sensors can be assigned to the increase of the fraction of embedded SiO_x particles with the accompanied increase of the uncovered oxide surface. Due to the increased uncovered oxide surface, the active sensor material can adsorb more water molecules in a shorter time, which results in a reduced response time and a larger variation of the relative capacity. The further improvement of the sensor performance will depend critically on the control of the uncovered oxide surface which can be precisely identified via decoration with appropriately functionalized fluorescent dye molecules.

References

Baumgärtel, T., and Graaf, H. (2014) Spectral shift of cresyl violet luminescence on charged silicon oxide nanostructures, *Physica Status Solidi A*, **211**, pp. 905–909.

Baumgärtel, T., Rehm, S, Würthner, F., von Borczyskowski, C., and Graaf, H. (2014) Functionalized bisimides dyes bound via electrostatic interactions to oxide nanostructures generated by AFM lithography, *Applied Surface Science*, **318**, p. 51.

Baumgärtel, T., von Borczyskowski, C., and Graaf, H. (2010) Fluorescence studies of Rhodamine 6G functionalized silicon oxide structures, *Nanotechnology*, **21**, p. 265306.

Chen, Z., and Lu, C. (2005) Humidity sensors: A review of materials and mechanisms, *Sensor Letters*, **3**, pp. 274–295.

Graaf, H., Vieluf, M., and von Borczyskowski, C. (2007) Selective binding of dye molecules and CdSe nanocrystals on nanstructures generated by AFM lithography of silicon surfaces, *Nanotechnology*, **18**, p. 265306.

Horcas, I., Fernándes, R., Gómez-Rodríguez, J. M., Colchero, J., Gómez-Herrero, J., and Baro, A. M. (2007) WSxM: a software for scanning probe microscopy and a tool for nanotechnology, *Review of Scientific Instruments*, **78**, p. 013705.

Jal, P. K., Patel, S., and Mishra, B. K. (2004) Chemical modification of silica surface by immobilization of functional groups for extractive concentration of metal ions, *Talanta*, **62**, pp. 1005–1028.

Kikteva, T., Star, D., Zhao, Z., Baisley, T. L., and Leach, G. W. (1999) Molecular orientation, aggregation, and order in rhodamine films at the fused silica/air interface, *The Journal of Physical Chemistry B,* **103**, pp. 1124–1133.

Li, Y, Yang, M. J., and She, Y. (2004) Humidity sensors using in situ synthesized sodium polystyrenesulfonate/ZnO nanocomposites, *Talanta*, **62**, pp. 77–712.

Martin, J., Piasta, D., Hammacher, J., Wegener, M., Dittrich, C., and Otto, O. (2010) Sensitive Feuchtigkeitssensoren auf der Basis von Nanokompositien, 2. *GMM Workshop Technologie and Werkstoffe der Mikrosystem- und Nanotechnik.*

Nafday, O., and Lenhert, S. (2011) High-throughput optical quality control of lipid multilayers fabricated by dip-pen nanolithography, *Nanotechnology*, **22**, p. 225301.

Tleugabulova, D., Sui, J., Ayers, P. W., and Brennan, J. D. (2005) Evidence for rigid binding of rhodamine 6G to silica surfaces in aqueous solution based on fluorescence anisotropy decay analysis, *The Journal of Physical Chemistry B,* **109**, pp. 7850–7858.

Wang, S., Kang, Y., Wang, L., Zhang, H, Wang, W., and Wang, Y. (2013) Organic/ inorganic hybrid sensors: A review, *Sensors and Actuators B*, **182**, pp. 467–481.

Zhuang, J., Friedel, J., and Terfort, A. (2012) The oriented and patterned growth of fluorescent metalorganic frameworks onto functionalized surfaces, *Beilstein Journal of Nanotechnology*, **3**, pp. 570–578.

Zhuravlev, L. T. (2000) The surface chemistry of amorphous silica. Zhuravlev model, *Collodis and Surfaces A*, **173**, pp. 1–38.

Index